P9-DHD-559

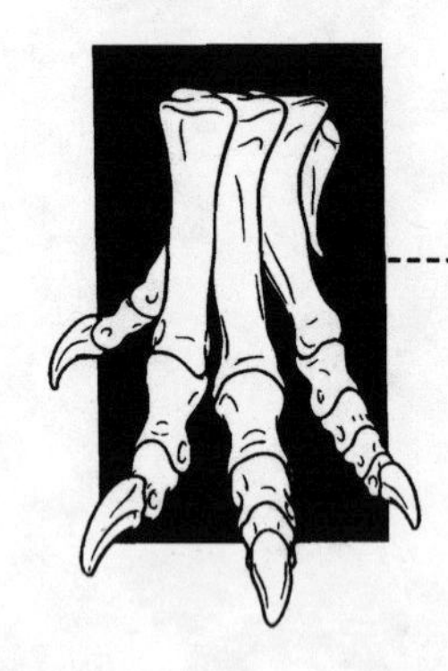

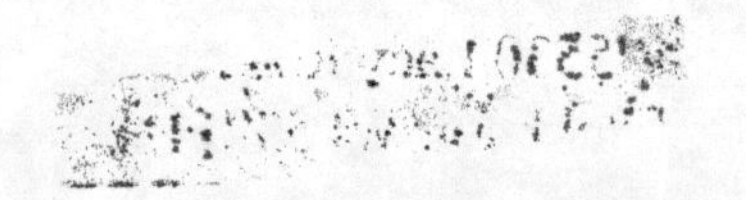

DINOSAURS
OF THE
LAMING CLIFFS

MICHAEL NOVACEK

VITH ILLUSTRATIONS BY ED HECK

NCHOR BOOKS
OUBLEDAY
EW YORK

To Vera, Julie, and my parents

ACKNOWLEDGMENTS

To write about our continuing exploration of the Gobi Desert—an adventure that is still unfolding—is rather odd, as if my reflection on the serendipity behind various events and discoveries had anything to do with predicting the future of our efforts. Nonetheless, enough has been experienced to justify my recollection. I also felt the sheer panic of loosing vivid feelings and images before I could flesh them out in any form beyond random notes in my field journals. As a result, I had the sheer pleasure of revisiting these experiences in a small room of an old eastern Atlantic Victorian house, about as far away from the land of camels, sand dunes, and dinosaur graveyards as one can imagine.

These thoughts embrace people as well as exotic places. Mark Norell and Malcolm McKenna were my fellow travelers from the start, sharing with me the years of the Gobi expedition from that first dreamlike roaming of the desert in the summer of 1990. This extraordinary opportunity was offered by the Mongolian Academy of Sciences, notably our colleagues Demberelyin Dashzeveg, Altangerel Perle, and Rinchen Barsbold. In addition, the Mongolian Academy–American Museum expeditions of the past six years have enlisted a large international team. Many of these people appear in the book as part of the story, but all of them contributed passionately and indispensably to the effort. Jim Clark, Lowell Dingus, Priscilla McKenna, Amy Davidson, and Louis Chiappe were key team members who participated in the expeditions over several seasons. Others who joined us were Bayersaichan Dashzeveg, Kevin Alexander, Batsuk, Temur, Tumur, Sota, Gunbold, John Lanns, Dan Bryant, Ganhuij ("Mad Max"), Mangal Jal, George Langdon, Boyin Tok-Tok, Carl Swisher, Christa Sadler, Andy Taylor, Minteg, Ines Horovitz, Argil, Mark Carrasco, Eungeul, Vera Novacek, Julie Novacek, Chimbald, Elizabeth Chapman, Ned Saandar, and Jim Carpenter. Journalists John Wilford and Donovan Webster, photographers Fred Conrad, Louis Psihoyos, and John Knoebber and filmmakers John Lynch, Jerry Pass, Tim Watts, Lisa Truitt, Nina Parmee, Ruben Aranson, Dick Kane, and Richard Confalone doubled, at times, as able field assistants. United States ambassador to Mongolia, Donald Johnson and his wife, Nelda, were gracious and helpful, demonstrating both a kinship for the expedition and a love for the strange and wild country where they are now in residence. Although they did not reach the field, several people contributed to the laboratory effort. Among these were Evan Smith, Jeanne Kelly, Marilyn Fox, Jane Shumsky, Bryn Mader, Ed Pederson, Bill Ameral, Lorraine Meeker, and Chester Tarka. Several saved the project from hopeless bureaucratic, budgetary, sociological, and logistical snarls. These included Dumaajavyn Baatar, president of the Mongolian Academy of Sciences, and General Secretary Tsagaany Boldsuch, my extraordinary assistant, Barbara Werscheck, Joan Davidson, Krystyna Mielczarek, Myra Biblowit, and Erdene Dashzeveg. My parents, as well as Malcolm McKenna's parents and Mark's father, willingly converted their Southern California homes to supply depots during the frenetic shipping months before each season. Museum President Ellen Futter, and her predecessor, George Langdon, were warmly encouraging and extraordinarily tolerant of my schizophrenic life as administrator and explorer. A number of generous people and organizations kept an expensive expedition financially healthy. These included the Frick Laboratory Endowment of the American Museum of Natural History, the Phillip McKenna Foundation, Richard Jaffe and the Jaffe Foundation, the National Geographic Society, the Eppley Foundation, the International Research and Exchange Program (IREX), and the National Science Foundation (NSF).

If I had any success in blending exploration of wild terrain with exploration of some central scientific questions, it is due not only to the flourishing literature in modern paleontology but to the input and inspiration of friends and colleagues. In addition to some team members, these individuals included John Wible, Guillermo Rougier, Niles Eldredge, Meng Jin, Andy Wyss, Zofia Kielan-Jaworowska, Jorn Hurum, Gao Keqin, Donald Phillips, Phil Currie, John Ostrom, Ian Tattersall, Don Lessem, Dick Tedford, Henry Gee, Sherri McGehee, Tim Rowe, David Archibald, and the late Lev Nessov. Of course, such productive interchange began years ago with my paleontological mentors, Peter Vaughn, Everett (Ole) Olson, Jason Lillegraven, Don Savage, and Bill Clemens.

I must also extend special thanks to those who helped me directly in the development of this book. Ed Heck is responsible for nearly all the artwork herein. His skills in depicting prehistoric beasts either as detailed renderings or with the economy of an ink sketch are admired and appreciated. The photographs include special contributions from Louis Psihoyos (the cover *Oviraptor* skeleton), Fred Conrad, Mark Norell, Dennis Finnin, Mick Ellison (who also contributed some excellent drawings), Chester Tarka, Amy Davidson, and the archival collections of the American Museum of Natural History. Mark Norell did a marathon reading of the text, highlighting errors and offering crucial corrections. Any errors of fact and concept are, of course, my own doing. Roger Scholl, my editor at Anchor/Doubleday, took on the challenge of my original draft with dedication, focus, and friendly encouragement. I learned much from him in the process. I am immensely grateful to my agent, Al Zuckermann of Writer's House, for adopting an untested client with the conviction that there was a story to be shaken out of the bones in the Gobi. Finally, I thank my wife, Vera, and my daughter, Julie, for showing me that love brings an enchantment to life that no expedition to the ends of the earth can match.

CONTENTS

Prologue: A Paleontological Paradise 3

Chapter 1: 1990—JOURNEY TO ELDORADO 18

Chapter 2: DINOSAUR DREAM TIME 49

Chapter 3: 1991—THE GREAT GOBI CIRCUIT 94

Chapter 4: THE TERRAIN OF EONS 131

Chapter 5: 1992—THE BIG EXPEDITION 160

Chapter 6: DINOSAUR LIVES—FROM EGG TO OLD AGE 184

Chapter 7: 1993—XANADU 224

Chapter 8: FLYING DINOSAURS AND HOPEFUL MONSTERS 243

Chapter 9: 1994—BACK TO THE BONANZA 269

Chapter 10: DISASTERS, VICTIMS, AND SURVIVORS 289

Chapter 11: THE SECRET HISTORY OF LIFE 323

Notes 333

Select Reading List 349

Index 351

Is it surprising that I was filled with regret as I looked for the last time at the Flaming Cliffs, gorgeous in the morning sunshine of that brilliant August day? I suppose I shall never see them again! Perhaps some day I may view the cliffs from the window of a trans Gobi train, but my caravan never again will fight its way across the long miles of desert to this treasure-house of Mongolian pre-history. Doubtless it will be the hunting-ground of other expeditions for years to come. We have but scratched the surface, and every season of blasting gales will expose more riches hidden in its rocks. Who can tell what will come from a place that has already given so much?

ROY CHAPMAN ANDREWS.
1932. *The New Conquest of Central Asia.*

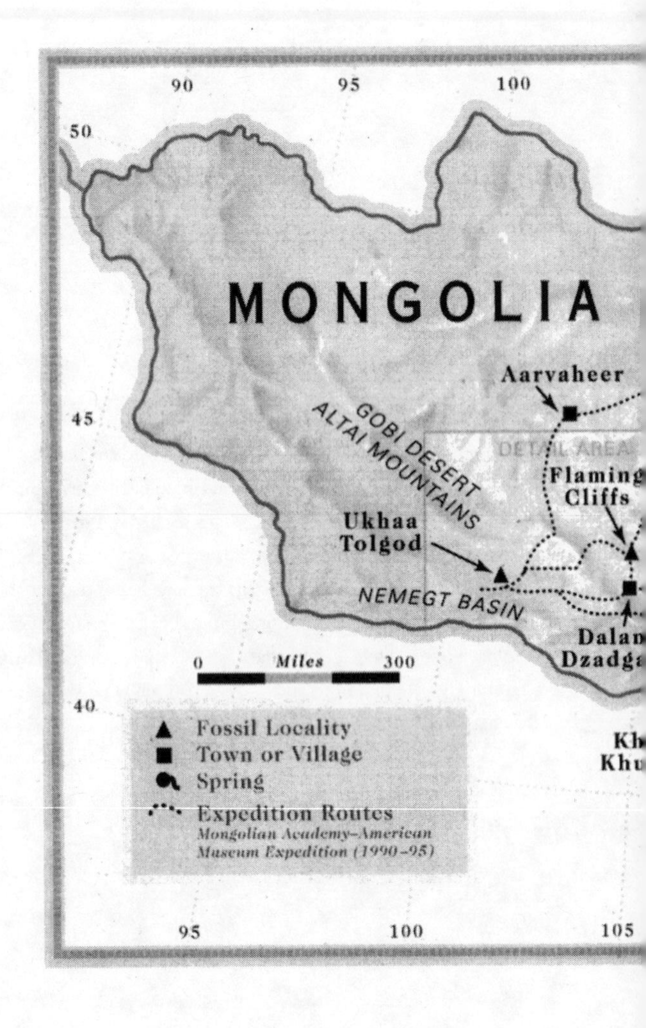

90
95
100
50
45
40
MONGOLIA
Aarvaheer
GOBI DESERT
ALTAI MOUNTAINS
DETAIL AREA
Ukhaa
Tolgod
NEMEGT BASIN
0
Miles
300
Fossil Locality
Town or Village
Spring
Expedition Routes
Mongolian Academy-American
Museum Expedition (1990-95)
95
100
105

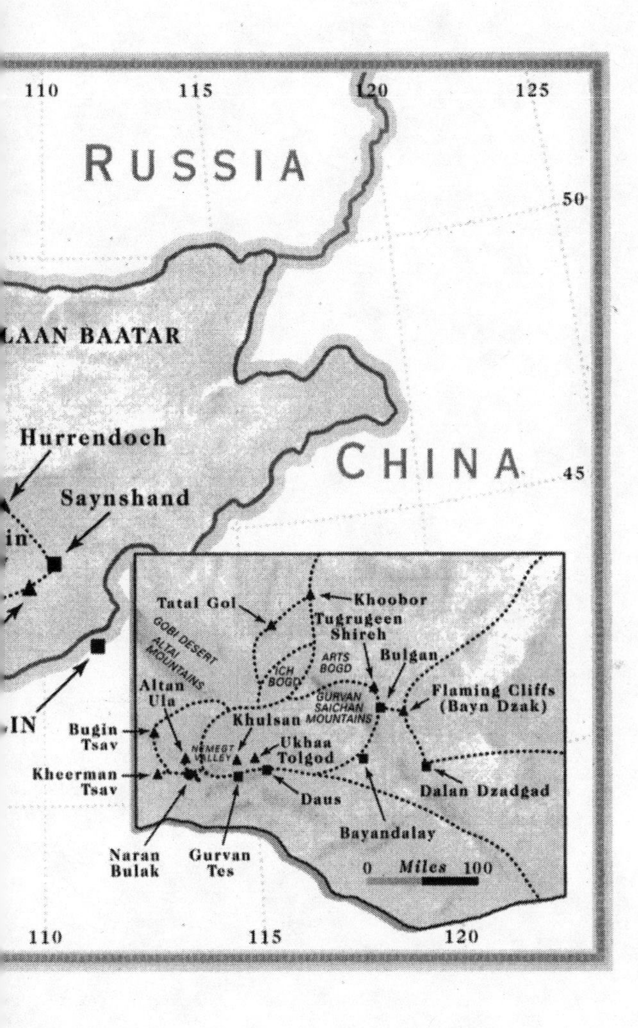

110
115
120
125
RUSSIA
50
LAAN BAATAR
Hurrendoch
Saynshand
CHINA
45
.IN
Tatal Gol
Khoobor
Tugrugeen Shireh
GOBI DESERT
ALTAI MOUNTAINS
ARTS BOGD
ICH BOGD
Bulgan
Altan Ula
GURVAN SAICHAN MOUNTAINS
Flaming Cliffs (Bayn Dzak)
Khulsan
Bugin Tsav
Ukhaa Tolgod
Kheerman Tsav
Daus
Dalan Dzadgad
Bayandalay
Naran Bulak
Gurvan Tes
0 Miles 100
110
115
120

PROLOGUE

A PALEONTOLOGICAL PARADISE

TIME AND DATE: 11 A.M., JUNE 24, 1990

LOCALITY: UNCERTAIN. SOMEWHERE IN THE GOBI DESERT, APPROXIMATELY THREE HUNDRED MILES SOUTH-SOUTHWEST OF ULAAN BAATAR, MONGOLIA

It was a dusty, windblown day. We were stopped once again, standing out of the vehicles on the dry mud-caked surface of an old lake. The great mountains of the Gurvan Saichan were only a shadow in the clouds on the southern horizon. To the west, a few camels stared at us, their spidery legs and necks dancing in the rising heat. The blowing sand stung our eyes as we squinted through our binoculars. But the wind was not strong enough to keep the flies from biting our sweating foreheads.

We were all tired and despondent. For the last seventy-two hours our wild Mongolian driver (whom we nicknamed Mad Max) had jockeyed his lumbering GAZ, a Soviet war truck, over the poor excuse for the Gobi highway. To avoid the ruts of the original road, travelers over the past seventy years had seemed bent on creating their own. The result was a multi-lane rake across the highland, each lane worse than the other.

Now there was no road at all. We were lost. Not even the expertise of Dr. Demberelyin Dashzeveg, a world-famous paleontologist and our Mongolian guide, who knew the Gobi Desert perhaps better than any person alive, was enough to keep us on course. Some miles back, impatient with the road, we had decided on a shortcut that would enter our destination from the poorly explored northern region. In the least densely populated country in the world such bushwhacking can be risky, if not downright dangerous. The Gobi is one of the great empty spaces on earth. Its nearly half million square miles of badlands, sand dunes, and saw-toothed mountains are alternately baked in summer's inferno and frozen in winter's Siberian winds. Only a smattering of grassy knolls and ancient springs sustain the rugged nomads with their horses and their dome-tented gers. Scientists like us, and the few tourists who take the guided circuit out of the northern Gobi town of Dalan Dzadgad, are uneasy intruders. Mongolia has little infrastructure and economy to sustain an expedition to such a frontier. Our old Russian trucks have somehow managed to transport us three hundred miles from the nearest nexus of civilization—the capital city of Ulaan Baatar. But the trucks have few spare parts and gasoline is severely rationed. We endure through the uncanny resources of our Mongolian mechanics (on more than one occasion they had to substitute makeshift replacements cut from our oatmeal boxes for rotten rubber gaskets in the engine block and transmission) and by our own ability to plot a route direct enough to conserve precious fuel.

Dashzeveg, a man in his late fifties, was tall and wiry, with the lean and hungry look of a Siberian wolf, the dark wrinkled skin of his face burnished by years of desert winds. His long back was slightly hunched, as if he had stooped under the low doorway of a ger once too often; his brown, sinewy arms resembled the twisted dark branches of an ironwood tree. The golf cap he constantly wore framed a pair of oversized ears that looked like sonar detectors capable of picking up the slightest sound in the dry desert air. Using his hand as a visor, he peered at the horizon, rotating slowly around nearly a full 360 degrees. He was looking for the tiny white speck of a ger in hopes a herdsman might help us get back on track. Having spent nearly thirty years in the Gobi, he was clearly at home in this desert. As a young graduate of Moscow's Paleontological Institute he accompanied the Poles in the late 1960s on several dinosaur expeditions. Since that time, in effect, he had never left the Gobi; every summer he had returned with one or two colleagues and a small platoon of assistants to stalk his favorite badlands. The limited funds and resources available from the Geological Institute in Ulaan Baatar limited the size of these forays. But I had the feeling, after knowing him only a week, that Dashzeveg was something of a loner, and he seemed to prefer these small expeditions.

Dashzeveg spoke little English. Now he only grunted, pointing his finger in a southwesterly direction. I could see nothing—there was only the deep brown scabland of volcanic hills, identical to the miles of terrain we had driven for hours.

"I guess we're heading for the dunes," Mark Norell said.

Dr. Mark Norell, my companion, friend, and colleague in the field for over a decade, was one of the best vertebrate paleontologists in the world. Like me, Mark grew up far from open fossil country in the urban sprawl of Los Angeles. But as a youth he and his friends were within three hours' drive of the great deserts of Mojave and Anza Borrego, affording him the opportunity for years of field apprenticeship. Now barely thirty, he was a new assistant curator in vertebrate paleontology at the American Museum of Natural History, where I was curator and dean of science. His career as a scientist was on a steep upward trajectory. He had come to the job with hefty academic laurels—a Ph.D. and postdoctoral fellowship from Yale University—but his long hair and casual L.A. demeanor belied his ivy-clad pedigree.

"Well, at least we're still heading south," I said.

"Yeah, but how is Mad Max going to cross those dunes?"

Mark's apprehension was well founded. An hour later we started plowing through the dune field. Through the tiny window slit we could see Max nervously tweaking the gearshift. The steering wheel whiplashed his arms. Ahead, Dashzeveg's truck had just conquered an impossibly steep dune slope. Mad Max followed, but too erratically. He quickly lost momentum and veered to the left. The tail end of the eight-ton truck began to drift.

"We're going to roll!" I shouted above the noise of pots, metal stoves, shovels, and tent poles careening off the inside of our canvas-covered compartment.

The truck continued to slide, its wheels no longer seemingly attached to the ground. I felt weightless. There was a violent crash and Mark and I were slammed together against the floor. The truck rocked and came to a dead stop. We were buried in maps and journal articles. But the truck was still upright, its nose pointing uphill, although its hulk tilted steeply to the port side. The flimsy door of our compartment creaked as it swung in the wind.

"I think I'll walk from here," Mark said.

"We may have to," I said.

Max jumped out and squatted under the truck, checking for damage. On top of the dune, Dashzeveg waited impatiently. Soon Max was standing in front of the truck, cranking the engine with an enormous metal shaft. At first it only shuddered, but eventually his efforts brought the motor to a rumble. The truck was nearer to the top of the dune than we had first realized. It crawled slowly forward, kicking up fountains of sand.

At last it leveled itself on the hard desert pavement beyond the top of the dune. We could see Dashzeveg standing on a high point near a dead zak tree. Smiling, he beckoned us out of our truck. We followed the direction he was pointing, to a low spot far south and east of the volcanic hills. There, beyond burnished grass, sand fields, and shimmering heat, was a thin orange line.

"Flamuung Cleeefs," he said.

From this great distance the line was neither ablaze nor clifflike. But after three days of tortuous roads and a near catastrophe in an old Russian truck, the Flaming Cliffs, our long-sought destination and perhaps the most famous dinosaur locality in the world, was within one or two hours from where we stood.

As we grew closer, the cliffs took on the shape I knew so well from dozens of archival photographs. I could see the great tower of sand rising from a smooth buttress at the end of the central section. We were approaching from the northeast, a profile familiar to me since I was seven years old, when I was captivated by an etching of the cliffs in one of my dinosaur books.

Within a mile of the main part of the cliffs we stopped again and Dashzeveg gave us a brief lecture in his broken English.

"Main localeety there," he said, pointing to the sand tower at the end of the main buttress. "Deenosaur eggs found high in seection, mammal skuuulls low."

Mark and I, both in ragged shorts and shirts, stood some yards away, staring at the sand castles forming one huge rampart that extended from us until it evaporated in the dusty horizon. Neon colored caps we wore sporting the legend "Lucy's Retired Surfer's Bar: New York" kept our long hair out of our eyes. We are unlikely successors to the 1920s Gobi expedition members in their ironed khaki and wide-brimmed hats.

Dashzeveg carried on a conversation about geological details with our more senior colleague, Dr. Malcolm McKenna. Malcolm, too, was a paleontologist at the American Museum in New York. An experienced explorer in his late fifties—nearly twenty years my senior—his determined, well-chiseled face was crowned by a healthy head of silvery hair. His usually light complexion had reddened from three days in the Mongolian sun. Malcolm shared with me an academic nurturing at the University of California at Berkeley, in their formidable Department of Paleontology, the largest graduate training program in that field anywhere in the country. As a graduate student and young professional he had a reputation as an *enfant terrible,* and indeed he still made his points with conviction and vigor. Malcolm, wearing a neatly pressed shirt and long pants and at this stage of the trip still well groomed, seemed a more fitting descendant of the impressive explorers of early decades.

With the assurance and authority expected from one of the world's leading paleontologists, Malcolm provided us with various geological facts and details about the cliffs. It was as if he had explored them many times before, although he had been there only once on a brief stop as an officially approved and carefully scrutinized tourist. We took several pictures and I shot a few minutes of 8mm video.

To my slight disappointment, it all seemed so undramatic. Yet Mark,

Malcolm, and I would be hard pressed to remember a more important day. We were living what for our colleagues and their predecessors was only a dream. For sixty years Mongolia, the most isolated pocket of the Soviet Empire, had closed its doors to Western scientists. In 1990 it had broken free, declaring its independence and opening its borders to scientists beyond the fragmenting Soviet sphere. We were the first team of paleontologists from the West to explore the Gobi Desert of Mongolia since the 1920s, when Roy Chapman Andrews, with Walter Granger and others from the American Museum of Natural History, led one of the most dramatic and successful scientific quests of this century. Appropriately, our initial target had been the Flaming Cliffs, where American Museum scientists in 1922 made the first discovery of dinosaur eggs from the Gobi Desert.

Our two trucks stopped at the ledge over the main section of the cliffs. A sinuous canyon marked the way to the bottom. The canyon had been christened by the Poles after Dashzeveg in honor of his discovery of a precious ancient mammal skull there. We unpacked, and carrying a few spare implements—a rock hammer, a day pack, and a one-liter water bottle—began a lazy but watchful stroll down the cliff face. Except for the camels and our dark-skinned Mongolian comrades, it could have been any hot afternoon in Wyoming's badlands.

Prospecting the narrow gullies, I soon found myself separated from the others. I stopped for a swig of lukewarm water. Malcolm was already far below, at the base of the cliffs, crawling on all fours in search of mouse-sized fossil mammal skulls. Mark was prospecting on the opposite canyon wall, perched on an absurdly steep slope, his white shirt glowing against the orange embers of the cliff face. The Mongolians shouted from the rim of the cliffs, "Come back for afternoon tea." But we were in no mood to comply. A little heat could hardly keep us off the trail we had started 13,000 miles away.

The hours passed uneventfully. I encountered a few scraps of a turtle shell, fragmentary evidence of life from eighty million years ago, but little else. Skiing down a steep wall of loose sandstone, my heavy climbing boots filling with Cretaceous pebbles, I suddenly stopped near the floor of the wash. At a small knob of the cliff was a streak of brownish-white bone, forming a curved surface like a bird beak. A few inches away more bone extended out into a kind of small shield. I carefully brushed the sand away from the edge of the jaw where some odd, comblike teeth were exposed. *Protoceratops andrewsi.* It was a scruffy specimen, and the American Museum of Natural History, in fact, already had the best collection in the world—more than one hundred skulls and skeletons retrieved by the 1920s expeditions. But at that moment I felt good. I had found my first Mongolian dinosaur.

BONES

Although they seem so fantastic as to be almost beyond imagination, Cretaceous dinosaurs are not phantoms. Nor are they flesh and blood, at least not to our eyes. But they were once as real as (and even related to, as we shall see) the robin plucking worms out of the great lawn in Central Park today. We know this because dinosaurs have left us with their cleaned corpses. The same skeleton that brings form and movement to an animal, the trusswork that cradles its vital organs and fastens its explosive muscles, endures when all else of what once made it alive decays to nonexistence. Although once in a while we do find other remarkable remnants—a footprint, a stomach stone, an egg, even impressions of skin—the legacy of the backboned creatures we call vertebrates is overwhelmingly a magical derivative of calcium phosphate—bones—tempered into hard though sometimes brittle rods, plates, and knobs. It is these bones that we extract from terra firma with wild eyes and trembling hands, what we labor over and argue about, what we reconstruct into their original glories, like some massive skeletal bridge, and plant at the threshold of a museum entrance. Bones tell us what these animals were, to whom they were related, how they spent their lives, and perhaps how they died. At the entrances to the town square in Jackson Hole, Wyoming, great heaps of elk antlers (another derivative of the bony skeleton) form archways; beckoning you to the village green and the flower gardens inside. Just like those archways, fossil dinosaur bones mark the threshold and the route to primeval swamps and forests.

Probing such lost worlds, recreating the action on an ancient stage, and plotting the transition from one prehistoric empire of creatures to another is the business of the science of paleontology, the study of the history and evolution of life. Paleontologists claim the whole world as their oyster. Yearly, they penetrate the arid deserts of every continent, scour the cliffs of thousands of miles of coastline, drill and extract rock cores from the ocean bottom, even scramble over the ice-encrusted precipices of Greenland and Antarctica. But this universal theater of operations has a few hot spots, a few regions that are more expansive, richer, and more interesting than any others.

One of these meccas is the Gobi Desert of Mongolia. Few if any places in the world rival the Gobi for the completeness, quality, and variety of fossils recording the Late Cretaceous Period, the final phase of dinosaur dominance before the great extinction event of sixty-five million years ago. The Cretaceous was the last glorious dinosaur regnum before the land was left to the mammals.

Why is the Gobi so enticing to paleontologists? To be sure, there are also spectacular Cretaceous fossils in other places. The Rocky Mountain region of the western United States and Canada is justly famous for its dinosaur specimens disinterred, studied, and eventually displayed in the American Museum of Natural History, the Tyrrell Museum of Alberta, the Smithsonian Natural History Museum of Washington, D.C., the Carnegie Museum of Pittsburgh, the Field Museum of Chicago, and other treasure vaults. There are also great collections from important sites in Europe, Argentina, South Africa, and western China. These have found their way to the major museums of London, Paris, Buenos Aires, Berlin, Stuttgart, Frankfurt, Beijing, and Zigong. But even by these lofty standards the fossil finds from the Gobi stand high. For one thing, preservation of fossils in this desert is extraordinary. Many of the skeletons, when first discovered, look more like a cleaned assembly of white bones from a recently dead carcass than the heavily mineralized, blackened "rock-bone" typical of such ancient fossils at many other locations. The Gobi bones still preserve tiny surface features—grooves and pits that mark the routes of blood vessels and nerves. They also remain connected by complex joints of the knee, elbow, or ankle. And the Gobi skeletons don't lose those delicate important bones—claws, finger bones, wishbones, and tiny rings of bone that surround the eardrum in almond-sized mammal skulls. These details are the feast to the eye of the paleontologist; they satisfy a hunger for anatomical minutiae, clues that are the essence of understanding possible function, adaptations, and the evolution of ancient species.

The remarkable preservation of the Gobi fossils relates to another superlative. This protective entombment produces a complete assembly of the skeletons for many species, regardless of size or durability. Although the larger dinosaurs are often magnificently preserved in other fossil-hunting grounds, smaller, more delicate, but equally important species are almost always represented by only a few bits of bones and teeth. The Gobi not only provides us with the big and the bulky: great long-necked sauropods; the theropods, like *Tarbosaurus,* a powerful, ferocious cousin of North America's *Tyrannosaurus;* and tanklike, armored ankylosaur dinosaurs. This desert also entombs the richest and most diverse array of complete skeletons of less imposing vertebrate land animals from the Cretaceous. These include small carnivorous theropods, like *Velociraptor,* the agile Cretaceous killer featured in the book and movie *Jurassic Park.* There are many other smallish theropods from the Gobi, including species yet to be named and fully described. These rocks also provide evidence of bizarre early birds that show a close resemblance to forms like *Velociraptor.* The birdlike dinosaurs offer evidence for one of the most fascinating theories in modern paleontology—that feathered birds arose from the theropod branch of the dinosaur tree. Complementing this bounty of diverse dinosaurs is one of the world's best sources for dinosaur eggs. Indeed, dinosaur eggs were discovered in the Gobi in the 1920s, a find that surely ranks as one of the great paleontological triumphs of the twentieth century. This was not the first such discovery. Some fossil eggs from the French Pyrenees were attributed to dinosaurs as early as 1877. But the abundance of eggs, and possibly nests, juxtaposed with numerous dinosaur skeletons at Flaming Cliffs far overshadowed these early findings.

Other creatures even smaller, more fragile, and less commonly found elsewhere are also a gift of the Gobi. These include more kinds and better skulls and skeletons than the rest of the world's collection of ancient mammals. Some of the mammals are small enough to curl up on a teaspoon, and none were larger than a gopher. But these inauspicious vermin are part of our own heritage. They are the first known high-quality skeletons of the great group of mammals, the placentals, to which we and the other primates belong. During the dinosaur reign, eighty million years ago in the Gobi and elsewhere, these placental mammals kept a low profile. Soon after all the dinosaurs, except for their bird relatives, went extinct, the placentals branched and diversified into a great array of more modern forms as divergent as aardvarks, anteaters, and antelopes. These were lineages that also returned to the sea as whales and seals and took flight as bats. Ancient tree-dwelling creatures not so distant from the Gobi placentals represented early primates that eventually diversified into branches that included our own species.

The Gobi then is so enticing because it reveals biological empires in transition. The diversity of theropod dinosaurs and delicate bird skeletons tells us about species straddling the line between the extinct dinosaur groups and their living bird relatives. The mixture of both very early primitive mammals with the first glimpses of our own progenitors records the last phases of ancient mammal evolution and the rise of the modern mammal groups. It is difficult to find another place in the world where such transitions are so magnificently revealed.

A final factor makes the Gobi especially appealing to paleontologists. In the heyday of dinosaur hunting a century ago in the American West, prospectors encountered valleys and canyons where myriad skeletons were exposed like corpses on a deserted battlefield. Feverish collecting was so successful that many regions of the Rockies are no longer so generous in revealing their treasures. We still find great fossils in North America, but we look back in envy to our predecessors who roamed that virgin territory. As my colleague Malcolm McKenna has said about a newly discovered fossil site, "It's never as good as the first day you find it."

The Gobi, by this standard, still has a lot to offer. Although the Andrews expeditions of the 1920s and more recent Mongolian, Polish, and Russian expeditions have successfully mined many Gobi localities, it is surprising to find a wealth of fossils continually eroded and exposed at these sites. Even the cumulative activities in Mongolia over the past seventy years cannot match the intensive paleontological effort in less remote parts of the Americas. Moreover, the very difficulties of travel and ignorance of terrain that characterize the Gobi increase the possibilities for discovery of rich and wholly unexplored pockets of badlands.

In making this claim about the lure of the Gobi, I don't mean to undervalue or discourage the chances for great discovery in North America or many other regions of the world. As this book will show, our own triumphs are testament to the joy and possibility of discovery in a remote valley, around the corner of an eroded outcrop of rock, or over the next sand dune, even in an age when digital images of the earth from outer space can discriminate rocks a few feet away from each other. All our technology, our satellite global positioning receivers, our aerial maps, and our temperature-controlled, stereo-equipped, four-wheel-drive vehicles are no substitute for a lot of walking and looking, and an appreciable amount of luck. And this strategy doesn't just apply to a Central Asian desert. The Gobi is a great wilderness for exploration, but it is not the only auspicious hunting ground in the world. As I write this, there is fresh news of discoveries of dinosaur skeletons from the African Sahara, human remains from 4.4-million-year-old rocks in the Awash region of Ethiopia, and a magnificent cave full of 20,000-year-old paintings by Paleolithic artists that rival the Sistine ceiling. The last was found in the beautiful, appreciably populated, and familiar region of the Ardèche Valley in southeast France. Despite some trite claims to the contrary, the age of discovery is not over.

RETURN TO THE GOBI

The age of paleontological discovery began in the Gobi in the 1920s with the American Museum team led by Roy Chapman Andrews. This enterprise gave the world evidence of dinosaur eggs, new kinds of dinosaurs and ancient mammals, and a vivid sense of the geologic history as well as the present terrain and wildlife of the great vastness known as the Gobi Desert. By the end of the 1920s, however, after the conclusion of five expeditions, the American Museum was not allowed to continue its research. Andrews and his team did not return, nor did any scientist from Western countries for the next sixty years. During those decades the Gobi was assaulted by important paleontological expeditions—the Sino-Swedish team between 1928 and 1935 and the Russians in the 1940s, the Polish-Mongolian teams in the late 1960s and early 1970s, and thereafter a series of expeditions by the Russians and Mongolians. These efforts resoundingly demonstrated the incredible fossil wealth of the Gobi, in ways that even exceeded the early collections mustered by the American Museum teams. The evidence could be scrutinized by Western scientists in the literature, or studied in collections in Moscow, Warsaw, and, with some skillful maneuvering through the Mongolian bureaucracy, in Ulaan Baatar. Unfortunately, this opportunity did not include clearance for actual exploration of the Gobi. Mongolia was under the shadow of the Soviet Union and was compelled to adopt a xenophobic policy that blocked such entry from the West.

But important political changes would eventually open the Gobi to the world beyond the Soviet hemisphere. By the end of the 1980s the Soviet Union was in its final paroxysms, like an old dinosaur writhing in the burning sun of a Cretaceous desert. In the midst of this tumult Mongolia, a country of cheery, strong-willed, and charmingly blunt people, in January 1990 was among the first of the Soviet satellites to declare its independence. Within weeks, a delegation of Mongolians paid a visit to the American Museum of Natural History. A series of spotty communications and hasty preparations led to our visit to Mongolia in the early summer of 1990. The reconnaissance that eventually got us to the Flaming Cliffs that late June day was only the prelude to our larger, more ambitious, and more costly expeditions to follow. Along with my Mongolian counterpart, Dashzeveg, I would play the role of administrative leader, striving (sometimes with difficulty) to ensure the forward momentum at home, abroad, and in the field of a complex international expedition. Mark and Malcolm as well would provide leadership in many operations that, through the years, have involved more than a score of scientists from the American Museum of Natural History and the Mongolian Academy of Sciences.

The 1990 reconnaissance played a crucial role here. Even this brief small-scale jaunt would bring back enough fossils to demonstrate that the international collaboration necessary was well worth pursuing. The reconnaissance culminated in the discovery of an impressive sampling of fossils, including a new, large lizard, one previously unknown to science, that resembled the living Komodo "dragon." The fossils were found in the Nemegt Valley, not far from the treasure house that the Polish-Mongolian expedition of the late 1960s had dubbed Eldorado. Upon our return to Ulaan Baatar that summer we drafted a protocol for future collaboration.

The mission of the Mongolian-American Museum Expeditions was in much the same vein as previous efforts. Its basic goal doesn't sound very technical—to find great fossils. But beyond that basic need to fulfill a prospector's greed we had more specific goals. We focused on certain choice items. As I mentioned earlier, the Gobi seems to be the world's most bounteous cradle of evolution for particular groups—the theropod dinosaurs, to which meat-eaters like *Velociraptor* belong, and the Cretaceous-aged mammals that represented both the ancient lineages of mammals that eventually died out and some key antecedents to the lineages of the modern mammals that still flourish today. These represented primary targets for the expedition. We were also set on providing a better geological description of the Gobi, one that would more properly position the isolated localities of fossil vertebrates in a logical sequence in time as well as in space. Such geological information would be critical to reconstructing more vividly and accurately the kind of environments in which the dinosaurs of the Gobi lived and died. Armed with new technology—like the global positioning systems, or GPS, that received longitude and latitude coordinates from satellites—we could more precisely plot fossil sites than could previous Gobi expeditions. These plots could be superposed on detailed images taken from outer space, where remote sensing from LANDSAT and SPOT satellites produced images with incredible definition.

We could also bring new technology—by this time well developed in the West but only poorly so in Russia and its neighboring countries—to the study of the fossils themselves. By the early 1990s, I and other members of our team were collaborating with Tim Rowe at the University of Texas in applying fancy machinery like Computerized Axial Tomography, or CAT scans, to detailed study of the anatomy of fossil skulls. We could also work at institutions like Yale University's Medical Center, shoving in a fossil skull under the CAT scanner while a distracted hospital staff looked on with amusement and curiosity. This procedure would be very handy for detailed observations of the internal features of the skull around the brain cavity in important fossils like *Velociraptor* or tiny mammals. Details could be described, downloaded, and analytically measured without damaging the skull by breaking it open to view its internal architecture.

Ultimately even our more sophisticated scientific goals hinged on one simple but formidable accomplishment. It would not be enough to find new fossils at some of the famous localities worked earlier by the Americans, the Russians, the Mongolians, and the Poles. To ensure the really dramatic findings that come with virgin outcrops we would have to stake our own claim on new rich valleys of fossil treasures, places not previously known to earlier expeditions. Otherwise the successes of the expedition would be overshadowed by its legacy. Even our discovery of spectacular fossils at such sites as the Flaming Cliffs would be muted by the rueful recognition that we had not worked this locality at the acme of its wealth.

We had a chance for success on this front. Much of those 450,000 square miles in the Gobi had not been thoroughly explored—there almost certainly must be hidden crypts of fossil-engorged rocks somewhere in that vast wasteland. Our strategy for maximizing the chances for this discovery was to be highly mobile. We assembled a caravan of reliable Japanese four-wheelers as well as Russian jeeps and heavier Russian GAZ trucks. We traveled relatively light, packing for mobility, rather than establishing elaborate and long-term camps that were so characteristic of earlier expeditions. Thus we resisted the temptation to settle down for the summer in a previously known spot to remove a giant *Tarbosaurus* skeleton, impeding our chances for finding our own dinosaur graveyards.

Accordingly, the Mongolian-American Museum expeditions have logged thousands of miles over the past six summers. Each of these expeditions was carefully planned according to the amount of gasoline carried, the availability of fresh spring water, and the location of important destinations. Many parts of our central route—from Ulaan Baatar to Flaming Cliffs to Nemegt and back—have become more familiar to me than areas of Wyoming and other Western states where I served my apprenticeship in field paleontology.

In the course of those journeys we could tick off various successes. Famous sites like the Flaming Cliffs, Khulsan, and Tugrugeen Shireh were generous with important new fossils of dinosaurs and mammals. There was even the remarkable discovery in 1992 of fossils of a strange stubby-armed flightless bird, *Mononykus,* whose portrait appeared on the cover of *Time* magazine. But the early seasons, though successful, were shy of the discoveries we yearned for. Our sweep across a good chunk of the Gobi terrain had not revealed a new "Valley of Kings" for the dinosaurs, one to rival the discoveries of earlier expeditions. By the end of 1992 we were beginning to reflect soberly on the future of the enterprise.

But on a mid-July day in 1993, the last year of our initial agreement, our dreams were realized. In a forgotten corner of the Nemegt Valley, an unassuming cluster of low red-brown hills known to locals as Ukhaa Tolgod, we found our Xanadu—an incredible mass of skeletons of dinosaurs, mammals, and lizards—the richest site in the world from the late age of the dinosaurs. But Ukhaa Tolgod was not only astounding for the wealth of its fossils. The fossils found therein were extraordinary, unlike anything found elsewhere in the Gobi or the world. The treasures included nests with eggs encasing the first known embryos of meat-eating dinosaurs, new delicate skeletons of birdlike theropod dinosaurs, and aggregates of exquisitely preserved skeletons of fossil mammals not before known to science. And, as the fieldwork of the 1994 season further demonstrated, the locality revealed more yet—the mode of preservation of these remains gave us clues to one of the most extraordinary cycles of life, death, and burial ever recorded in the fossil record. Rinchen Barsbold, one of the Mongolian scientists who had spent many decades in the Gobi, remarked that the discovery of Ukhaa Tolgod was the greatest in dinosaur paleontology since the American Museum team found dinosaur eggs at the Flaming Cliffs over seventy years before.

What follows is the story of our journey to the Gobi, and our fitful route to the discovery of the dinosaur Xanadu. The story enfolds the paleontological lessons that can be appreciated en route. Ultimately, it is an account of adventures, difficulties, and more than a little luck—and how an unexpected triumph affected our own lives and altered an important chapter in the history of life.

CHAPTER 1

1990—JOURNEY TO ELDORADO

Then in the early twenties of this century, the Asiatic dinosaur rush began, shifting the attention of dinosaur hunters (at least for the time being) from the Western to the Eastern Hemisphere. And it began by accident. For the initiation of this new phase of dinosaurian discovery and research was a side effect of other activities—something unexpected, something that came as a very pleasant surprise indeed, and something that led to later expeditions and studies, the ends of which are still in the future.

EDWIN H. COLBERT.
1968. *Men and Dinosaurs.*

The evening of our first day at the Flaming Cliffs, we took the time to count our take. My first "proto" was not the only thing I found that day. I later discovered another pretty skull frill in the shadows of a gully; the jawline of the beast was highlighted by the afternoon sun, and the yellow enamel of the wrinkled cheek teeth glinted, as if this animal were appointed with a full set of gold caps. Bits and pieces of his foretoes stuck out of the sand one or two feet away. "Protos" were a common species in the eighty-million-year-old Gobi. These sheep-sized dinosaurs likely roamed in large groups. The squat proportions of their skeletons suggest an animal with lots of fat-laced flesh—probably a delectable culinary staple for meat-eaters like *Velociraptor.*

Mark found several protos too, as did Malcolm and Dashzeveg. We even found bits of dinosaur eggs. The dust in the Gobi air had dissipated late in the day, and by eight o'clock the cliffs were bathed in wondrous sunlight and etched by deep blue sky. It was probably not unlike the warm evening in 1922 when the cliffs were first encountered. They remain beautiful and brilliantly colored, their contours essentially unchanged. At the foot of the cliffs, in the center of the basin, is a small lake bordered by sand dunes, just as Roy Chapman Andrews described it. The green valley is spotted with a few brilliant white gers. On some afternoons the wind blows unremittingly and tornadoes of sand, as terrifying as those recounted in the 1920s expeditions, assault our exposed camp.

Yet some things have changed since the first American Museum team entered this valley nearly seven decades ago. The Flaming Cliffs lie only about forty-five miles west of Dalan Dzadgad, a respectably large town on the edge of the Gobi, complete with an airstrip. For some years an occasional Zhulchin bus, full mainly of Russian and Eastern European tourists, has bounced along on the flat, two-track road leading from Dalan Dzadgad to the rim of the cliffs. The tourists unload themselves, stand dazed in the heat, snap a photo or two, find some gully to use as a latrine, and return to the bus without much lingering. When prospecting the base of the cliffs in 1990, I occasionally spied bits of newspaper sticking out from under a pile of sandy rocks. *Pravda* had become the toilet paper of the Gobi.

By 1992 the Russian tourists had been largely succeeded by Japanese. The tourist camp where former U.S. Secretary of State James Baker stopped in 1991 had installed a karaoke machine in its nightclub. Nonetheless, the decussation of roads leading from the tourist camp to the Flaming Cliffs is effectively navigated only with considerable experience or a remote satellite receiver. Maps of the Gobi are still deficient. Roads and trails multiply, trying to track the nomadic settlements that drift through the years like shifting sand dunes. Elevations given on aeronautic charts, which we've double-checked with GPS readings, are often dangerously inaccurate. Large swaths of southern Mongolia are marked on these charts as "data not available." When Dashzeveg wrote me for a few items from the States—a rock hammer, a field book, a compass, and sundry—he ended the list, intending no irony, with a request for "a good map of Mongolia." Even as the twenty-first century approaches, getting lost is an acceptable and frequent event for a Gobi expedition.

Our second day in 1990 at Flaming Cliffs was uneventful—we found a few more protos, a few egg shards, but no treasured *Velociraptor* or gemlike mammal skulls. The place through the years had been scoured by the American Museum teams, then the Russians, Mongolians, and Poles. There were even a few plunderers, who, with tourist groups, sneaked surreptitiously away with bits of proto or dinosaur eggs (Mongolia, like many countries, has strict laws and harsh punishments for the unauthorized exportation of fossils). We call a site like this "played out," though by North American standards this is an absurd assessment. Every year of our expedition since 1990 has demonstrated that a stop, however brief, is well worthwhile. Yet the good dinosaur skeletons, egg sites, and mammal skulls are scarce; they certainly aren't anything like the cornucopia spread before the 1920s expedition. As Malcolm reminded us, "It's never as good as the first day you find it."

Dashzeveg agreed. Through thirty years of work in the Gobi, he had visited the Flaming Cliffs more than any other human alive, and was anxious to move on. The drivers had done their miracles with their Russian trucks. Our damaged transmission and head gaskets had been repaired with cereal box tops precisely cut and sealed with rubberized gasket compound, and the transmissions had been reassembled. Early in this second afternoon we struck out due south for the great jagged escarpment of the Gurvan Saichan Uul (Uul is the Mongolian word for mountain range). By that evening it would be fifty degrees colder in those mountains.

As we advanced through the village of Bulgan, Dashzeveg's truck, barreling ahead of ours, disappeared from view. After some desperate searching, the only fresh tracks we found led back in the direction we had come, toward the Flaming Cliffs. Damnit, I thought: the Gobi was not the proper place for a car rally.

I looked up and saw the northern horizon frothing with a tremendous dust cloud. One of those legendary Gobi sandstorms, our first, was hurtling in our direction. It was, in its terrible power, magnificent. The cloud itself had an orange glow, like the fiery columns rising from a battlefield. It stretched from Arts Bogd, the blue mountain fifty miles to the west, to the small volcanic outcrops some twenty miles east of the Flaming Cliffs, casting a great shadow, a curtain, of brilliant green.

The brunt of the storm soon hit us, but it was mercifully brief, having spent most of its force out on the plain. When the storm passed, the slopes south of Bulgan were highlighted by a late afternoon sun; every wet, mica-laced pebble glistened. We could see a shimmer of a windshield on the alluvial plain near the entrance to a canyon. Before long we were rejoined with Dashzeveg.

The afternoon warmth released the pervasive smell of sheep fat, which left a congealed layer on the canvas of the truck, clothes, even skin surfaces of our Mongolian companions. A piece of bacon in Mongolia is a brick of pork fat with one or two fibers of muscle tissue. Fat is never cleaned from a sheep carcass, it is instead simply thrown in the pot along with muscle, bone, and various entrails. The resultant mutton stew is the most common meal in the Gobi. Not surprisingly, Mongolian food, I discovered, was not to my taste. This is not the stew so delectably barbecued in "Mongolian hot pots" in various restaurants in the United States. That cuisine is derived from Nee, or Inner, Mongolia, the part of the Gobi and environs that lies in northern China. The food of Mongolia proper more closely reflects the influence of the hardy but primitive fare of Siberia than the great, flavorful cuisine to the south. Its main components are fat and grease. As a food review in the *New York Times* once stated, there is no word in Mongolian for cholesterol.

We slowly climbed up the long slope toward the canyon entrance. Beyond the mountains, still farther south, stretched huge tracts of desert cast in oranges and fiery reds, canyonlands of sandstones—dinosaur country—much of it never explored by paleontologists. These were places outside the navigational range of Andrews' party, with their Dodge motorcars and their fuel cans ferried by camels. They were places whose existence Andrews could only have imagined. As we approached the Gurvan

Saichan, I felt the thrill of first encounter and anticipation. Looking back from where we had come, the Flaming Cliffs had receded to a thin orange line.

The maze of Gurvan Saichan challenged both drivers. A number of passageways led to blind pockets. We stopped to ask directions from the few nomads we came across, only to end up in another dead-end canyon. Yet we slowly made progress through the maze. Each time we regained the main canyon we were a little closer to the crest of the range. We were gradually gaining elevation; the fields had become less broken by shrubs and took on the smooth silvery sheen of short alpine grasses and mosses. Magnificent steppe eagles exploded from the grass, making great circles in the sky as they searched for marmots and pikas.

We talked little among ourselves, enjoying the vista and the clear mountain air. Bayersaichan, or Bayer, the twenty-three-year-old son of Dashzeveg and our companion in the compartment of the truck, occupied himself with a book I had brought along, Stephen J. Gould's *Wonderful Life,* the story of earliest phases of multicellular life, complex animals and plants, as seen through the remarkable fossils from a mountain range in British Columbia.

At one point I asked Bayer, "What is Mad Max's real name?"

Bayer laughed, "It is Ganhuijag, or Huij for short."

The actual spelling of Max's name is something of a mystery to me. The truth is, however, that translating Mongolian names requires some creative liberties in spelling. Dashzeveg's prename (which in Mongolian is traditionally rarely used, even in casual conversation) is Demberelyin, but he himself has spelled it variously in his own scientific papers. Ulaan Baatar is also commonly listed on maps as Ulaan Bataar, Ulan Bator, or Ulan Baatar. One longs for the return of the original name of the capital, Urga, which seems less susceptible to variant spelling. Names on the few maps available for the Gobi are often not just misspelled; they are dead wrong. On one occasion, it dawned on us that our destination, Guchin Us, was mistakenly labeled on the map as Bogd. Had we asked a herdsman for directions to Bogd, we would have ended up a hundred miles farther away from where we intended to go, and across an impassable river from our next stop for fuel and fresh water.

Bayer continued, "Huij [I could now hear the *j* as a very soft *shh*] in Mongolian is a very good word, it means strength. It also is the word for the little sandstorms [those mini-tornadoes we often call dust devils]. But *huij* in Russian sounds like a word that means penis. Since Huij is a very common name in Mongolia it is embarrassing for us sometimes. I remember Mongolian students at Moscow University at roll call on the first day. When a Mongolian had to say this name, everybody laughed."

As we climbed, we cut our way through great walls of blackened volcanic cliffs. Although the cliffs looked charred, like an old sequoia tree from a long-ago forest fire, their deep black comes from the high basalt content, which represents fossilized lavas from archaic eruptions. As we cruised past the basalt ramparts, it was impossible to tell how many eruptions had occurred. Was this mountain range the result of one gigantic outflowing of lava? Probably not. There were places where the rocks had a jumbled, red cast, embracing sharp brown-colored fragments as well as bits of basalt. These seemed to be andesitic layers, products of particularly violent eruptions, like those of explosive volcanoes, such as Mount Pelé of Martinique, or Krakatoa west of Java, or the Vesuvius that smothered the town of Pompeii. The andesites were sandwiched in between the exposures of basalts—the "gentler" effusions of broad, low-shield volcanoes, producing great rivers of lava—the Mauna Loas of ages past.

The details of this history of fire and brimstone were not apparent to me as I bounced along in the back of Max's truck. Complex rock sequences like those in the Gurvan Saichan are a mélange and a mess. Faults cut through one rock unit and raise it much higher, above a section of the same age on the other side of the valley. Sometimes rocks are perversely overturned, so that the young ones are actually *below* the older ones. The Gurvan Saichan had inscribed in its steep walls a calendar of eons, but one whose dates and events were obscured by the tortured changes in the rocks.

Roy of the Flaming Cliffs

Our expeditions to the Gobi in June 1990 began, in many respects, nearly seventy years before, with the audacious expeditions of the American Museum of Natural History in the 1920s headed by Roy Chapman Andrews. The five Gobi expeditions, officially called the Central Asiatic Expeditions (C.A.E.), were carried out in 1922, 1923, 1925, 1928, and 1930. These are chronicled in Andrews' remarkable narrative, imperialistically titled *The New Conquest of Central Asia.*

Andrews was a serious scientist but not without the frontier flair that many allege made him the inspiration for the movie character Indiana Jones. As George Andrews remarked of his father, "That man knew how to shoot a gun!" Indeed, Roy Chapman Andrews claimed to have shot and (probably) killed from a great distance one or two brigands who threatened his field party near the Mongolian-Chinese border.

Andrews became a zoologist in his boyhood home in southern Wisconsin, scrambling among the grassy hills and thick deciduous forests, on the prowl for frogs and salamanders. When, as a young man, he came to the American Museum of Natural History in New York, he was hardly an individual of great reputation. He started low in the ranks, scrubbing floors of collection rooms and polishing tables. His more interesting assignments were trips to the eastern seaboard to deflesh the huge stinking carcasses of beached whales for the museum's skeletal collection. He was soon studying cetaceans (the name for the order of whales, dolphins, and porpoise) for the museum along the Asiatic coast of the Pacific. As his intrigue with Asia grew, he shifted his focus inland. Andrews claimed that his dreams of "conquering" the heart of Asia took form in 1912, inspired by Henry Fairfield Osborn's theories, published in 1900, on the role of Asia as the dispersal center for all mammalian life, including the roots of our own human species. Coincidentally, he was inspired by a building collection of strange new species of living mammals coming back from China to the British Museum of Natural History through the efforts of expeditions under the Duke of Bedford. Andrews' desires soon attracted the attention of the powerful Henry Fairfield Osborn, conveniently then president of the American Museum. With Osborn's blessing, Andrews launched the museum's venture to Asia in 1916–17, a project dubbed the First Asiatic Expedition. Small teams under Andrews penetrated Yunnan, southwest China, and the borders of Burma and Tibet in search of specimens. Important zoological collections were made, but Andrews chafed at the lack of time to do more and the absence from the team of important experts in fields like paleontology and geology.

In 1919 he launched the Second Asiatic Expedition for the museum. This time Andrews roamed far north into the steppes and mountains north of Ulaan Baatar (then called Urga). He brought back large collections of mammals representing three different faunal areas, but became particularly impressed with the potentials of Mongolia for broad-based study in botany, zoology, geology, and especially paleontology. Andrews returned to New York with the intention of mounting a large expedition to Central Asia with many different experts, an unprecedented reliance on motorcars, camel caravans of supplies, and a well-established jumping-off place from the populous city of Beijing. Osborn committed museum funding at about $5,000 as well as continued salary support for the museum staff involved. The America-Asiatic Association contributed $30,000 to the expeditions. Andrews was expected, with the help and advice of Osborn and other men of wealth, to raise the balance of funds. That effort was considerable; the initial price tag of $250,000 Andrews intended for the expeditions inflated to about $600,000 by the end of the eight-year span of the project.

Especially in his earlier professional years, Andrews' career was fueled by no little element of luck and good timing. But this should not detract from his talents. He was distinctly gifted in transforming his enthusiasm for nature into ambitious and expensive expeditions. The team he assembled by this time included specialists of greater scholarship than he—paleontologists such as the brilliant Walter Granger, who was really the scientific leader of the expeditions, and the geologist Charles Berkey; even Osborn himself contributed to the research of fossils from the Gobi. But Andrews was able to draw effectively on the collective power of those egos. Before the American Museum team entered the Gobi with their caravans of camels and Dodge motorcars, we had no notion that great dinosaurs once thrived in Central Asia; we didn't even know with absolute certainty that dinosaurs laid eggs (although such a possibility, based on fossil eggs from Europe, had been reported some years earlier). The Andrews team unearthed hundreds of dinosaur fossils, as well as the jewellike skeletons of small mammals, fossils that offer important clues to the early history of the group to which we belong. In Cenozoic rocks some millions of years younger, they uncovered the remains of a huge rhinolike beast, *Indricotherium,* the largest land mammal that ever lived.

It's hard to decide whether Andrews was driven more by scientific curiosity or by wanderlust. His passion for Asia and his talents for forming strong alliances with the residents are much in evidence. In the summer of 1990, a year after the tragedy of Tiananmen Square, I toured Beijing's Forbidden City with Mark Norell. At the top of the steps, near one of the great temples, we could see a large maze of tile roofs and walled courtyards stretching before us. Our Chinese student guide pointed to an impressive cluster of buildings beyond the city walls. "Andrews there," she said. With the blessings of the royal house, Andrews had set himself up comfortably in a coveted corner near the Forbidden City. Among the opulence of jade, gilt, and wood carvings, pampered by a large staff of servants, he prepared for the formidable journey to the gates of Mongolia. A more inviting base camp for an expedition could not be imagined.

The Central Asiatic Expeditions developed in ways not at all foreseen by its participants. Andrews and Osborn dreamed about finding multimillion-year-old humans, and the cradle of human evolution, somewhere in the secret crypts of Central Asia. This of course did not happen, even though the quest occupied much of the energies of the team. The failure to turn up ancient fossil humans recently prompted a historian to characterize the C.A.E. as a failure, despite their monumental, albeit unexpected, discoveries. Using this kind of argument, James Watson and Francis Crick, the codiscoverers of DNA, one of the most important findings in the recent history of science, would be discredited because the series of accidents and coincidences that set them off on a new pathway distracted them from their initial thesis work. This kind of revisionist nonsense violates both history and the essence of scientific enterprise. Anyone involved in the process of discovery knows that success is not always a matter of fulfilling expectations. Andrews and his team were not expecting to find a clutch of dinosaur eggs. Yet this does not diminish the importance of a discovery that shed light on the reproductive biology of some of the most fascinating creatures that ever lived.

On the first day of September 1922, Andrews and his colleagues were roaming indecisively on a vast, apparently unbroken plain just north of the Gurvan Saichan Mountains. Andrews stopped at a ger to ask some frontier soldiers for directions. The expedition photographer, J. B. Shackleford, hung back with the caravan of spindly-wheeled Dodge motorcars. To pass the time, Shackleford took a brief walk. Far to the north on the horizon he could see some volcanic hills that looked like islands floating in a sea of pink sands. As he walked in this direction, he saw an abrupt edge to the burnished grass, and a thin orange line beyond. He walked to the edge of the plateau. There below him extended a fantasy land of orange-red cliffs and spires. As Andrews later wrote, "Almost as though led by an invisible hand, he [Shackleford] walked straight to a small pinnacle of rock on top of which rested a white fossil bone." This was a skull of a parrot-beaked, frill-headed dinosaur, a year later named *Protoceratops andrewsi.*

What happened next was fairly predictable. "The tents were pitched on the very edge of the escarpment, and every available man of the expedition, native and foreign, went down into the badlands," Andrews wrote. "Quantities of white bones were exposed in the red sandstones, and at dark we had a sizable collection. However, Shackleford's skull still remained the best specimen, with the possible exception of the skull and jaws of a small reptile, found by [the team geologist, Charles] Berkey."

The place of this momentous event in the history of exploration and science was originally known as Shabarakh Usu, referring to an ancient name for the spring-fed valley with a shallow lake, long recognized by wandering herdsmen as a dependable source of water for their sheep, horses, and camels *(shabarakh* is the Mongolian word for mud; *us,* for water). The area is also called Bayn Dzak, which means "many zaks." Zaks or zaksaul, scientific name *Haloxylon,* are stubby, contorted trees that thrive in sand dunes near the lake. But the region is more notable for the chromatic sand cliffs that rim the basin on the south. In the early morning and late afternoon this escarpment truly burns with sunlight and Andrews' name for the place, "Flaming Cliffs," is widely used even by local paleontologists like Dashzeveg.

The specimens on that first day of collecting at Flaming Cliffs in 1922 included an enigma. Walter Granger found an egg which the team suspected belonged to a fossil bird. But this and other matters would not be clarified until the next year. One of the great ironies of the discovery of what Andrews was later to claim as "the most important deposit in Asia, if not the entire world," was the abbreviated time devoted to it. The team worked for only three or four hours on the evening of Friday, September 1, 1922. The next day they pushed northward over sandy, zak-choked terrain. Andrews was anxious to depart; the crisp mornings of autumn in the Gobi were already heralding the bitter winters of Mongolia. Not until July 1923 would the American Museum team return to this hidden valley of orange-red cliffs to find nests of dinosaur eggs and scores of *Protoceratops*.

The successes of the Andrews team were not limited to a plethora of "protos" and their supposed eggs. In 1923, on top of a "nest" of bits of dinosaur eggs, the team found the skeleton of a bizarre toothless dinosaur, which Osborn subsequently named *Oviraptor philoceratops*. That name is quite evocative: *Ovi* meaning egg; *raptor* meaning robber; *philoceratops*, which translates to a fondness for ceratopsians (the dinosaur groups to which *Protoceratops* belongs). The name was given because the skull was found lying directly over a nest of dinosaur eggs. As Osborn himself wrote, "This immediately put the animal under suspicion of having been taken over by a sandstorm in the very act of robbing the dinosaur egg nest."

The 1923 expeditions also uncovered the skull and jaws, one front claw, some associated phalanges (finger bones), and a hind foot of a small theropod dinosaur never seen before. The skull was no longer than seven inches, but it was bristling with sharp, recurved teeth. The expanded cranial vault suggested a large brain, and the distinctly capacious eye sockets were meant to cradle large and presumably alert eyes. The claw of the first digit of the hand was sharp and flattened like that of a falcon. The fossil was named *Velociraptor*, "the swift-robber," an agile, lean killer of the Cretaceous. But the nature of the beast would not be better known until many decades after the American Museum left the Gobi.

One of the most important fossils found at the Flaming Cliffs by the 1923 expedition was also one of the smallest, a tiny skull in a sandstone nodule simply catalogued by Granger as an "unidentified reptile." Preparation of this specimen back in New York revealed that the specimen was actually the skull of what was at that time one of the oldest known mammals. The new mammal did not cause quite the stir on the

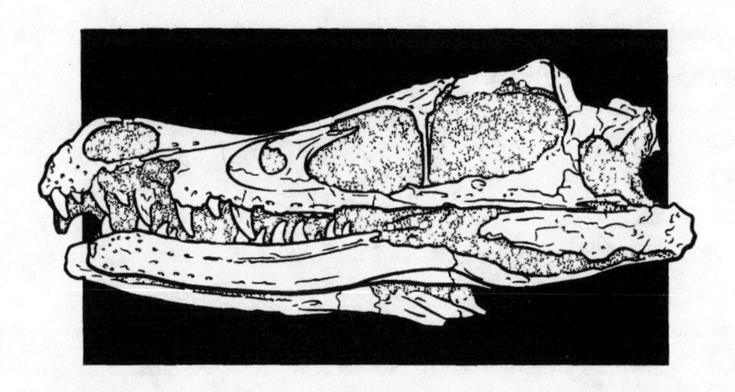

The skull of *Velociraptor* found in 1923 at Flaming Cliffs by the Central Asiatic Expeditions *(Ed Heck)*

publicity front as the dinosaur eggs, but it was, to many of the scientists back at the American Museum, more valued than any of the dinosaur finds. To this point in time only one other ancient skull of a supposed mammal was known *(Tritylodon,* housed in the London Natural History Museum, is now actually thought to represent a lineage closely related to, but outside of, mammals). These forms were of great importance to science because they told the beginnings of the story behind a world later dominated by large land mammals and ultimately the story behind our own origins. During the 1925 return to the Flaming Cliffs, Granger and three other members scoured the outcrops for a week and came up with seven mammal skulls, several of these with associated skeletal remains. Andrews, with characteristic fanfare, asserted: "It was possibly the most valuable seven days of work in the whole history of paleontology up to date. Those skulls were the most precious of all the remarkable specimens that we obtained in Mongolia."

The accumulated riches and accomplishments of the 1920s expeditions are extraordinary: dinosaur skeletons, dinosaur eggs, mammals, and lizards from the Cretaceous; a wealth of mammals from the early Cenozoic; unprecedented geological, zoological, and botanical surveys of the Gobi; and important archeological evidence for "dune dwellers," people inhabiting Central Asia thousands of years ago. Andrews and his team had successfully carried out one of the greatest (and, for its time, one of the most costly) scientific ventures of the twentieth century.

The Way of the Winds

When Roy Chapman Andrews and the rest of the American Museum team left Mongolia in 1928 they did not intend it to be their last trip. But troubles plagued the thorny negotiations for future expeditions. Permission to continue exploring was at last granted with strings attached. Andrews was allowed only to explore China's Inner Mongolia during the 1930 season. He had hoped that the demands made by Chinese officials who controlled entry to Mongolia would eventually be lifted. This hope for a renewed spirit of cooperation was in vain. Neither Andrews nor any other scientists from Western countries were to return until our expeditions some six decades later. Increasing xenophobia in China, political unrest, wars, and the suturing of Mongolia to the Soviet Empire blocked all such scientific opportunities. A few expeditions, like the Sino-Swedish team in the 1940s, probed the Gobi, but their activities were confined to the desert south of the Mongolian-Chinese border.

There was, however, one country with an important paleontological investment that did gain entry to Mongolia's Gobi—the Soviet Union itself. In 1941 the Scientific Commission of Mongolia issued an invitation to the Soviet Academy of Sciences for paleontological work. Following World War II, a crack Soviet team in 1946 began their reconnaissance in the Gobi. The principals included I. Orlov, a distinguished academician and authority on fossil mammals and reptiles, and Ivan Efremov, a paleontologist specializing in assorted tetrapods, particularly the synapsid line leading eventually to the ancestry of mammals. Efremov, who was assigned the responsibility of actually leading the Gobi expedition, had another distinction: he was a best-selling author of science fiction in the Soviet Union. His writings capture evocatively the adventures of the Soviet campaign in the 1956 book, *The Way of the Winds.*

The Soviets audaciously penetrated much farther into the white-hot core of the Gobi than Andrews was in a position to attempt, places where the winds blew harder and the sun was more scorching. This was the "outback," even to nomads. But the expedition was up to the demands of the terrain. The mobile caravan was appointed with some of the accouterments of the Soviet war machine, heavy-duty ZIL military transports (the acronym of the Zavod Imieni Likhacheva truck and limousine factory) and the ancestors of our expedition's beloved all-terrain GAZ trucks. Mounted on these monsters, Efremov's explorers ranged along nearly the whole length of the Gobi. After finding the important Cretaceous locality of Bayn Shireh in the eastern Gobi near the old route between Ulaan Baatar and the Chinese border, they revisited Bayn Dzak and the Flaming Cliffs. Then they turned southwest, finally entering the broiling isolated depression known as the Nemegt Valley. There the Soviets found a great wonderland of fossil vertebrates on a scale far beyond anything Andrews encountered. The Nemegt runs about a hundred and ten miles east and west and between twenty-five and thirty miles north and south. It is flanked on its northern and southern borders by mountain ranges reaching nearly 10,000 feet in elevation. As they moved slowly on the tough route along the long axis of the valley, Efremov and his team were overwhelmed by the garden of earthly delights: massive white, brilliant red, and in some places candy-striped badlands of fossil-bearing sandstones filled the valley floor and opened in huge amphitheaters against the flanks of the great Nemegt Uul and the blackened hump of Altan Ula. It was the grand canyon of fossil beds. The three-mile exposure of the Flaming Cliffs would be only a modest blemish on this red world.

The enticing beds of Nemegt proved to have a high yield. And the results, based on the work of three expeditions (1946, 1948, and 1949), were spectacular. They included nearly a dozen excellent skeletons of a *Tyrannosaurus* relative *Tarbosaurus,* massive skulls and ribs and limbs of hadrosaurs, duck-billed dinosaurs, and an armory of tanklike, spike-tailed ankylosaurs. Some of these were prepared and mounted at Mongolia's Municipal Museum in Ulaan Baatar, where they can still be admired (this institution changed its name after 1990 to the apolitical Natural History Museum). The Soviet team put the Nemegt on the map; it would be the sanctum sanctorum for future expeditions: Mongolians, Poles, and, much later, a new generation of Russians and Americans.

ZOFIA'S ELDORADO

In 1961 a new spirit of cooperation arose out of the Warsaw Convention, which brought together many Soviet satellites, including the Eastern Bloc countries and Mongolia. During the next year, the famous Polish paleontologist Roman Kozlowzki and other delegates from Poland met with Mongolians from the Academy of Sciences to establish a protocol for joint expeditions to the Gobi. Professor Kozlowzki, upon returning to Poland, asked a young energetic paleontologist named Zofia Kielan-Jaworowska if she wouldn't mind interrupting her studies of fossil marine invertebrates to head and organize the expeditions.

Invitation translated efficiently into action. The collaboration of Polish and Mongolian scientists, which included our colleague Demberelyin Dashzeveg, was a formidable record of achievement. In all, the joint expeditions comprised eight seasons in the Gobi: the initial reconnaissance of 1963, the full-scale expeditions of 1964, 1965, 1970, 1971, and the less elaborate but equally intensive scouring of the Flaming Cliffs in 1967, 1968, and 1969. Their explorations centered on an impressive list of old and new localities for Late Cretaceous vertebrates: Bayn Dzak and the nearby Tugrugeen Shireh, the Nemegt series, and farther west the beds of Tsagan Khushu, Altan Ula (the "Gold Range"), Kheerman Tsav, and the Sheeregeen Gashoon basin north of the Nemegt. The expedition even ventured far west to Trans Altai Gobi, near the lost mountains of Ederingian Nuruu. In its journeys, the expedition retrieved *Tarbosaurus*, hadrosaurs and sauropods, theropods like the dromaeosaur *Velociraptor*, a plethora of protoceratopsids, and spike-studded ankylosaurs. In addition they found numerous lizard skulls, and the *pièce de résistance* for Kielan-Jaworowska, scores of mammal specimens from Bayn Dzak, Tugrugeen, and a rich badlands the expeditions christened Eldorado located in the Nemegt Valley. It was a trove that even overshadowed the findings of the American Museum expeditions of the 1920s.

At the beginning the Poles, even with the collaboration of Mongolians, seemed an unlikely team for such triumph. Their leader lacked formal graduate training in vertebrate paleontology; Zofia Kielan-Jaworowska's early research work centered on Paleozoic trilobites and polychaetes. The assembled team was, by international standards, poorly equipped for fieldwork in the hot, arid fossil country of Central Asia. Team members with experience had concentrated on invertebrate fossil sites closer to home. As Kielan-Jaworowska herself observed, eight of the fifteen people on the 1965 expedition had never even been to a desert before. But these deficiencies were decisively overcome, as demonstrated by Zofia's scholarly work on Cretaceous mammals as well as the important papers on dinosaurs and lizards by other team members. By the time the expeditions were drawing to a close, the names of Kielan-Jaworowska and her Polish colleagues as well as those of the Mongolians, Dashzeveg and Rinchen Barsbold, were well known to the scientific world.

The Russians and Mongolians continued to explore the Gobi through the 1970s and 1980s, extracting many marvelous vertebrate fossils. Yet the closest a Western team came to the Mongolian sites was a Sino-Canadian expedition, which made important discoveries in parts of the Gobi Desert in China known as Inner Mongolia. The party, however, did not cross the frontier into Mongolia itself. Mongolia's Gobi remained forbidden territory to scientists outside the Soviet sphere.

By the late 1980s, however, the wall began to crack. In 1988, I was paid a visit by the distinguished president of the Mongolian Academy of Sciences, Dr. Sodnam, who cordially expressed his wishes that someday in the near future we might come to his country to work jointly with Mongolian scientists. A year later, in the summer of 1989, Barbara Werscheck, my assistant at the museum, relayed a strange message to the hotel I was staying at in Buenos Aires. "The Mongolians have sent a telex via the consulate in New York, wondering where you are?!" Apparently our Mongolian colleagues expected us to arrive suddenly at their doorstep, though there had been no messages, plans, or commitments.

Then, at the beginning of 1990, an unexpected turn of events transformed the politics of much of Europe and Asia, accelerated our communications with the Mongolians, and changed rather decisively the direction of my scientific career.

THE INVITATION

In January 1990, the first month of Mongolian democracy, a delegation of Mongolians visited the American Museum of Natural History to invite a scientific team to their country. Upon my return from a fossil-hunting trip to the Chilean Andes, I dispatched a message hand-carried by an oil geologist en route to Ulaan Baatar, the capital of the new democracy, to confirm details. (Despite the recent arrival of fax machines in Mongolia, a hand-carried letter is still the most reliable and often the fastest way to communicate with our distant colleagues.) The Mongolian Academy wanted us to come in March, now barely a month away. Because of the complexity of organizing such a trip, I suggested June instead, and asked for details on costs, schedule, and transportation.

My message finally received a response in May. A short trip to the Gobi could be planned and it would cost about $3,000 for food and supplies, transportation, and lodging for three or four of us. Naively, I sent flurries of telexes and letters by courier with the names of our team, suggested itineraries, and questions about available supplies. These were greeted with silence. Since then I have learned that, to a Mongolian, beauty exists in economy. A single reply will do; the rest will fall into place. We did, through the efforts of the Mongolian embassy, manage to get visas, and we proceeded as if all was in order. However, as Malcolm McKenna, Mark Norell, and I left New York for Central Asia in early June, it wasn't clear there would be any expedition at all. Our last message from the Mongolian consulate in New York was disquieting: "Sorry but we were unable to secure tickets for you on the Trans-Siberian Railway from Beijing to Ulaan Baatar. Good luck."

We did, through the help of our Chinese colleagues, manage to catch the Trans-Siberian Express. The train followed the Andrews route to Mongolia, over the steep mountains north of Beijing, through the Great Wall near Kalgan, where Andrews had also launched his caravan of motorcars. North of the mountains we struck out over the barren and dusty plains surrounding the sprawling industrial city of Datang. The cities were surrounded by low hills pockmarked with the graves and ruins of centuries. The train kept to a northward bearing, crossing tawny steppes before entering the Gobi itself. But there was nothing to see from our car—just some red sands and dust clouds. In the heat of the afternoon the train itself created its own dust whirlwind which penetrated the cracks at the edges of windows and doors. For the first time I experienced the parched, choking sensation which I would experience over the next several years of Gobi exploration. Fortunately, the heat of the Gobi Desert evaporated with the dying light of the sun.

The sun had long set when we at last reached Erlian, an unruly, bustling town on the Chinese-Mongolian border, where, over the next two hours, the cars of the train were elevated off the tracks by a system of huge cranes and pulleys in a dark hangar. An energetic Chinese crew, mostly women, noisily removed the Chinese wheel sets and replaced them with wheel sets that fit the Russian gauge of the rails in Mongolia. There are means other than a Great Wall to strengthen international security.

The next morning we awoke virtually out of the Gobi. The train took broad curves between grassy hills. There were a few stops at lonely, dilapidated towns that had the appearance of Siberian outposts, with old stone buildings that looked as if they were built by a tsar. By midday the Trans-Siberian Express reached Ulaan Baatar.

We were greeted at the station by an unfamiliar delegation headed by Davasambu, director of the Mongolian State Museum. Davasambu was a portly man, a member of the waning establishment, who resembled in appearance and manner the stereotype of a bureaucrat of the Soviet Empire. In his capacious but sparsely furnished office, a portrait of Lenin was set on the floor, propped against the wall. Our first morning meeting went unusually badly; we talked only of exchanging exhibits and protocols between museums. When I finally asked, after several hours, when we were going to the desert, Davasambu looked at me with the somber eyes of an old bloodhound. "But we have no responsibility for or knowledge of scientific expeditions!" he exclaimed. It was difficult to suppress our agitation. Across the table, I saw Malcolm's face redden as he broke a pencil in half.

It was not until the next morning that all was clarified. Our *real* colleagues from the Mongolian Academy of Sciences had also waited for us at the train station but were thwarted by the greeting delegation of museum bureaucrats. Apparently our flurry of messages, which had been choked up in the primeval communication lines for months, had at last burst into Ulaan Baatar. As it turned out, the belated news of our arrival had spread far and wide to several Mongolian institutes. A number had jockeyed for the prestigious role of collaborating on the first American scientific venture in over sixty years. In any case, before long matters were resolved. On the afternoon of June 22, five days after our arrival in Ulaan Baatar, we found ourselves in the back of a Russian truck, entrusted to the uncertain helmsmanship of Mad Max, bouncing over the rutted roads toward the Flaming Cliffs and beyond.

THE BITTER ROAD

Bayandalay, a small village just south of the Gurvan Saichan, had none of the charms of Bulgan, the spring-fed village to the north of the range near the Flaming Cliffs. The buildings of decaying plaster were pockmarked by sand and wind, like Cretaceous cliffs. The few people looked equally sandblasted and rather sullen, with more than the usual drunks on the streets, given the late morning hour. A *Lonely Planet* travel book on Mongolia includes the observation that some towns have so many drunks staggering about by midday that they look like the aftermath of a germ warfare attack. Bayandalay easily qualified for this unflattering distinction. Only the dogs seemed energetic and purposeful; they swept through town in packs, barking menacingly at the dust-caked children. Diverted by our two drab green Russian trucks, the curs descended upon us like a platoon of *Velociraptor*. As in many countries, it is best to greet a dog in Mongolia with your cab window rolled up. Roy Chapman Andrews, describing their savagery in the 1920s, warned that they would attack, kill, and even eat a human with little hesitation. This may be a bit of hyperbole to add to the drama of his tale of the Central Asiatic Expeditions, but there is a common and relevant salutation issued by nomads who come upon another ger. It translates roughly as "Hold back your dog."

"Life here hard. Town not too good," Dashzeveg said.

He and Max exchanged a few words.

"Father thinks it best if we wait outside of this town," Bayer informed me. A phalanx of street urchins came running up to us, yelling and waving their arms as we pulled away.

Dashzeveg stopped long enough for a run on fresh bread, which, not surprisingly, ended in failure. In Bayandalay, the bakery was only sporadically supplied with flour. Life was a bitter, hand-to-mouth existence. By contrast, the lives of the nomads—with their tentlike gers, their camels, sheep, plentiful meats, homemade bread, "milk vodka," and camel cheese (called "hrud" and pronounced just the way it looks)—seemed well endowed. I was later told by another Mongolian colleague, "Mongolians are nomadic people; we don't know how to live in towns or cities."

The road was now worse than ever, deeply rutted by rain, strewn with large sharp volcanic rocks, and veneered with a washboard surface that bounced our truck like a canoe in a cataract. The sharp lines of mountain ranges faded in the blowing dust. It was fiercely hot. We could see the shimmer of giant faux lakes on all horizons, the mirages of the rising heat in the inferno of Central Asia. The only objects of attention were the great columns of dust tornadoes, or huijs, phallic expressions of the forces of nature, which materialized down in the valley and drifted aggressively toward us up the valley flanks.

It was now my turn to serve in shotgun position next to Max in the cab. As the hours went by, I noticed Max grew very quiet, his eyes fixed straight ahead with an occasional glance down to the gear box. Streams of sweat ran down his brown cheeks. Clenching his teeth, he wrenched at the steering wheel with every hairpin turn in the canyon we were driving through. Then I heard, below the thunder of the truck engine, a high-pitched noise like the sound of a chisel dropping on a granite floor. Max stopped the truck and jumped out of the vehicle. He squatted below the front differential of the truck. Then he slowly got up and started sauntering back down the road.

After about fifty yards Max started bending down to pick up a bolt here and a small spring there. Farther on he spied a greasy black rod of carbon steel, brilliantly mirrored in the early afternoon sun.

"The steering stabilizer. We've had it now," Mark said.

Max stood there a moment, looking at the ground in some kind of deep reverie, like someone standing over the grave of a loved one. Then he walked back with the broken pieces to the truck.

"No way that can be put back," Mark said with authority.

Max looked rather wistfully under the truck, stuffed the stabilizer and components into his toolbox, and mounted the cab again. Mark, Malcolm, and I looked at each other and shrugged, following suit. Max revved up the engine and we pushed on. The truck could only be steered now with Promethean effort. Whenever Max loosened his grasp, the wheel spun wildly, like a ship's helm in a typhoon. Yet Max struggled dutifully forward, in hopes of catching Dashzeveg's vehicle, which had surged unmercifully ahead. Despite my frustrations with Max the last few days, I felt some admiration for his fortitude.

At last we reached the crest of a hill where the other truck waited. Dashzeveg frowned as Max related our problem. The scientist shook his head, uttering, "*Moh* [bad]."

We drove a little farther and stopped for a late lunch where the wash opened into a small valley. The drivers seemed cheery, almost as if the steering stabilizer had cured itself. But Max kept rubbing his arms in pain. No one ventured a question concerning the impact of the destroyed stabilizer on our plans. We all knew it was a long walk back to Ulaan Baatar.

"Russian trucks *moh.* Next year you bring Japanese *machina* [the Mongolian word for car or truck], okay?" Dashzeveg laughed.

The two drivers were soon under one truck and then the other, replacing bolts or wrapping cloths around leaky hoses. I was beginning to understand the weak link of the Soviet army's invasion of Afghanistan.

As we drove over Severy Mountain into the Nemegt Valley we passed a camp of mining geologists. The geologists had a GAZ 66 in dry dock, stripped down to its skeleton. Now there are probably no more ravenous scavengers of car corpses than Mongolians, and there seemed little chance that this piece of junk would have our part. But, to Max's joy, it did still have a steering stabilizer, and one in decent condition. A business transaction was swiftly concluded. Max danced back to his truck, stashed the newly acquired part for later installation, and we were on our way.

By the time we reached the town of Daus in the late afternoon the conditions were reasonably pleasant, warm but not hot, breezy, not windy, a few huijs in the distance. Daus (which also translated to Dabs or Davs) is the Mongolian word for salt. Appropriately, the town sits at the edge of a great salt flat in the lowest and most sizzling sector of the Nemegt Valley. The salt-mining operations had been shut down in recent years and the town seemed to be on the withering end of existence, just waiting for one of those huge sand dunes nearby to drift over and bury it. But the few people we met seemed gracious and reasonably upbeat. Dashzeveg indeed had old friends here, people he had known over twenty-five years. On the outskirts of town, near a well, we congregated with a lively bunch of locals, taking Polaroid photos of a regal elder man in his black bowler and his dehl, the traditional cloak with the satin waist sash. He smiled broadly, an impressive battery of gold caps, even on the front incisors, shining in the sunset, as he cradled his lovely granddaughter, a girl of four or five with an enormous pink bow in her hair. He looked as old as the Gurvan Saichan. Bayer remarked that he was indeed old in Mongolian terms, perhaps fifty-one or fifty-two. As Dashzeveg would say with some redundancy, "Life is hard here."

The visibility at the Daus well was excellent, and the landscape surrounding us was sublime. To the south was a corrugated range of dark purple and maroon rocks that were swirled into folds and arches, like a child's finger painting that becomes overloaded with primary colors. Malcolm noted that those beds were laid down in the Triassic Period, about 200 million years ago. A Russian team had reported some fossils from those mountains. Beyond this range and out of view was a large basin of Jurassic-aged rocks, about 160 million years old. As yet, this area was unexplored for fossils. Between those Triassic mountains and Jurassic basins lay more than a few miles of jeep-destroying terrain. During the forty million years or so separating the birth of the two rock layers, "Grand Central Asia" had been profoundly altered. In the Triassic, the Gobi was part of a single mega-landmass, a giant world continent. By the Jurassic, it was no longer part of such a cosmopolitan sweep, having been separated from the continental mass that dominated the Southern Hemisphere by a great equatorial ocean. In that span, the world had also changed in other ways. It had seen the end of a major ice age, the development of a warmer climate, the planting of stately pine-coned conifers, and the rise (both in height and the diversity of species) of long-necked sauropod dinosaurs.

After pondering this rock record a moment we shifted our eyes farther west, where the blue shadows of the Nemegt range marked the gateway to the red dinosaur-rich rocks of the Gobi Cretaceous, badlands too distant and too low in profile to pick out from Daus. To the north, however, there were a few pimples of orange-red sandstone, only a few miles distant, nicely lit in the evening light against a stark mountain range of black volcanics, the Gilvent (or Gilbent) Uul. I asked Dashzeveg about these outcrops.

"*Mitqua* [I don't know]," he shrugged. "Maybe next year we see, okay?"

Tomorrow we intended to prospect the maze of dinosaur badlands that the Polish-Mongolian expedition had named Eldorado. So we turned our backs on the inauspicious-looking outcrops north of town, which the locals referred to as Ukhaa Tolgod, or, plaintively, "brown hills."

It would be three years before we realized our mistake.

Two Red Hills

The next morning I felt a little uneasy. It was Wednesday, June 27, 1990, about midway through our brief reconnaissance of the Gobi. We had found some dinosaur and other fossils at Flaming Cliffs, seen some desert wonders, and experienced the excitement of travel, Mongolian style, in decaying Soviet trucks. But we would need to find something to register this adventure as more than an odd and very primitive tourist trip. We would have to find fossils, something unusual or impressive. Money for real future expeditions would be impossible without them. Dashzeveg seemed to understand this well and he felt the pressure. He was confident that success would come. Yet, even in the Gobi, a two-week jaunt hardly gave us a sporting chance. He knew as well as we did that those unreliable trucks could break down again and delay us, or that taking the wrong route through a valley or a mountain pass could eat into our precious time. He had an added concern. Dashzeveg was familiar with our scientific publications and our work in the field on other continents. But how would we do on his home turf? Would we have the right instincts, the right search image, to make a significant fossil find here?

Toward the Nemegt were some small hills of pink-orange. In the far western reaches of this terrain was a splotch of rainbow-colored rock. Cutting across the labyrinth toward this target took us hours; it was noon before the first pink outcrops looked within reach. But Dashzeveg for some unexpressed reason turned west straight through a formidable set of sand dunes. Eventually this route brought us to a gravelly plain. Not far from us at the base of the mountains was a small but impressive set of pink-orange badlands.

Upon reaching these sandstones, we gleefully descended the trucks for our first prospecting in two and a half days. The sands were friable—breaking into small chunks and dusty to the touch—and there were no small concretions. Dashzeveg warned us that these rocks were not optimal Eldorado quality: they looked somewhat different in the texture and color, what geologists call *lithology*. Nevertheless, Dashzeveg rightly felt the place was worthy of our attention. It was probably virgin outcrop, not previously visited by paleontological teams. It was not unlike some sandstones in Wyoming where I have found an abundance of bone.

It was also completely barren. After an hour of looking, we found not one fragment of a shiny tooth or a *Protoceratops* finger bone. Perhaps a few days of undaunted searching might yield something, but it was not worth the effort. Dashzeveg and I looked at each other and simultaneously nodded toward the truck.

We headed west again, toward the rainbow badlands I had seen several hours before. These exposures would pop in and out of view, as we descended into or climbed out of the myriad gullies draining the mountains. Finally, we spotted a series of likely outcrops. In my field notes of that day I had written: "Fine-grained, pale-red sandstones with intercalated brown siltstones. SS [sandstones] are massive [that is, smooth and unbroken in surface texture] or horizontally- or cross-bedded. Cross bedding is tabular in the thicker SS. Siltstones vary in thickness from several to 50cm. Calcareous cementation represented by ovoid, resistant sandstone clasts within the softer sandstone matrix. Small pink or white, nodular, calcareous concretions more common."

My notes on this rock formation matched published descriptions of the notorious but ill-named "blank series" first discovered by the Russians, who failed for some mysterious reason to find much of anything in these red sands. But the Polish-Mongolian team found these beds rich with fossils. They named the sequence the Barun Goyot Formation. The name Goyot derives from the goyo *(Cynomorium sangaricum)*, a strange, phallic-looking plant that grows in the Nemegt Valley and other areas of the Gobi. The goyo is rooted erect in the soil, standing several inches to a foot tall; it resembles a cucumber covered in brown velvet. The general area of these Barun Goyot exposures was called Khulsan, whose northern and most intricately sculpted badlands were Eldorado itself. Dashzeveg kept heading south, away from them, toward two red hills, where we finally stopped.

Not more than ten feet from the truck we found a flat sandy surface littered with tiny bones, lizard jaws, fish vertebrae, and small finger bones of mammals. It was a delight to our eyes, even for the Gobi-jaded Dashzeveg.

After a brief rain and hailstorm, which turned the sandy substrate into an unwalkable sponge of mud, we were soon back out on the wet sand, swathing lizard jaws and finger bones in toilet paper and a few strips of masking tape. These small tape-wrapped nodules started to fill our cloth sample bags. Mark headed over into the adjoining gully, where he found a gorgeous little proto skull. He and I spent two hours carefully trenching the soft sand around the skull, before applying wet copies of *Pravda* to the bone surface. We then wrapped the skull and its newspaper cap in water-soaked plaster bandages, the kind used to set a broken leg, slapping extra globs of plaster on the smooth surface of our Cretaceous sculpture. Once the plaster was semidry, it was easily undercut with small chisels. Finally, the pivotal moment came. We overturned the plaster jacket so its soft sandstone underside was now on top. We covered the upturned side in plaster and moved on in the evening light toward the more northeastern red hill.

Heading up the slope, we could see Dashzeveg's footprints winding fitfully toward the more impressive crest of the red hill. The tracks disappeared in the broad wash that lay at the outskirts of Eldorado. Content to stay near our objective, we circled the red sand gullies and small pinnacles.

It wasn't long before Mark turned up a beautiful nest of several dinosaur eggs. I nabbed a delicate set of limbs of a small theropod in the friable sandstone. Bayer came around the side of the hill to tell us about a "really beeg old camel" whose scattered bones were bleaching on the hillside. It turned out to be a complete ankylosaur. The weathering suggested that this fifteen-foot creature had been exposed on the surface for many years; in more pristine condition, the skeleton would have been greedily removed by any field party that happened to pass this way before us. The top part of the skeleton had been shattered into small shards of armor plating and bone. But the limbs and feet were exposed at the surface and at least a large part of the skull seemed to be in good shape.

There are several species of ankylosaurs named from skeletons discovered by the Russians, Poles, and Mongolians on earlier expeditions. You could hardly distinguish these beasts by their size; they were all rather big and bulky, with adults probably ranging between fifteen and twenty-three feet long and weighing several tons. But they would be distinguishable by the patterns of their armor. One, *Tarchia,* has a very heavy and spiked-appointed skull, with many short spines in parallel rows over the back and larger spines sticking straight out laterally from the flanks of the body. *Saichania* ("the beautiful one") has intricate cobblestone armor plates that encircle large spikes on the back. The top of the head is plated but its skull spikes are relatively small and laterally directed. There is a bizarre set of studded collars in back of the head, and there is a thick layer of ventral (belly) armor. *Talarurus,* a deep-bellied ankylosaur, is not as well known from its armor. It likely had rather prominent spines on the back facing both up and to the side, and its expanded gut was probably not protected by ventral armor. *Pinacosaurus* was a more "slender" animal, which held its body rather high off the ground on erect solid limbs. Its armor plates were relatively small, pointed, and widely separated from each other. Each of these ankylosaurs had a tail ending in a blunt bony club that was powered by massive muscles extending from the hind limbs and pelvic region.

The teeth of ankylosaurs are rather blunt and low, with a fan-shaped profile. They were not predators. These animals probably fed on a variety of plants and perhaps insects and other small invertebrates. The expanded cavities in the snout indicate extreme development of the olfactory

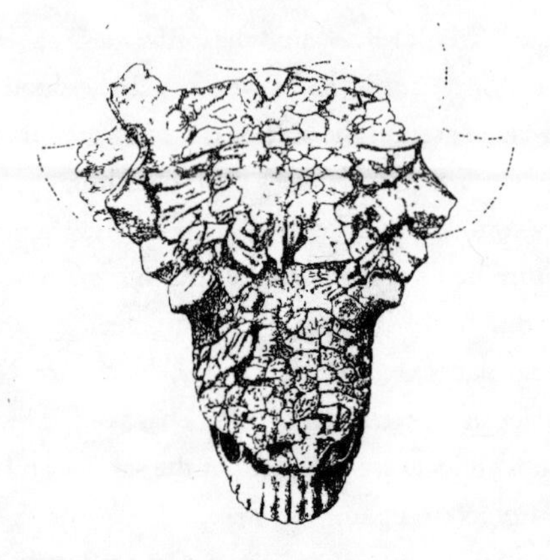

The skull of the ankylosaur *Pinacosaurus* *(reprinted with permission, AMNH)*

(smelling) centers. Jacobson's organs, pouchlike structures connecting the nasal passages to openings in the roof of the mouth, as in snakes, were probably highly developed. The hyoid bones which support the muscles for the tongue are also quite large. The animal may have protruded its long tongue, flicking it in and out, picking up tactile sensations, smell, and taste. Indeed, ankylosaurs may have foraged the way anteaters do today, although their menu must have encompassed a wide range of plants and "invertebrates." It is hard to imagine the outrageous decimation necessary to sustain one of these monsters on a strict diet of ants and termites!

Ankylosaurs were hardly defenseless. Restagings of Cretaceous times in North America often show a bull *Triceratops,* its horns pointed forward and its head shield slung low, staring down a charging *Tyrannosaurus.* But the large horned and shielded dinosaurs (ceratopsians) like *Triceratops* are not found in the Cretaceous sequence in Mongolia. In the Cretaceous Gobi, a similar standoff would pit *Tarbosaurus,* the Mongolian cousin of *Tyrannosaurus,* against an ankylosaur like *Tarchia,* except that it is likely that the encounter was head to tail rather than face to face. That tail club on *Tarchia* and other ankylosaurs could do serious damage. Moreover, this boilerplate backside did not offer an easy entry, even for a set of *Tarbosaurus* teeth. The ankylosaurs could simply crouch down to the ground, legs folded in, bellies protected, and become an impregnable pillbox (*Saichania*, with its extensive ventral armor, wouldn't even have to crouch). Finally, ankylosaurs could probably charge with their spikes bristling; trackways of North American species show that the legs were probably held directly under the body, instead of splayed out to the sides. This means that ankylosaurs could counterattack with momentum and speed. Certainly this kind of heavy tank defense strategy has proved its mettle. A strikingly similar approach is seen, millions of years later, in the glyptodonts, massive extinct mammals resembling (and related to) living armadillos. Glyptodonts mimic ankylosaurs in their heavy armor plating, from their heads down to their nasty, spiked tails. This meant a terrific defense against saber-toothed cats.

Mark and I carefully brushed away on the surface of Bayer's ankylosaur skeleton, exposing an array of back plates and the stubby toe and finger bones of the splayed-out hands. The generic identity of the skeleton was not apparent to us from the bits of skeleton we could see on the surface. As it was nearly dark, we resolved to jacket out the skull the next day. Our best day yet in the Gobi slipped away. Camp that evening was a relaxed and pleasurable affair: a small zak fire, some tea laced with pure Mongolian alcohol from Bayer's molecular biology lab, enthusiastic talk about future days of good prospecting. A brilliant orange-red sky mirrored the orange-red sands around us like a gigantic inverted lake. The color lingered long after the sun slipped behind Altan Ula. It was an evening worth savoring. Even the Mongolian mutton stew tasted pretty good.

THE KHULSAN DRAGON

Breakfast the next morning was brief. My field journal notes listed taking out Bayer's ankylosaur skull, removing Mark's egg nest, and jacketing my small theropod limb as things to do before hiking north to prospect further. But Mark and I had barely begun trenching around the ankylosaur skull before Malcolm came running with the news that he had found a theropod skull.

We dropped our trenching tools and ran after him.

Pinacosaurus (*Ed Heck, reprinted with permission, AMNH*)

Not far from Mark's egg nest, Malcolm had discovered an exquisite, eight-inch-long skull with a series of knife-edged teeth, exposed in a vertical slab of sandstone. Skilled preparation of the specimen at the American Museum two months later revealed it to be the skull of a wholly new kind of large predaceous lizard. We named the lizard *Estesia,* after our friend and colleague Richard Estes, the world's foremost authority on fossil lizards, whose death a year earlier was a great professional and personal loss to us.

Estesia is an important addition for understanding the evolution of the lizards. It bears resemblance to the varanids, the diverse group of predaceous lizards that includes the Komodo dragon from Indonesia. Interestingly, *Estesia* shows a series of canals at the base of the teeth that suggest this lizard was capable of injecting poison into its prey. This lethal weapon is not common to living varanids but is found in the Gila monster of southwestern United States and northern Mexico. Gila monsters belong to another lizard family, the helodermatids, and *Estesia* may be much closer to this group. Since our initial discovery, we have found fragments of *Estesia* in other sites where smaller lizards, tiny mammals, and dinosaur egg shells are common. Just as modern Komodos are noted for their voracious and wide-ranging appetites, we speculate that *Estesia*'s diverse menu included smaller vertebrates as well as small dinosaurs and possibly dinosaur eggs.

By noon, the jacketed *Estesia* was drying in the Nemegt sun. Mark and I took a brisk hike over the eastern wash to the outlying pinnacles of Eldorado, wandering about the ridges hoping to spot a dinosaur skeleton exposed on the surface. Our meanderings, however, were only mildly successful, although we enjoyed the spectacular maze of Eldorado, whose canyons stretched away from us in routes that would require a more extended future field season. Malcolm, though, was on a hot streak. On the cliffs opposite Red Hill East he found a partial, but impressive, skeleton of a fair-sized Cretaceous lizard. By 2 P.M. Mark and I had carried the partially jacketed ankylosaur skull down slope and finished the job with some poor-quality Russian plaster. Two hours later the trucks were loaded and we set off for Naran Bulak ("Sun Spring"), some forty miles away.

Dashzeveg felt that Red Hill East and Red Hill West, where we had made our discoveries, were probably new localities. We could not be completely sure of this, but no spot matching these precise coordinates and description had previously been published. In any event, *Estesia* had helped us make our mark on our first foray to the Gobi. The trip was no

Estesia (Ed Heck, reprinted with permission, AMNH)

longer just a pilgrimage to the shrines made famous by Andrews, Efremov, Kielan-Jaworowska, and Dashzeveg himself. We had retrieved something of value for paleontology. And in this accomplishment we had reached the outer walls of Eldorado. We would come back another year to take the city.

CHAPTER 2

DINOSAUR DREAM TIME

TIME AND DATE: AN AFTERNOON, THE LATE CRETACEOUS PERIOD; SOME EIGHTY MILLION YEARS AGO.

LOCALITY: A SMALL VALLEY, IN WHAT IS NOW CALLED FLAMING CLIFFS, IN CENTRAL ASIA.

The landscape is, on first inspection, familiar. A meandering stream empties into a pool of stagnant water. A cluster of small shrubs, alive with perennial flowers, dots a nearby hill. In the distance are a line of pink dunes that form the apron of a set of cliffs. The water surface is broken from time to time with the smooth profile of a turtle's back and, with less frequency, a crocodile's snout. Away from the water's edge, bees and wasps industriously work the shrubs and flowers. The grass rustles with the movement of small birds and tiny shrewlike mammals. There are lizards everywhere: small spindly-legged lizards with long tails and sharp teeth, blunt-nosed lizards with large crests over their eyebrows, and

(Overleaf) *Velociraptor* attack on a group of *Protoceratops* *(Ed Heck)*

massive wrinkled-skinned creatures looking very much like Komodo dragons. One of the big lizards breaks an oscillating trail toward the nearest dune as its sprawling body follows its flicking tongue.

A herd of four-legged Protoceratops *drifts ponderously in the shadows of the dune fields. The size of large sheep, their bodies are stout and square, like those of small, compact rhinos. Their oversized heads are appointed with a prominent snout ending in a hooklike beak and jaws and a distinctive shield that rises from the top and the back of the skull roof. There are thirty in the herd, migrating toward a drinking pool below a steep sand dune. Two of the larger individuals prance in circles, suddenly turn toward each other, and collide. There is the sound of scraping bone as the combatants butt heads. Amidst a choked chorus of snorting and howling from the herd, the dust rises as the two large males carry on the battle.*

But the scuffle between the shield-headed Protoceratops *is only the prelude to real disaster.*

The oasis in the dunes is a sanctuary, a place where the frilled creatures can congregate around a pool of water away from the heat and the aridity of the sand. They are not the only denizens seeking sustenance. Foraging clans of spike-tailed armored ankylosaurs hang close to various groves of trees. Several small-headed ostrichlike creatures speed across the flats near the pond. Beyond this lush vale, the beige and pink dunes stretch endlessly in the distance, broken in the north only by the shadows of blue mountains. The oasis in the sand is voluptuous with life, a seemingly safe haven from the rigors of the desert.

Distracted by the commotion of the battling Protoceratops, *the guards of the herd leave their posts. A line of black figures stand unnoticed at the top of the dunes, etched by the sun behind, completely absorbed by the riot of noise and writhing beak-headed bodies below. Taut and erect, they are perched on thin muscular hind limbs, their forelimbs set free off the ground. Their eyes betray an extraordinary single-minded concentration. The dragonlike heads of these* Velociraptor *are propped on serpentine necks that bend from their trunks in a short S-curve. Their one-inch recurved teeth, sharp enough to tear a floating piece of silk, make them formidable predators to virtually all but the largest and most imposing creatures. Traveling in a pack of five, they are unparalleled killers.*

The predators had been attracted by the noise of the protos. Suddenly they explode into motion, their huge leaps carrying them in an instant to the foot of the dune. They form a wedge flanking the largest pack member and cut a blurring swath through the herd, moving so fast that the frilled creatures barely have a moment to react. One of the hunters lands almost accidentally on top of a frilled creature. It rakes a long, cavernous incision with the enormous claws on its hind foot, virtually disembowelling its victim, but it does not stop. With the others, it converges on a hapless parent and young at the edge of the herd.

There is little struggle. The frilled Protoceratops *snaps its beak at the largest of the pack, but its flank is exposed to the slicing hind claw of another attacker. The victim is soon a carcass of blood, muscle, and bowels, ripped and shredded from five directions. The other frilled protos stampede out into the dunes, helplessly bleating cries of panic. The* Velociraptor *hiss and snap viciously at each other as they divide their feast.*

An Oasis in the Dunes

The stuff of dreams? Or nightmares? Perhaps. But this day in life and death, this vignette from some lost time and place, is captured with extraordinary detail by some of the fossil bone beds of the Gobi. The Flaming Cliffs of Shabarakh Usu appear to be the remnants of a varied landscape some eighty million years in age, not unlike parts of the Gobi today. There is evidence in the cliff face of ancient sand dunes. There are also places where the sculpturing of sandstones indicates what was once a meandering stream. Bits of turtle shells (even sometimes complete turtle skeletons) as well as crocodiles suggest the existence of ponds and lakes. The small shrewlike skeletons found at the foot of the Flaming Cliffs are among the most perfect and precious scientific clues to the origins of more modern mammals.

And, as the above story goes, there are other strange things here as well: the tanklike, spike-tailed ankylosaurs and the sheep-sized frilled *Protoceratops.* Though not known elsewhere in the world, *Protoceratops* is extraordinarily abundant in the Gobi Cretaceous. We call them Cretaceous sheep, because, as Mark Norell puts it, "they're what everything else ate."

But do accumulations of *Protoceratops* fossils really document the "social tendencies" of these animals? It is impossible to tell whether or not protos traveled in herds. It is true that their skeletons are found at the same rock horizon, indicating that a number of these beasts lived and certainly died at the same time and the same place. But were they social creatures? Did they sweetly graze in gathered herds? The abundance of juvenile skeletons representing various growth stages in proximity to adults suggests that families may have stayed together. Again, the matter requires some speculation, but the circumstantial evidence here is helpful.

Protoceratops is related to an impressive assortment of frill-headed creatures known as the ceratopsians. A number of paleontologists have suggested that the horns and frills were in many cases more important for sexual rituals and mating hierarchies than for defense. Many of the specialized ornaments would have been virtually useless against the attack of a ferocious predator, but they might have been very good for attracting mates or scaring away challengers of the same species. Moreover, fossil skulls of juveniles often lack horns and head frills; if such appointments were so crucial to defense they might be expected to develop earlier even in younger animals. Finally, some preliminary findings suggest that males and females might be distinguished by heads and horns. This possibility of sexual dimorphism—the blatant difference in form and size between males and females—also suggests an organized and aggregated social life for many ceratopsians.

Does this necessarily mean horns or frills had no purpose against predators? Not at all. In fact, there is dramatic evidence that *Protoceratops* were not mere sheeplike creatures when it came to defending themselves. As in many mysteries of the past, one or the other possibility—that frills and horns were primarily defensive or were ritualistically employed in highly organized mating behavior—cannot be cleanly excluded. Protos may have gathered in large herds to graze. The adults of one sex may have gathered harems of the other sex. They may have fought with each other or issued dominance signals in competition for mates. Or they may not have done any of these social things.

Less speculation is necessary to say something about the feeding behavior of protos and other ceratopsians. These animals had cheek teeth that were closely packed into tooth batteries. During chewing, the vertical outer surface of the lower teeth slid against the vertical inner surface of the

Protoceratops (Ed Heck, reprinted with permission, AMNH)

upper teeth. Chewing involved cutting and shearing rather than grinding. In the Cretaceous Gobi, the chopping was probably reserved for some tough xerophytic (arid-adapted) plants. On this count, as well, protos were certainly not sheep. They snapped and sheared great quantities of food, probably uprooting and cutting up tough shrubs or stunted trees. They were browsers, like rhinos and elephants, rather than grazers, like sheep. Some authors have postulated that the frill in *Protoceratops* and other ceratopsians enlarged the area of attachment and thus allowed the development of massive muscles for snapping the jaws against woody branches. This theory is controversial, because the surface features of the bone in the frill area suggest a skin covering highly infiltrated with small blood vessels, a situation not expected in areas of bone that attach massive musculature. It is likely that such muscles extended to the base of the shield. Nonetheless, the biting force of a *Triceratops,* a North American relative of *Protoceratops* with a head seven feet in length and a jaw nearly as long, must have been stupendous. A more awesome threshing machine cannot be imagined.

Velociraptor, in the imagined scenario above, is clearly the showcase fossil of the Gobi. It has been immortalized (and scaled up) as the infamous child-chaser of the movie and book *Jurassic Park. Velociraptor* and its dromaeosaur relatives, like *Deinonychus* from North America, were the apotheosis of the agile killing machine. In *Velociraptor* the skull is elongated and rather like a stock reconstruction of a dragon with razor-sharp, recurved (backwardly curved) teeth, each serrated fore and aft. Large eye sockets indicate keen eyes. The arms and hands, complete with three elongate fingers that end in formidable, hooklike talons, are designed for grasping, clutching, and slicing prey. Behind the shoulder girdle and the forelimb, a rigid horizontal backbone suspends a shallow and doubtless taut belly. The tail extends in a horizontal line with the back; it is reinforced by thin bony rods, extensions of articulations between the vertebrae. These were attached by muscles both at the top of the pelvic region and at the caudal vertebrae, keeping the tail stiff and straight while the animal was on the run. The hind limbs were, as one might expect, elongate and powerful, with a pronounced knob (called the greater trochanter) at the top of the upper leg bone, or femur, which may have served as a site for the attachment of a muscle useful in a "leg-kicking" action.

Perhaps the deadliest aspect of *Velociraptor* pertains to the hind feet. The foot has four toes, but the inner or first digit is small and spurlike and probably, as indicated by trackways of theropod relatives of *Velociraptor*, did not reach the ground. But it is the second toe that is the devastating weapon of *Velociraptor* and other dromaeosaurs. It is long and powerful with a nasty hook. The individual bones on this second digit are specially designed for flexibility. The toe can be retracted upward to avoid impeding the animal or damaging the toe while the animal is on the run. One can also imagine a leaping dromaeosaur, its killing claw coiled upward, ready to spring down and into the flank of a victim.

The bloodbath at the oasis witnessed above is not far off the mark as a recreation of the past, as powerfully demonstrated by a remarkable fossil.

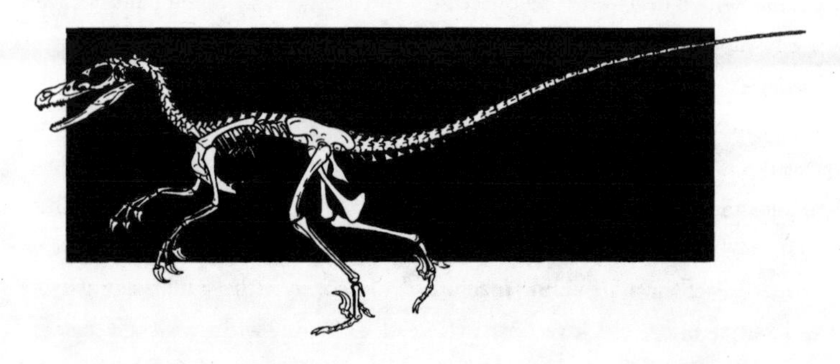

Velociraptor skeleton *(Ed Heck)*

Velociraptor hind foot *(Ed Heck)*

In August 1971 the Polish-Mongolian expedition found at Tugrugeen Shireh, a locality only thirty miles west of the Flaming Cliffs, lying in white sandstone, as if frozen in time, two nearly complete skeletons—one *Protoceratops* and the other, *Velociraptor*—locked in mortal combat. *Velociraptor* was preserved with its right arm in the mouth of the proto, desperately grasping at the face of the defender. The killing hooks of its hind claws were raised high against the neck of the proto, as if to slice open the skin and the vital carotid artery. This seems to be the case of a solitary *Velociraptor* attacking its prey. *Velociraptor* and other dromaeosaurs, however, are often depicted hunting in packs like African hunting dogs—a reasonable strategy, especially when the prey is a medium to large dinosaur. But the reality of pack hunting cannot be directly established. Nonetheless, the "fighting dinosaurs" from Tugrugeen show that their taste for *Protoceratops* is more than a matter of speculation. The rather poorly lit display of the "fighting dinosaurs" at the Natural History Museum in Ulaan Baatar belies the fact that it is one of the greatest of all prizes in vertebrate paleontology.

Velociraptor, Protoceratops, and other Cretaceous creatures seemed to thrive side by side despite the apparent rigors of their habitat. It is not likely that the Late Cretaceous Gobi represented by the rocks at Flaming Cliffs and other localities matched the extreme desert conditions known today in places like the Sahara. The Flaming Cliffs and other outcrops have a striking predominance of sandstones with distinct curved grooves, called cross-beds. These lines represent the surfaces of migrating sand dunes and windblown (aeolian) dune deposits. There are also clays, shales, and fluvial beds—indications of ponds or streams—sandwiched in between the aeolian deposits.

But there is also a frustrating absence of a critical piece of evidence. Despite its wealth of vertebrate fossils, there are essentially no plant fossils from the red sandstones of Flaming Cliffs or other closely similar sites. Although there are faint trace burrows of worms and other soil invertebrates, there is also a notable rarity of other ancillary evidence of life, such as trackways and fossil feces, or copralites (a footprint, probably of a small lizard, was identified by the Sino-Canadian team from the red sandstones of Bayan Mandahu, a locality in northern China). For the moment, we can only infer that the plants inhabiting these ecosystems were, with the exception of those populating the riverbanks and marshes, hardy enough to withstand appreciable aridity and seasonality. The trees and shrubs were perhaps more like the cottonwoods, creosote, and thorn scrub that opportunistically plant themselves in the water-deprived washes of dry regions in North America today.

Thus the fossils of the Flaming Cliffs and other Gobi localities do not perfectly mirror the recreated scenario of eighty million years ago that opens this chapter, but they do reveal an impressive amount. In succeeding chapters I will describe how the clues our expedition came upon enhanced this image of life and death in the Cretaceous Gobi. My efforts, though, are not simply a matter of painting gothic (or in this case Cretaceous) landscapes full of dark shadows and fearsome dinosaurs. Rather, it is important to relate the ancient echoes we discovered in the Gobi to the history and organization of life as a whole. To do this, we must put dinosaurs and other prehistoric animals in a proper perspective—in terms of both the creatures to which they are related and the time of their reign on earth.

Dinosaurs Rule

Exhibits in natural history museums have much to lure visitors—lost artifacts from Mayan and other ancient civilizations, meteorites, gems, and magnificent dioramas. But it is hard to imagine an item more popular than the dinosaurs. Dinosaurs are such a draw that museum directors chafe that their institutions, so enriched with great artifacts of all kinds for all ages, giant-screened IMAX theaters, and responsible exhibits about environments, floras and faunas, are often known by the public as "dinosaur museums." The label has some unfortunate connotations—a mausoleumlike edifice with old, bulky, extinct creatures, represented by dusty skeletons—a metaphor for obsolescence itself. Yet dinosaurs also signify grandeur, strength, ferocity, and the wonder of lost worlds. They have sex appeal. At the American Museum, a good proportion of its three million annual visitors ask guards and clerks, "How do I get to the dinosaur hall?" On the first weekend in June 1995 the museum opened its newly renovated dinosaur halls—embracing the world's greatest collection—after a five-year hiatus. Dinomania erupted. Over thirty thousand fossil-deprived fans stormed the museum that weekend, and the crowds continue to be massive. The onslaught was not totally unexpected. After all, one of the most popular Hollywood epics of all time, Steven Spielberg's *Jurassic Park,* based on the popular book by Michael Crichton, was not about horticulture, courtroom drama, or romantic intrigue.

Dinosaur fossils, like other echoes of ancient life, are discoveries of the science of paleontology. But dinosaurs have a special status that transcends their importance to science—people, especially children, love them.

Because I'm a paleontologist, people often ask me, why are dinosaurs so popular? The truth is, I have no idea. Perhaps they vindicate the wild, disturbing fantasies of childhood. Perhaps those great trusses of bony vertebrae, skulls, ribs, and limbs, like some filigree of massive bridgework, fire our early curiosity about the elegance and beauty of nature. Perhaps we are captivated by the epic sweep of this 150-million-year history, from its lowly beginnings to great dominance to dramatic demise (though this picture of mass extinction of the dinosaurs is not fully accurate: one branch of the theropod dinosaurs—the birds—survives today). But my favorite hunch is that some things are simply too big and too bizarre to be ignored.

This kind of answer does not always please my interrogators. Unfortunately, my passion for paleontology doesn't seem to endow me with insights into popular culture. I can say why *I* like dinosaurs. Like most kids, my passion for dinosaurs was kindled despite a limited access to them. Aside from the odd trip to the Los Angeles County Museum of Natural History (whose major attraction was the collection of mammoths, ground sloths, and saber-toothed cats from the La Brea tar pits), my ritual meeting with the dinosaurs were through books. The faded red cloth cover and green and black inkings of Roy Chapman Andrews' book *All About Dinosaurs* was the first of many thumb-worn titles. I vividly remember the sketches of skeletons in the American Museum of Natural History that were in the book, and the etchings showing paleontologists lifting huge dinosaur bones on pulleys, and piloting a caravan of motorcars through herds of wild asses on the Mongolian steppes. I remembered as well, of course, the scenes of life-and-death struggles among dinosaurs and other prehistoric beasts. My indulgences in these readings were not always appreciated by those around me. When a nun caught me with a book about fossils in class, I was dispatched to a sector of the room for mystics and dreamers she dubbed the "spaceman row." But the fossil world I had discovered more than made up for this indignity.

Like many others then, I was nurtured on dinosaur books and museum exhibits as a child, but my passions really now pertain not "just" to dinosaurs, but to the greatness and complexity of history revealed by many different kinds of fossils. Indeed, although I am still a dinosaur hunter in the field, my lab research largely concerns the precious remains of fossil mammals—the blueprints of our own heritage—that lived during the time of the dinosaurs and beyond. Dinosaurs are, true enough, one of the grandest phases of life's history, but it is hard to deny the sheer intellectual exhilaration of trying to disentangle the complex branching of myriad products of evolution, whether they be gargantuan dinosaurs or the tiny denizens of a 300-million-year-old coral reef. Indeed, many paleontologists bristle at the attention lavished on just a few chapters of this fossil history; paleontology isn't strictly a matter of dinosaurs, mammoths, saber-toothed cats, and early apelike humans.

In the spirit of fair treatment, and by way of general introduction, I would like to consider dinosaurs and other fawned-over creatures in relation to their actual places in the pageantry of life. When I invite candidates to join our Gobi expedition team, I assume they are familiar not only with the tools of the trade but also with the basic concepts that bring meaning to the fossil record. Readers, too, need a handle on some essential terms and concepts that anchor the science of paleontology.

THE NAME OF THE ROSE

To begin with, let me turn to the second most frequently asked question I get (next to "Why are dinosaurs so popular?"). Namely, when a paleontologist finds a bone or a skeleton in the field, how does he know whether or not he's found a new dinosaur? Conversely, how does he know if the bone belongs to a dinosaur already named? Of course, very often such a determination can't be made in the field at all. The fossil may be only partially exposed in the rock, failing to reveal the clues to its identity. In the field one must resist the temptation to poke around at the specimen before it can be protected, removed, and carried back to the laboratory for preparation. Eventually, whether in the field or in the lab, a paleontologist can venture an identification with some level of precision. Sometimes the fossil may be exposed enough in the field for a paleontologist to muster a pretty good identification. When I found that dinosaur skeleton at Flaming Cliffs, I suspected I had a *Protoceratops andrewsi,* even though only parts of the skeleton were exposed, because I could see several diagnostic traits—a parrotlike beak formed by a bone in the snout and the upward curvature of a plate—the telltale shield—at the back of the skull.

But in a way I was being rather glib about this identification. There are a couple of different species of Protoceratopsians. There are also a couple of different kinds of shielded, beaked dinosaurs, such as the Cretaceous Gobi forms *Bagaceratops* and *Udanoceratops,* that closely resemble and are closely related to *Protoceratops.* These animals all shared a characteristic look. They had a squat, stout body with short legs and a rather short, thick tail. Larger adults were about six or seven feet long and probably weighed roughly seven to eight hundred pounds, about as heavy as an adult tapir or a young rhino. The head was notably large in relation to the body and was appointed with a skin-covered (in the living animal, anyway) bony frill at the back and top. There is at least one rather incompletely known species, *Protoceratops kozlowskii*, closely related to *Protoceratops andrewsi*. The two species can be distinguished for the most part by subtle features of the bones that form part of the eye socket (orbit) and cheek region, although the specimens referred to as *P. kozlowskii* are juvenile and some of the differences might merely reflect growth stages of *P. andrewsi*.

In the field I didn't even think about mistaking *Protoceratops andrewsi* for *Protoceratops kozlowskii*. The latter is very rare—it is known only from a couple of skulls at a locality in the Nemegt Valley—and, as noted above, its validity as a distinct species is even questioned. In contrast, hundreds of skeletons of *P. andrewsi* have been found at the Flaming Cliffs. But this is not sufficient grounds for proof positive. Only a more careful study of the skull, taking notice of those diagnostic details of the eye socket, will confirm its true identity. The place of occurrence and abundance of a particular species are merely circumstantial evidence for separating it from other species. In the end, only the anatomical evidence is admissible. I was, however, on safer ground when it came to identifying my Flaming Cliffs skeleton as *some* species of *Protoceratops*. The closely related *Bagaceratops* has a much smaller frill and a more prominent bump, or "horn," on the snout. In the lab, the identification of *Protoceratops* might be refined, because this form can be distinguished by some curious small teeth that extend from the roof of the mouth in the premaxilla, a small bone near the front of the mouth cavity and the snout. The point is that zeroing in on a species may call for some rather exacting anatomical study.

We often amaze untrained companions with field identifications of isolated bits of bone. Mark will pick up a single recurved claw with a diagnostic groove running along its length and sing out, "Theropod!" Or maybe, if the specimen warrants, a more precise "Dromaeosaur!" or even "*Velociraptor!*" I might confidently identify one pectinlike tooth as *Protoceratops* or a small splint of jawbone with teeth no larger than peppercorns as

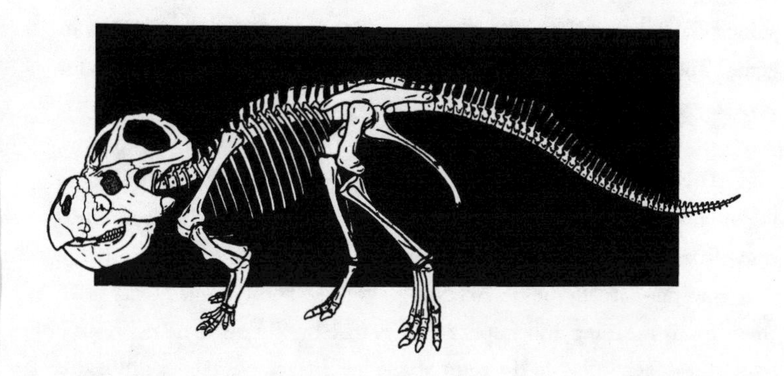

Protoceratops skeleton *(Ed Heck)*

a particular species of ancient mammal. "Look, Mark, I think it's *Zalambdalestes!*" But such identifications are only tentative—based on a bit of experience and a keen eye. There are also tricks to the trade. Mammals have highly differentiated teeth (think of your incisors, canines, and molars); most other things don't. Protos have those frills and beaks. Theropods have those recurved claws, and so forth. One sometimes needs very little to tell what a fossil is, if one has the right anatomical part.

In sciences like paleontology this anatomical know-how is combined with an understanding of the kinds of organisms the parts belong to. The process has to do with the rigorous, but not always perfect, application of the rules involved in naming things. The first lesson then—one that concerns not only dinosaurs but all present and past life—must involve our effort to appropriately name all the millions of biological entities, the things we call species, and put them into some kind of organizational chart. The fruit of this labor is biological classification. Building such a classification started many centuries ago with the Greeks, and took on new formality in the seventeenth and eighteenth centuries when people like John Ray of England and Carolus Linnaeus of Sweden started arranging life forms into lists based on very explicit characteristics from every part of the anatomy. Today the effort continues, fueled by new, more sophisticated theory, wondrous machines like electron microscopes, and computer imagers which probe and depict anatomical detail. Clues to the order of life can even be gained from the code of inheritance revealed by the structure of DNA in genes. Today we have some kind of foundation, some superstructure, for myriad named species—about 1.6 million names (mostly insects, by the way).

Although a noble activity and achievement, isn't this all rather boring? Who wants to pigeonhole a bunch of organisms and clutter our minds up with millions of arcane names? But what is often portrayed in such a dull way can actually be a fascinating venture. Those names carry with them much meaning and implication. A rose by any other name is *not* a rose, at least according to the rules of science. In fact, our names of various species of roses powerfully symbolize our understanding of a complex story of evolution, one that interweaves intricately with human history and horticulture. Each one of the names of the *Rosa* group tells us something different about its namesake; perhaps about its color, habitat, hardiness, and certainly about its family tree or genealogy. Likewise, the evocative *Tyrannosaurus rex,* a simple conjugate of some Latin words which translate literally as "king of the tyrant lizards," provides an instant picture, replete with images of skeletal structure, tooth size, evolutionary position, and predatory behavior. Names like *Tyrannosaurus rex* and *Rosa gallica* (the French rose) are given to an important unit of life, the species. Species, like atoms in the physical world, are the basic elements of life's diversity. They comprise individuals in reproducing populations that form distinct things—things that are not like other things.

Naming species, then, may seem straightforward; *Tyrannosaurus rex* is not likely to be confused with *Rosa gallica.* Oddly enough, biologists have a hard time dealing with species as a general phenomenon. There is a continuing debate over which are the best criteria for recognizing species. Some biologists favor the notion that species are best recognized as populations that are reproductively isolated (do not interbreed with members of other populations to produce fertile offspring). Unfortunately, most objects in the biological world are not like our familiar roses; we cannot examine directly their capacity for interbreeding. When it comes to fossils, that is certainly true. We cannot mate a *T. rex* with its Mongolian relative *Tarbosaurus bataar* in order to see whether it is the same or a different species. There are other complications. Some species—selected plants, marine creatures, bacteria, and one-celled protozoans—do not reproduce sexually. They simply clone copies of themselves. Yet they can be recognized for what they are: individuals of a distinct species, whether or not they choose to breed with other individuals of their kind.

How do we deal with this problem? At a practical level, we recognize different species not only by their failure to interbreed. We also base this recognition on our ability to discriminate them by any number of clues in a way that is consistent and explainable. Different anatomical parts, or genes, or behaviors all imply different histories and ultimately different species. For this reason, some biologists simply accept the self-fulfilling concept that species are kinds of organisms with their own distinct histories (they are unique lineages), regardless of how we recognize such divergent histories. Sometimes this spells trouble. One person's species is not another's. A species may be recognized for a trait that varies so much in its populations that it blends with such a trait in other species. We call those who get carried away with naming species on questionable differences "splitters." Alternatively, a species may be ignored because it has been only superficially studied. Some species, known as sibling species, are virtually identical in form but differ in the code of DNA that constitutes their genes. Such genetic measures have been used fruitfully in the cases of sibling species of birds, mammals, plants, and insects. In such cases, the boundary of a species is more distinct than originally recognized. We call those who tend to "overhomogenize" different species "lumpers."

The science of describing species and applying such names is called *taxonomy.* Taxonomists are in dire fear of being either lumpers or splitters. In reality they can be both, at least some of the time. No one is perfect. For instance, our difficulty in distinguishing *Protoceratops andrewsi* from *Protoceratops kozwalskii* may be a false problem. If the latter is only a juvenile form it may not be a different species at all, just one artificially "split off" from *P. andrewsi.* Scientists thrive on scrutinizing the conclusions of others. A taxonomist not only makes new designations but sees to it that other taxonomists get the names right. Out of this process of careful description, naming, and cross-examination has come some appreciation for life's organization and its incredibly baroque diversity. Thus there is some foundation for identifying a new bone as *Tyrannosaurus rex* or, alternatively, a new dinosaur. We also can be confident that those 1.6 million names in the taxonomic dictionary are a fair representation of life as we know it. Importantly, these named *taxa* are orders of magnitude short of the actual numbers of species past and present yet to be discovered—taxonomists suggest there may be as many as 20 million species, as yet unrecognized, hidden in tropical rain forests, soils, ocean bottoms or, as undiscovered fossils, in ancient rocks. Despite a few centuries of work, taxonomy continues to face an enormous task.

DISPATCHING DINOSAURS

According to some treatises, there are about 410 named species of dinosaurs. This is far short of all the dinosaur names that have stuffed the literature. Dinosaurs have been outrageously oversplit. A recent volume on dinosaur taxonomy edited by David Weishampel, Peter Dodson, and Halska Osmólska listed, in addition to valid species, about 209 *nomina dubia,* that is, species of dubious distinction. These rejects were originally described from inadequate bits of the skeleton, a part of a rib, a limb bone, or an isolated tooth. In a more formal sense, *nomina dubia* are cases in which the name is so poorly based that additional specimens are not likely to be assigned to them. Perhaps the most notorious of the dinosaur *nomina dubia* is *Scrotum humanum* Brookes, 1763 (in taxonomic papers, the author of the description and the date of the publication often follow the taxon name in this manner). *S. humanum* was earlier described by Richard Plot in 1677 from the distal end of the femur of what is probably a large meat-eating theropod. But the rounded ends of the paired femoral condyles, as the scientific name later bestowed evocatively indicates, were mistaken for the petrified genitalia of a giant race of antediluvian humans.

In addition to these *nomina dubia* there are numerous names that have been invalidated simply because they were preempted by earlier names for the same species. We call these rejects junior synonyms. The most famous of these is perhaps the name *Brontosaurus.* It is a shame that such a wonderful name—it translates literally to "thunder lizard"—is a casualty of taxonomic progress. The bones of *Brontosaurus* were found to be identical to those of a not so poetically but earlier named animal *Apatosaurus* (the Latin translates to "deceptive lizard"). But taxonomy can be a grim reaper. In total, about 500 *nomina dubia* and junior synonyms have been eliminated through the efforts of many dinosaur specialists, a great house-cleaning service to the profession.

A list of 410 species may seem like an impressive tally, but dinosaurs are certainly not a group of notable diversity. This is less than half the number of species of all living bats, for example. Even so, a list of a few hundred species without any more parceling and organization would be a chaotic mess. Fortunately, through the efforts of several valiant dinosaur taxonomists, these names have been sorted out into more inclusive groups. Identifying and dispatching these species to their groups is in a sense even more important than naming them. Group membership conveys many important facts, just as our surnames relate us to our families and clans that have a particular past and a particular geography and culture.

To appreciate the principles and practice of grouping things, consider the major lines of dinosaur evolution, in other words the major branches of the dinosaur tree. The tree has at the base of its trunk a mighty bifurcation. One of these massive branches twists and divides itself through the dinosaurs which include plate-backed *Stegosaurus,* armored ankylosaurs, and *Protoceratops* itself. The most telling feature of this group, known as the ORNITHISCHIA, is in the hip girdle, or pelvic bones. The relevant anatomy here is rather complex, but one familiar to many museum visitors. One of the three major bones of the pelvis, the pubis, is a long, strutlike element that extends downward from the region of the hip joint. In ornithischians, the pubis has one process projecting forward and one deflected backward, so the latter lies just ahead of the other ventral strut, the ischium. The pelvis is quite different in the other great group of dinosaurs, the SAURISCHIA, which include the flesh-eating theropods and the long-necked, massive sauropods. Saurischians have a pelvis in which the pubis projects forward and downward, so the lower pelvis has the shape of an inverted V. The differences are simplified in the accompanying figure.

These differences in the dinosaur hip region have to do with muscle power, posture, and gait. Instead of having splayed-out legs and a sprawling gait, like lizards, dinosaurs are capable of a more weight-bearing pos-

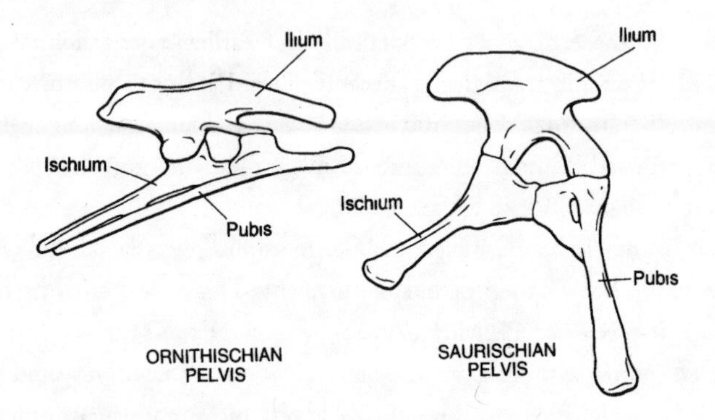

The hip (pelvis) in ornithischians and saurischians *(Ed Heck)*

---

ture, with the legs tucked in under their bodies. This means that the important and massive upper bone of the hind limb, the femur, comes into closer position with the hip, especially the pubic bone. Such an arrangement creates problems because it greatly reduces the space for big muscles that have to power the upper leg. Thus dinosaurs must somehow solve the problem of having a femur too close to the pubis. They seemed to have confronted this in different ways as demonstrated by the design of their hipbones.

The evolutionary story of dinosaur hipbones is then rather complicated, but not impenetrable. Remember that saurischians have a basic plan of the pelvic girdle that contrasts strongly with ornithischians: the pubis extends well forward and is broadly separated from the backward-extending splint of another upper pelvic bone, the ischium. This is not, however, a specialized and useful character for recognizing saurischians as a distinct group; it is likely the primitive condition found in the ancestor of all dinosaur lineages. For a useful derived character of the saurischians we must look to the forefoot. Here, the ancestral condition for the group is one wherein the foot is designed for grasping: the thumb is offset, the second digit is the longest of the hand. It may be hard to visualize this as the index character of saurischians; so many members—long-necked sauropods and flesh-eating carnosaurs, not to mention birds—have drastically modified forelimbs. Yet the blueprint for these redesigned hands and appendages can be traced back to the condition described here and is well exemplified in very primitive early dinosaurs like *Plateosaurus.* Another good saurischian feature is the elongate, highly flexible, and S-shaped neck that gives many of these animals their serpentine or dragonesque appearance. The sinew and length of this neck of course go to extremes in sauropods, but the neck can be radically and secondarily compressed as in *Tyrannosaurus* and *Tarbosaurus.* Saurischia contain a number of groups, the most familiar of which are:

The SAUROPODA—the gigantic defoliating "snake-necked" dinosaurs that we hold in such awe. The diagnostic sauropod characters include a string of twelve or more cervical vertebrae, each excavated with large cavities and openings, large nostrils well elevated on top of the skull, and several other traits. These features may seem rather esoteric for something as readily recognized as a hulking gargantuan. Yet these anatomical details are very reliable and useful for explicitly grouping the greatest of all land creatures.

The THEROPODA—one of the most fascinating and most passionately studied of all dinosaur groups. Theropods are, by lifestyle, primarily carnivorous, but their anatomical emblems are actually in the hind foot ("theropod" is derived from the Latin *ther,* beast; and *podes,* foot). The theropod hind foot differs from that of other dinosaurs. The first internal toe, if present, is small and stubby and generally does not touch the ground. Digits 2, 3, and 4 are elongate and "functional" toes, of which the central (third) toe is usually the longest and most robust. This tripod forms telltale footprints, useful for recognizing fossilized track impressions of theropods as well as tracks of their present-day members, birds. Theropoda contains an impressive sweep of taxa (the term for any entity, whether species or a group containing species, studied by taxonomists). *Tarbosaurus, Tyrannosaurus,* and the newly discovered *Giganotosaurus* are the heavyweights. An important theropod category is the MANIRAPTORA, the key group for the radiation of smallish, agile theropods that comprise forms like *Velociraptor, Oviraptor,* and the only group of living dinosaurs, the birds.

Starting back down at the lower trunk of the dino tree, we can climb the other limb, represented by the ornithischian dinosaurs. Ornithischia

Hind foot of *Tyrannosaurus (Ed Heck)*

make up a very distinct group recognized not only by the specializations of the pelvic region but also by a large number of advanced features in the teeth, skull, and other parts of the skeleton shared by its members. At this broad level, the group includes the armored ankylosaurs and their distant relatives, the plate-backed stegosaurs, as well as the hadrosaurian duck-billed beasts.

At one node of the ornithischian branches we at last reach the evolutionary neighborhood that contains *Protoceratops* and its relatives, like *Bagaceratops,* along with forms like *Triceratops,* some of the great rhinolike beasts of all time. These are the NEOCERATOPSIA. There is no doubt that neoceratopsians are a "good" evolutionary group. All neoceratopsians have dramatically large skulls in comparison to body length. The skulls of *Protoceratops* were larger on a relative scale, but some neoceratopsians are much larger in body size (up to 25 feet long). In absolute terms, ceratopsids had the most gargantuan skulls of any creatures that walked the earth (the skull of *Torosaurus latus* is 8.5 feet long!). In addition to big heads, this group has a distinctive emblem. There is some degree of expansion of frills at the back of the skull, giving the head the appearance of those helmets worn by Greek warriors.

There are also some dinosaurs, like *Psittacosaurus,* that seem to be closely related to neoceratopsians but lack a head frill. These forms can, however, be allied with neoceratopsians within the group CERATOPSIA on the basis of another important condition. In all ceratopsians the front of the snout ends in the curious downward-directed hook—formed from a special bone called the rostral—that is so reminiscent of a parrot's beak. One can see that in classifying groups we move quickly into a new language enriched with anatomical terms. The terms are not user friendly, and scientists as well as general readers rely heavily on pictures, which are worth thousands of words. I hope the accompanying illustrations will help elucidate our journey through the categories of dinosaurs and other life forms.

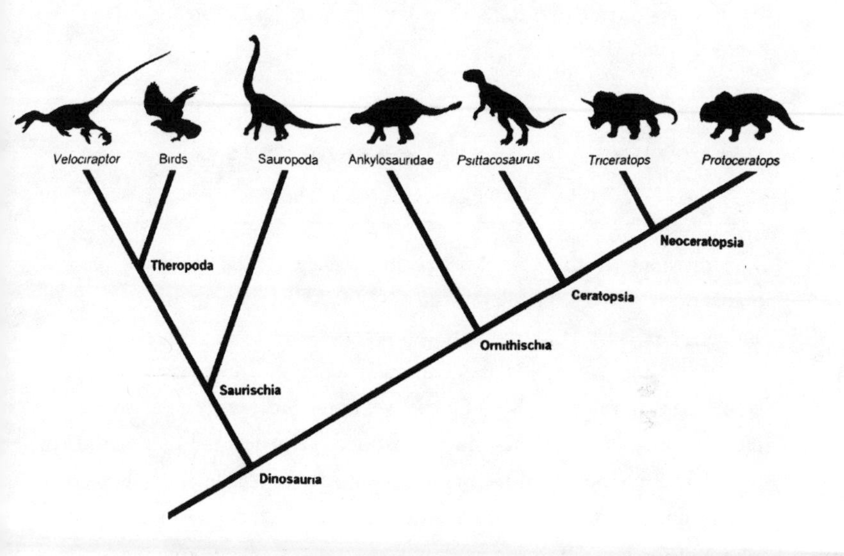

Cladogram for the major dinosaur groups *(Ed Heck)*

It is also important to understand the proper use of these anatomical bits of evidence. Earlier, we considered the problem of distinguishing species through any clue. Now we have a new problem. Grouping things together requires that we not simply recognize what distinguishes the group from others—we must also identify the traits that indicate new steps in evolution, or specializations. These traits evolved in the common ancestor of all the species assigned to the group. We might not even have found or identified this common ancestor, but we can predict that, if we did find it, it would have certain important traits shared by all its descendants. For example, we might say that the ancestor of all the species of Neoceratopsia had some kind of expanded frill at the back of the skull. These shared traits are extremely important. They indicate that a group is unique and has a single origin from a common ancestor, i.e., that a group is *monophyletic (mono,* single; *phyla,* branch). Relating species and their inclusive groups in this manner is a method whose formal practice is rather new to taxonomy, called *cladistics* (from the Greek *klados,* a branch). The product of this practice is a *cladogram,* like the one shown in the figure for *Protoceratops* and its dinosaur kin.

Dinosaurs . . . Are Vertebrates . . . Are Animals . . . Are Life

The divisions of the Ornithischia and Saurischia are the two major branches of that familiar level, the DINOSAURIA itself. Interestingly enough, dinosaurs as a group are less distinctive than the two major dinosaur divisions, as well as many dinosaur subgroups. The reason for this is the reality of evolution. Certain relatives of the dinosaurs very closely approach the basic conditions which must have existed in the dinosaur ancestor. The monophyly of Dinosauria is only demonstrable on the basis of a few subtle characters, the most important of which relates to the interaction of the femur and the hind limb. As noted above, all dinosaurs (including birds) have upright and erect posture with the legs shifted under the body for movement directly fore and aft. The legs are not splayed out from the trunk as in living lizards, crocodiles, and turtles. The hip region or pelvis is a vertical blade of bone. The articulation of the upper leg or thigh bone or femur is effected by a bony ridge at the top of the hip socket or acetabulum. The dinosaur acetabulum has a large hole, in unusual contrast to other land vertebrates. The latter have an acetabulum that is completely floored with bone and accepts the articulation of the femur. The pelvises of dinosaurs with the hole in the hip socket and the upper bony shelf for contact with the femur are therefore unique signatures of dinosaurs.

One may naturally ask, how does this "hole in the hip socket" help us identify a dinosaur when out in the field, say on a Gobi expedition? The answer is not very often. We can identify certain bones as dinosaurs by default. We might not have the diagnostic hipbones of a fossil but we can still recognize a frill as belonging to *Protoceratops,* and we know that *Protoceratops* is a dinosaur. If we found a more complete skeleton, we might note the ornithischian design of the hip, and ultimately note the giveaway "hole in the hip socket." More often, we don't have all the relevant parts of the skeleton. But the fact that an anatomical feature may not always be identified at the outcrop does not make it any less important. These characters, sometimes subtle and difficult to find, are the beacons for the pattern of evolution. They allow us to say a diverse assemblage of creatures arose from a common ancestor in the past, and that this event in the history of life was unique. They also allow us to place groups like dinosaurs or roses or humans and their primate relatives in their proper places on the tree of life.

That tree of course is so big and complicated that dinosaurs occupy only a small sector of its crown. Dinosaurs (including birds) along with crocodiles and fossil-winged pterosaurs belong to the ARCHOSAURIA. Archosaurs, lizards, snakes, turtles, mammals, amphibians (salamanders, frogs, caecilians, and many extinct lineages) are members of the TETRAPODA, referring to the origin of four limbs (later lost or grossly modified in some tetrapods, like snakes and caecilians) for walking about on land. Tetrapods along with fishes are VERTEBRATA, for those wondrous backboned elements that serve as the bridgework for our skulls, shoulder girdles, hip girdles, and their various muscles. Vertebrates, along with some curious little creatures like *Amphioxus,* share several basic features: gill slits (later lost in some vertebrate adults), a long strutlike back splint called the notochord, and an elongated nerve tube along the back. These are enough to recognize the supergroup (often denoted a phylum) CHORDATA. Chordates are just one of many lineages of mobile, consuming, and reproducing creatures that belong to ANIMALIA.

But animals are of course just one main branch of the tree of life.

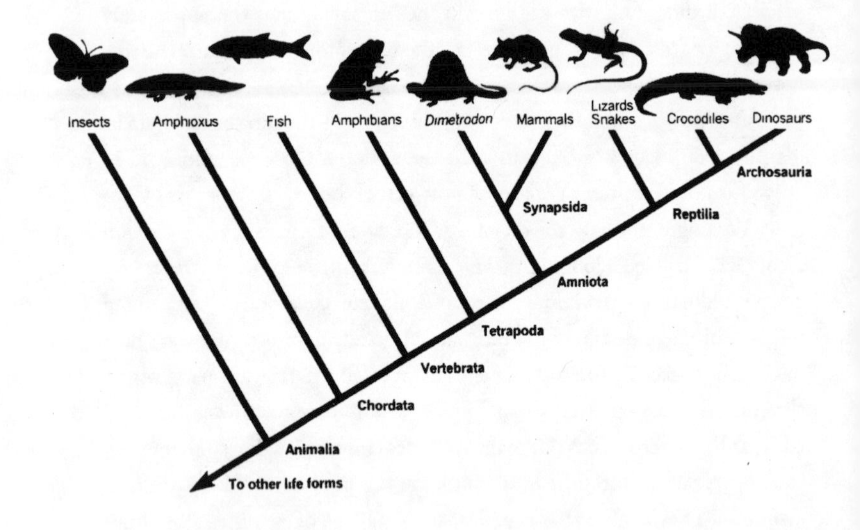

Cladogram for dinosaurs and other animals *(Ed Heck)*

Others include green, red, and brown algae, vascular plants, the fungi (mushrooms, slime molds, yeast, and lichen), one-celled protozoans, like the shoe-print-shaped *Paramecium* and the bloblike *Amoeba,* bacteria, and blue-green algae. These are the basic members of the largest group of organisms—LIFE.

The diagrams—cladograms—on these pages show how these major groups are arranged along evolutionary branches down to the smaller branches representing the dinosaur subgroups. Alternatively, we can spit back out the categories (I've actually skipped quite a few), *sans* the explanatory details, in the form of a biological classification. To do this we simply make a list from the "bottom down"; that is, we list the most inclusive group first and indent each group that represents its contents. Starting with the most inclusive group, LIFE, we can follow the branching pathway to *Protoceratops andrewsi.* In this particular list we will ignore, for the purposes of simplicity, the side branches—like amphibians, or saurischian dinosaurs—that do not contain *Protoceratops.*

LIFE
 ANIMALIA
  CHORDATA
   VERTEBRATA
    TETRAPODA
     ARCHOSAURIA
      DINOSAURIA
       ORNITHISCHIA
        CERATOPSIA
         NEOCERATOPSIA
          Protoceratops
           Protoceratops andrewsi

THE HIERARCHY OF LIFE

What I have related above in rank-order litany conforms to a venerable taxonomic tradition. We use the impenetrable phrase "Linnaean hierarchy" to describe this offset arrangement of groups within groups. It dates back to the great biologist Carolus Linnaeus, whose passion for and knowledge of the natural world, especially plants, was so acute and so precocious that at eight years old he was nicknamed "the little botanist." Linnaeus spent much of his productive career studying at the University of Uppsala in Sweden, where the garden of his delights is still maintained. He also cultivated a scientific following, and he spread the word about his organizational scheme for life, embodied in such great works as the *Systema Naturae,* published in 1735. The sheer enormity of his coverage had enduring influence. There is a Swedish saying which goes "God made the plants and animals, Linnaeus named them." We find the emergence of many group names in the works of Linnaeus; it would be hard to think about the divisions of life without them—Mammalia, for example. There were also some real charmers, anachronisms abandoned in light of a better understanding of real, monophyletic groups. Linnaeus' order *Beastiae* was an unlikely assortment—armadillos, anteaters, pigs, hedgehogs, and opossums—characterized by the great taxonomist as having long snouts and a tendency to grub for worms. We have long since put these unrelated forms in their proper and separate groups. Doubtless Linnaeus would not have liked some other recent developments in taxonomy. These ideas are not "anti-Linnaean" to the core but do seek to modify the use of names for categories in a way that gives one a better sense of the genealogy of related species. In a later chapter, I'll discuss these controversies, particularly with respect to some important problems, like the origin of birds.

Why this dustup over names and categories? Simply because we attach ever greater importance to their proper use. We seem to have a growing appreciation for phenomena—namely the true history of life—that the names represent. Taxonomy is the language we apply to the study and description of such patterns. But the actual investigation of the pattern, process, and history of life's diversity is called the science of *systematics.* Taxonomy is then the language of systematics. When paleontologists attempt to uncover the family histories of dinosaurs and other extinct creatures they are being systematists. When behaviorists recreate the evolutionary steps leading to the complex animal societies—for example, the caste systems of workers, slaves, and queens in different ant species—they are being systematists. When microbiologists ascertain which kind of bacteria was likely to represent the earliest form of life on earth they are being systematists. When molecular biologists reconstruct the patterns of descent of human populations based on changes in DNA they are being systematists.

Systematics is the magnet for these activities; it defines the mission for understanding our planet's life. It has a long and venerable tradition to the point that some have unfairly judged systematics to be old-fashioned. On the contrary, it's hard to think of another area of biology (including *paleo*-biology) that has undergone a more dramatic renaissance during the last decade. We now know how to unlock the secrets of the genes in 30-million-year-old fossils, and relate these blueprints to the evolutionary relationships of living species. The theory and goals that guide these studies are the same as those of a paleontologist hammering away at some exposed bone in the corner of a canyonland. Methods and tools are bringing us closer together. In the next millennium, as sciences hybridize like garden roses, we may have to rethink our labels for scientists and their pursuits.

THE EARTH CALENDAR

Like systematic biologists and ecologists who work on living things, paleontologists are passionate about the organized fabric of life. But we also have another dimension to reckon with: the axis of time. This can create some sticky problems, forcing us to untangle confusing stacks of rocks and geologic structures. The killer course in many geology departments is structural geology. One such infamous offering at San Diego State University was taught by a charming but sadistic Dr. Threet. Students were sent to a God-awful place, simply called in mock honor "Threetland," where the spur lines of the San Andreas fault system had done everything imaginable to the country rock.

To make matters worse, many of these rocks are hardly useful at all to paleontologists. Igneous rocks, those that form from the cooling of molten activity—whether the volcanics of the Gurvan Saichan or the massive granites of Yosemite Valley—hardly ever preserve fossils (though they are sometimes, as discussed below, very useful for dating rocks by measuring their radioactive minerals). Another class of rocks, the metamorphics, represent the recooked, squeezed, twisted, and deformed versions of older rocks. These are the byproducts of geological invasions by nearby molten rocks or the stretching or compression caused by mountains rising and continents slamming into each other. Metamorphics, like the marbles that once were limestones, make beautiful coffee tables, but their resident fossils become abstractions, wispy textures and colors misshapen by the forces of nature. The passage of time as seen through fossils is thus not accountable for a considerable chunk of the earth's terrain.

Fortunately there still remains a good deal of fossil territory because of the exposure of a third kind of rock. We watch these form, though very slowly, as the days pass by. Mud builds up on the spongy bottom of a lake. A beach is eroded by the action of the waves. A spit of sand forms off a coastline or in the point bar of the river. Occasionally violent floods create massive mudflows that carry with them great chunks of rock. The accumulation of these processes and their products result in sedimentary rocks.

These are the layers of rocks that are often stacked on top of each other, fine-grained limestones, dark, thin-bedded shales, clays, silts, and multi-hued, grittier sandstones. To get a layer of sedimentary rock thick enough to form a cliff face often takes significant time—sometimes thousands, even millions of years. Human experience is merely a blink in this timescape. Yet we can appreciate the process: "I'd swear, that sand dune was closer to the road when I was here ten years ago." "Wow, that last flood really gouged out this canyon!" And people who ignore the processes that reshape sedimentary rocks can end up looking either foolish or tragic: "That poor guy's house went right off the cliff in that last landslide!"

Paleontologists are scientists with time on their side. They must get into the habit of accepting the bizarre fact that a coral fossil in a road cut of limestone is actually millions of years older than some snails and clams a few feet higher in the same rock face. The Late Cretaceous Period that represents the age of the great bestiary of the Gobi Desert is a flourishing phase of the lost dinosaur empire. But the Cretaceous is just another day on the stone calendar. In order to appreciate this, at least in an abstract way, we must also take into account something nearly impossible to appreciate in any tangible way—the immense span of time that embraces life on earth.

THE TIME MACHINA

Imagine the Cretaceous Period—a span of time extending from 140 million years ago to about 65 million years—as just another twenty-four-hour day. Under this super-condensed calendar dinosaurs came and went over a period of roughly two and a half days. Life first came out on land a little more than five days ago, first populated the sea with diverse multiple-appendaged creatures a little more than a week ago, but first appeared as single-celled, bacteria-like bags of DNA appreciably earlier, more than a month and a half ago. On this scale, the 500,000-year history of our own species, *Homo sapiens,* is puny indeed; it is represented by about ten minutes. The actual, real-time numbers and historical highlights are as follows. The earth and its neighboring planets in the solar system date back to 4.5

billion years. Its bubbling and incandescent surface eventually started steaming; outgassing of volcanoes and molten rock produced atmospheric water through condensation. But this crust was devoid of life for roughly one billion years. The first life coalesced in some soupy pond enriched with amino acids (the chemical building blocks of proteins and all life) about 3.5 billion years ago. Most of life's early history is enshrouded in uncertainty; its clues are scattered and sporadic. A group of persistent paleontologists search for these clues in the odd outcrops of extremely ancient rocks in the heart of Australia, southern Africa, Canada, and elsewhere.

The first three billion years of life are a vast timescape called the Precambrian Era. Arguably most of the real action occurred during this span—the derivation of energy from nitrogen and sulfur compounds, then the harnessing of sunlight in the energy-producing process called photosynthesis, then the use of hydrogen in water for energy. This last process entailed the release of potentially destructive oxygen molecules—the oxygen that inflames a sulfur-loaded match head can essentially burn unprotected organisms, like some stinking, sulfur-using bacteria that lurk in the black ooze at the bottom of a swamp, a place essentially devoid of oxygen. Once the oxygen freed by organisms had saturated the iron-bearing rocks of the earth, it had to go somewhere else. It went up, infusing the earth's changing atmosphere. Once the air was oxygen-rich, cells evolved that could not only tolerate oxygen but could directly use it for respiration. The result was a great improvement in the machinery for metabolism—energy production and maintenance—that was necessary for the evolution of more sophisticated creatures, including, roughly 1.5 billion years later, the evolution of ourselves.

Now life had all its equipment in order: protective enzymes, photosynthesis, and aerobic respiration. Cells evolved more intricate and elegant architecture. Their DNA was enclosed in a membrane-bound nucleus, and their various components, energy batteries (mitochondria or chloroplasts) and sites for protein construction (ribosomes), floated in the cell plasm outside the nucleus. The whole system of soggy bags within bags was surrounded by a protective cell membrane. These were the eukaryotes *(eu,* Gr., "true"; *kary,* "nut" or "kernel")—our ancestors. From the start, eukaryotes were fully adapted to oxygen-rich environments. The infusion of oxygen in the atmosphere and the explosive radiation of new forms of life that followed together represent the crux pitch of the ascent of life, second only to the origin of life itself.

More mystery, more blank gaps in the record, and more sporadic clues. Finally, nearly a full three billion years after the emergence of life and one billion years after the emergence of an atmosphere with oxygen, we see the first evidence of a much more complex array of organisms, still all confined to oceans or marine seas. This baroque assemblage included species made up of many cells that arranged themselves into body parts—filmy or jelly-like umbrellas, eyestalks, jointed legs, tentacles, wormlike bodies, and harder plates of body armor. The event records, even in remote depths of our past, the bounty and diversity of the earth's biota, what paleontologist Stephen Jay Gould calls a "Wonderful Life," so well documented in the Burgess Shale of British Columbia's Rocky Mountains.

The explosion of multicellular creatures marks the beginning of a sizable chapter in earth history, the Paleozoic Era, that extends forward to a time line of 250 million years ago. The Paleozoic is subdivided into smaller time intervals called periods, beginning with that first pivotal phase called the Cambrian Period, and extending through several intervals—the Ordovician, Devonian, Silurian, Carboniferous, and Permian. The Burgess Shale may be the most exquisite, but it is not the first record of the Cambrian; it is dated at about 530 million years, a couple of geochronologic winks after the beginning of the Cambrian. The commonly accepted time for the onset of the Cambrian is 570 million years before present, although new evidence may push that date back a few million years.

Dinosaurs Emerge

The Paleozoic Era roared in like a lion and went out with a bang. The Cambrian explosion at the beginning of this span heralded the impressive radiations of the marine biosphere. The final period of the Paleozoic, the Permian, saw the sweeping expansion of coral reef systems, polar ice sheets, and continental deserts. The synapsids, like the fin-backed creature *Dimetrodon,* appeared. These represent the branches along the bough toward mammals. But the end of the Permian, circa 250 million years before present, was a catastrophe. Huge numbers of species of corals, moss animals (bryozoans), sea "lilies" (actually animals called crinoids related to living starfish), shelled clamlike brachiopods, and microscopic plankton were completely wiped out. Life, nonetheless, is remarkably resilient, as is documented by the recolonization of the blighted earth's surface during the next major interval of the earth calendar, the Mesozoic Era—the great span of time from 250 to 65 million years ago.

The first phase of the Mesozoic is called the Triassic Period, spanning a time roughly from 250 million to 210 million years before present. This is a crucial chapter—it offers the first glimpse of our targets of obsessive searching in the Gobi, the dinosaurs and mammals. In the case of dinosaurs, the actual emergence event is somewhat mysterious. Dinosaurs are part of a whole range of forms called archosaurs, where familiar lineages like crocodiles also eventually branched off. But the details of this story—namely which kinds of other archosaurs are clearly the closest kin of the dinosaurs—are not decisively known. It has been suggested that the nearest relatives of dinosaurs may have been some early forms of the winged "flying reptiles," the pterosaurs. Thus dinosaurs might be rooted in the unknown ancestor that also gave rise to the pterosaurs.

The first dinosaurs were less dramatic creatures than their successors—smallish, bipedal theropods like *Herrerasaurus,* described by the paleontologist Paul Sereno, or the ornithischian *Pisanosaurus.* These occur in rocks about 228 million years old, in the later phase of the Triassic, from the canyonlands of Argentina. A few million years later we encounter some interesting Triassic dinosaurs in North America—an early branch of theropods, the ceratosaurs, are again rather generalized creatures. These include the famous Triassic animal *Coelophysis,* whose abundant remains have been collected by American Museum teams under Dr. Edwin (Ned) Colbert. Ceratosaurs may not represent a true clade, although experts on the group contend that they can be recognized by the considerable scouring of the cervical vertebrae, the fused ankle bones, and sundry features of the trunk vertebrae and pelvis. Other Triassic dinosaurs include the long-necked prosauropods, like *Plateosaurus.* The footprint record also demonstrates that Triassic dinosaurs were numerous and diverse.

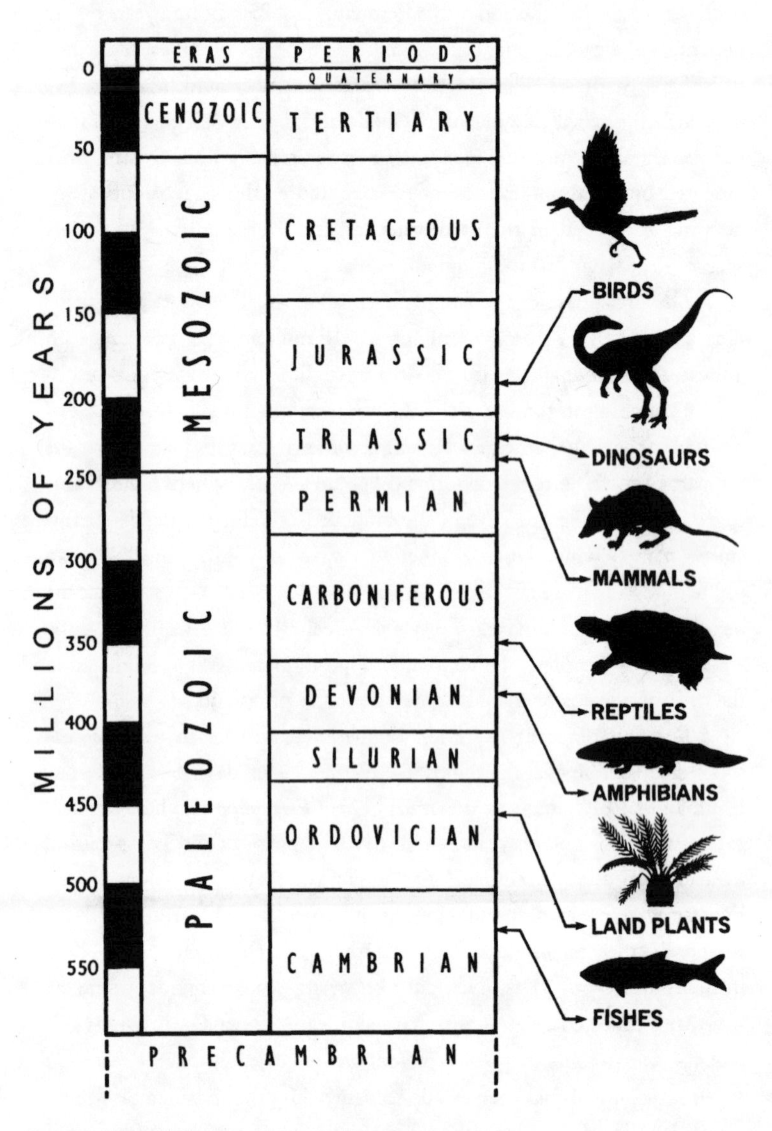

Earth Calendar calibrated in millions of years *(Ed Heck)*

Coelophysis (Ed Heck, reprinted with permission, AMNH)

Meanwhile, in an even less auspicious fashion, the relatives of our own group, the mammals, were emerging. Unlike the case of dinosaurs, a fairly precise array of branches leads up to the mammalian line. These are taxa known as synapsids, forms that superficially resemble dinosaurs but evolved special features of the skull and skeleton in a drastically different fashion. As noted, the fin-backed *Dimetrodon,* a denizen of the Permian, is an early synapsid. These were rather large creatures, several feet in length, with an imposing back fin and powerful jaws armed with large, sharp teeth. But the first near relatives of mammals were very small. The earliest such animals were tiny shrewlike forms, the triconodonts, that appeared some 200 million years ago, during the Triassic Period, when prosauropods were roaming the earth. The triconodonts were augmented by a number of later mammalian clades during the succeeding phases of the Mesozoic. Many of these Mesozoic "experiments" waned and died out before or at the time when the non-bird members of the Dinosauria went extinct. So mammals and their forerunners go way back; a fact that surprises most. They are indeed virtually as old as the dinosaurs.

The next chapter of the Mesozoic is marked by the beginning of the Jurassic Period, at about 210 million years ago and lasting for about seventy million years to the beginning of the Cretaceous. The Jurassic marks a fundamental change in the dinosaur environment. Seed ferns that dominated the Triassic landscape were on the wane, and the lavish fern gardens of the Late Paleozoic coal forests were no longer so baroquely flourishing. The dominant Jurassic plants were the gymnosperms, plants that carry their seeds exposed in cones to be fertilized by windblown pollen. They began as the palmlike cycads and the heart-leaved ginkgoes, which endure today in greatly diminished numbers of species. The gymnosperms culminated in the conifers—the group of pines, larches, firs, and sequoias—whose dominance was secured by later Jurassic times.

Perhaps this transition in gymnosperm evolution had something to do with changing reproduction. Cycads and ginkgoes carry their seeds and pollen cones on separate plants; even if the pollen lands close enough to the female ovules, the sperm are forced to "swim" in water that surrounds different plants. Conifers, by contrast, build pollen and seeds into two different kinds of cones on the same tree. In many species, the "male" (staminate) cones are carried lower on the tree than the "female" (ovulate) cones. The pollen is strongly encased and aerodynamically designed. It is transferred by wind, not water. Moreover, the pollen grains grow a tube that brings the male sperm straight to the ovule. Conifers thus efficiently expanded to a variety of habitats. Cycads, on the other hand, became more restricted through time to tropical and subtropical areas, where water tends to be more abundant and seasonality less marked.

Thus the Triassic scene was radically altered by Jurassic times. Jurassic forests had big trees—pine, spruce, redwoods that stretched to the sun—and big animals to match, the largest of all dinosaurs, the earth-shaking long-necked sauropods, like *Apatosaurus, Barosaurus, Seismosaurus,* and *Brachiosaurus.* These were of ultimate mass for load-bearing landlubbers—adults may have ranged between twenty and ninety tons. Among the branches of the trees were early experiments in flight, dinosaurs that took on the features that anticipated their evolution into feathered birds. Not to be confused with these fliers are a whole array of winged pterosaurs which, as mentioned above, represent a lineage of archosaurs possibly closely related to the dinosaurs.

Despite these dramatic changes in geography and environments, the Jurassic world still held on to some archaic trappings. Flowers were not true flowers. There were no angiosperms, no hardwood forests, no grasses. Beetles may have been major plant pollinators but some cycad cones look

Apatosaurus, a Jurassic sauropod *(Ed Heck, reprinted with permission, AMNH)*

as if they were constructed to repel such insects. For a number of uncertain reasons, the final period of the Mesozoic, the Cretaceous, which began some 140 million years ago, marks a dramatic change in all this. Although we are riveted by Cretaceous dinosaur life, perhaps the most important environmental and evolutionary event of this time was the emergence of modern land flora—one that still persists today. Conifers were complemented by deciduous leaf forests and magnolias and other members of the richly varied angiosperms. The insect group Hymenoptera—the ants, wasps, and bees—dramatically diversified. If not challenging beetles in species numbers, the Hymenoptera staked their claim as the dominant pollinators, and developed ecological roles and social systems of unprecedented complexity.

The transition to the Cretaceous is marked by other shifts in dominance. At the beginning of the Cretaceous there was a great rise in diversity of the coiled ammonites and a variety of new kinds of microscopic sea algae and other creatures, as recorded in the hard tests in the Cretaceous limestones from ancient ocean bottoms. On land, long-necked sauropods persisted, but in some regions, like North America, they diminished in numbers as several other dinosaur groups diversified. These interlopers included the shielded and horned ceratopsians, tanklike ankylosaurs, duck-billed hadrosaurs, and the varied agile theropods that culminated in killers big and small, like *Tyrannosaurus* and *Velociraptor.* The ground-hugging mammals, which had thrived throughout the Mesozoic, remained small and rather limited in adaptations. But the Cretaceous was a particularly critical time for our own heritage; tiny shrewlike forms in the Gobi and elsewhere first appear, the forerunners of the modern mammal groups, including primates—and us.

THE DAY AFTER

The Cretaceous Period ended appropriately 65 million years ago in perhaps the most famous of extinction events. The non-bird (non-avian) dinosaurs disappeared. So did the ammonites and marine reptiles and many other marine organisms. The nature and possible cause of this catastrophe will be explored later. But whatever the cause, the result was the extermination of many dominant Cretaceous creatures. The interesting facet of the next phase of geological history, the Cenozoic—the day after the stroke of the Cretaceous midnight—is that this mosaic of extinction and survival leaves a world, despite some notable absentees, much the same. Oceans still have their coral reefs and marine communities. The land vegetation is still ruled by the angiosperms. Lizards abound, with the added increasing diversity of their legless members, the snakes—yet another survivor of the Mesozoic. The real mark of the Cenozoic, at least primarily on land, is the rise of the more modern mammal groups.

The Cenozoic—the age of mammals (it could just as well be called the age of birds, snakes, or angiosperms)—is subdivided into the Tertiary Period, which extends from 65 million to about 1.8 million years before present (the origin of the human lineage, for example, dated in the fossil record at about 4.4 million years, is a phenomenon of the Tertiary), and the Quaternary, the second and last period which began about 1.8 million years ago. The Quaternary marks a phase when huge icecaps in the polar regions spread and retreated through several cycles called the ice ages. It is, in essence, only a thin slice of time, but one wherein our own species, *Homo sapiens,* emerged (at least 500,000 years ago). The Cenozoic is actually comparatively better known throughout its duration. Hence the Tertiary and Quaternary are more finely dissected into various epochs. For some chauvinistic reason the second epoch of the Quaternary, our current epoch called the Holocene, is distinguished from the penultimate Pleistocene. The Holocene, beginning about 10,000 years ago, is merely coincident (and very roughly so at that) with signs of "higher achievement"—more advanced toolmaking, cave art or symbolic pictographs that foreshadow writing and historical records. There are more dramatic ecological and evolutionary changes to animal and plant communities for the few million years preceding this fuzzy boundary. I prefer to think of us as creatures of the Pleistocene.

NAMES FOR TIME

With this strobelike time-lapse review of life's history, I must admit I am doing the fossil record and the work of hundreds of colleagues and predecessors great injustice. Myriad evolutions and revolutions, many of them of critical importance, are ignored here. I have, I hope, at least put the Cretaceous Period in its proper place on the calendar. There are some other general distinguishing features of the earth calendar. Note that the Precambrian is unfairly simplified because it is such a featureless blank for most of its eons. Its recently proposed subdivisions do not rescue us from our paltry knowledge of the first four fifths of life history. With the onset of the Cambrian the available fossil record vastly improves for paleontologists. Fossils become larger, more complex, and more readily preserved. The younger rocks entombing them are more commonly exposed on the earth's surface and less deformed, recooked, and dissected by erosion, mountain building, and other earth processes. It is a basic truth of paleontology that, as we get closer to the present day, our knowledge of fossils, the rock record, and the history of life becomes richer, and our sense of time becomes more acute. Thus, as noted above for the Cenozoic, there are more intervals representing shorter amounts of time.

And what is the purpose of the various names of the time intervals? We could, after all, simply call these divisions of the earth calendar Age II a, II b, and so forth. Or, more economically, we could simply calibrate our hourglass in millions of years without any references to intervals at all. In fact, many scientists fume that the Cretaceous means something different depending on whether we study fossil plants, or dinosaurs, or marine snails. They exhort us to abandon this cumbersome calendar. The names endure, however, because of a curious history, delightfully chronicled in William Berry's book, *Growth of a Prehistoric Time Scale,* and many other reviews. Because of this legacy, the names mean something to paleontologists. The word "Cretaceous" conjures up a world of dinosaurs, with a particular flourish of birdlike theropods and horned, shielded dinosaurs, like the Gobi's *Protoceratops,* that far exceeds the previous Jurassic Period. The Cretaceous is also a world that shows a striking amount of mountain-building activity, a good deal of crustal movement, an abundance of shallow seas and coastlines, a flourish of the strange coiled ammonites in the sea, and the appearance of flowering plants. The word "Cretaceous" is therefore economic and evocative; certainly it has more meaning than "an interval extending from 140 million years ago to 65 million years ago." In much the same way, the word "June" tells us much more about seasonal change, school graduation, and the surge of matrimonial ceremonies than simply "the interval between the one hundred and fifty-first and one hundred and eighty-first days of the year."

The meanings behind the names of these time intervals are, in themselves, not so profound. In most cases the names mean nothing more than the place where rocks containing a certain group of fossils were first described. The word "Cretaceous" comes from the German word *kreide,* for chalk. It refers to the chalky cliffs that align the English Channel on the coastlines of both the British Isles and continental Europe which admirably represent the marine animals and their environments of that time. Other names are more tied to specific locales. Cambrian refers to rocks of that age in Cambria, the ancient Roman name for Wales in western Britain. Devonian is for Devonshire. Permian is derived from rocks sticking out in the vicinity of the town of Perm in Russia. Indeed, most of the names of the major subdivisions of geologic time refer to European places. The calendar was already well divided by the time paleontologists struck out for the American West. By then places like central France and southern England were exquisitely mapped to the pebble. In John Fowles' novel, *The French Lieutenant's Woman,* Charles, the Victorian gentleman paleontologist, reveled in the bounty of the Jurassic rocks—actually described by Conybeare and Philips—rimming the coast near his hometown of Lyme Regis (he later revels in other bounties). And why was Charles so entranced by all this rock?

> Stonebarrow, Black Ven, Ware Cliffs—these names may mean very little to you. But Lyme is situated in the center of one of the rare outcrops of stone known as the blue lias. To the mere landscape enthusiast this stone is not very attractive. An exceedingly gloomy grey in color, a petrified mud in texture, it is a good deal more forbidding than it is picturesque. It is also treacherous, since its strata are brittle and have a tendency to slide, with the consequence that this little stretch of twelve miles or so of blue lias coast has lost more land to the sea in the course of history than almost any other in England. But its highly fossiliferous nature and its mobility make it a Mecca for the British paleontologist. These last hundred years or more the commonest animal on its shores has been man—wielding a geologist's hammer.

SQUEEZING TIME FROM STONES

By this point you may have noticed something rather odd about all this reference to geological time units. I have referred to these eras, periods, and epochs and the problems of their recognition and varying duration, and all the while I have applied to them some rather precise, clocklike dates in thousands or millions of years. What allows me to be so glib? Are some rocks conveniently labeled "I am 310 million years old"? The answer to this last question is really no and yes. Rocks don't reveal their ages so easily, but at least some of them, through modern technology and a lot of careful analysis, can provide a remarkably precise hourglass. Ironically, with one exception (the younger sediments that preserve remains that can be "carbon-dated"), these are the very rocks that usually don't preserve fossils. Igneous rocks contain certain elements, such as uranium, argon, and potassium, that occur in two or more forms, or isotopes. The more unstable isotope will tend to degrade over time—that is, its atoms will lose electrons or otherwise be reshaped in increasing numbers. This degradation also releases radioactivity, and the process of degradation is known as radioactive decay. The process can be precisely measured; in fact, the rate of decay is constant over time.

This decay constant is the key to success in radiometric dating. It allows a geologist to determine the age of the rock by measuring the proportion of the product isotope to the proportion of the element in its original form. The decay curve never reaches zero—in other words, never reaches a point where all of the original isotope has been transformed. Accordingly, the decay rates of elements are often described in terms of "half-life," the time necessary for half of the sample of a particular isotope to decay to the product isotope.

Sometimes igneous rocks good for dating occur in perfectly convenient places. In the late 1980s my colleagues and I found some superb mammal skulls entombed in dense, hard volcanic mudstone high up in the Andes Mountains of Chile. Such "Pompeii" situations are rare. In this case, they offered wonderful opportunities for precise dating. Above and below the "Pompeii layer" were basalt units, as dark as those lacing the walls of the Gurvan Saichan. These ancient volcanic flows were sampled and dated by Carl Swisher, a scientist at the Berkeley Geochronologic Laboratory. His results, based on potassium-argon decay, gave us some very reliable dates of around thirty-eight million years before present. The dates helped us in proposing a stage of mammal evolution that was not represented elsewhere in South America.

Carl does not always have this good fortune. In the Cretaceous Gobi, where Carl accompanied us in 1994, we were virtually skunked for any volcanic rocks that might give us isotopic dates. At this point I should admit that the eighty-million-year date for our sites in the Gobi is rather fuzzy; it is based on comparisons of the vertebrates from these sites to sites on other continents having similar organisms where the age of the rocks is more confidently known. On this basis, we could say that the Gobi Cretaceous fossils are perhaps not really eighty million years old; they could be ninety million years old or seventy million years old but almost certainly are not more or less old than this range. So our ball park figure is not too bad, at least with respect to the 3.5-billion-year history of life. We can narrow down our interval in the Gobi to the evening of the Cretaceous day.

On Life, Death, and the Pursuit of Science

Sciences like systematics, geology, and paleontology require a massive effort to capture data about the natural world through description and observation. This is not only hard work; it can be risky. Curiosity has killed the scientist. The great Roman naturalist Pliny the Elder is a case in point. As those ancient times called for, Pliny had an enormous range of scientific expertise. He published volumes on plant and animal taxonomy, lending firm diagnoses to many of the living organisms named by Aristotle. He was also interested in geology, atmospherics, and volcanology (the study of volcanoes). In late August, A.D. 79, Pliny went ashore near the volcano, Mount Vesuvius, to ascertain the nature of mysterious dark clouds. In the midst of this field research he was killed by the fumes in the clouds, which turned out to be the evil vapors of erupting Vesuvius.

On a less tragic note, the famous Harvard naturalist, Alexander Agassiz, paid dearly for his curiosity over the function of the tentacles of the sea anemone. Small fish swimming in or about these tentacles were stunned and consumed by the anemone. But when Agassiz stuck his finger in one of these crowns of waving arms he barely felt a tinge. Why? Agassiz got his answer when he promptly stuck his tongue in the animal. He demonstrated that the anemone triggers its stinging cells in response to the texture of various skin surfaces. In doing so, however, Agassiz suffered excruciating and prolonged pain, as well as a tongue almost too swollen to retract.

Sometimes one's excitement with nature is simply impulsive and rather stupid. Several years ago I had the pleasure of visiting the environs of Uppsala. Some friends and I stood in the florescent garden of Linnaeus. It was a lovely August day and we watched fat, pollen-encrusted bees, many with Linnaean names, pollinate Linnaean flowers. Then we went on a brief, guided excursion in the forests nearby. The mushrooms there were bounteous, and our guide, a mycologist (a mushroom or fungi expert) from the university, was brimming with fascinating observations and facts. At one point we stopped to admire one of the enumerable species of *Lactarius,* the group of milk cap mushrooms. "Very good to eat," exclaimed our guide. For some unknown reason, I immediately retrieved one of many individuals and bit out a chunk. I was treated to an Agassiz-like sensation on the tongue, a thousand pinpricks of peppery pain. Our guide finished his lesson, "but edible only after cooking." I have an indelible (and inedible!) recollection for the taxonomic name *Lactarius.*

Another encounter with nature was more serious. In 1985 we were working in Baja California, prospecting a very isolated set of badlands in the middle of the peninsula quaintly known on Mexican maps as "Lomas las Tetas de Cabra" (Hills of the Goat's Teats). It was sweltering hot. I slept in my underwear outside on top of my sleeping bag. About six o'clock one sultry morning I awoke to two points of sharp stinging pain high in the crotch, on either side of my genitals. I reached down and brushed aside two small scorpions, killing them in the process. What ensued was a taxonomic exercise, one not pertaining to dinosaurs or ancient mammals, but one of more interest to me than any I have undertaken before or since. We put the scorpions under a dissecting microscope and got out the field keys for desert arthropods. I rather impetuously stated after a brief glance at the books and a view of a barblike hook at the end of the tail of the specimen, "I think it's *Centruroides sculpturatus.*" Now this species is the deadliest scorpion in the New World, and its favorite locale is Baja and nearby deserts of the Southwest. It has killed more people over the last few decades than rattlesnakes. One of my field mates, Dr. John Flynn, put my careless taxonomy to the test. He said something like

"Don't be melodramatic" and proceeded to examine the materials with more deliberation.

Meanwhile I wasn't feeling too well. After some time John issued the peer review that taxonomists live for but don't necessarily want to die for. "I think you are right," he said with a deliberately soothing tone. We read on about treatment. The nearest place for antivenom was Phoenix, Arizona, but to get there would require a drive sixty miles south to Guerrero Negro, where we might be able to radio a small plane or hitch an emergency ride. It didn't look too good for evacuation. We both read another entry, "In case of serious and life-threatening symptoms and the unlikelihood of evacuation remove the appendage nearest to the point of the sting." I remember declaring, "I choose death," though we didn't laugh too much at the time. We decided to go about our business of prospecting. Mark Norell, my comrade on many subsequent expeditions I lived to experience, agreed to hang back in camp to "observe" me. If things got bad, he would throw me into a vehicle and speed down to Guerrero Negro. After all, the risks seemed limited. Our systematic analysis told us that these *Centruroides sculpturatus* were younger individuals, with less than a full dose of poison. Besides, I was a robust adult, in the prime of life; and *C. sculpturatus* usually only killed small children or elderly people. At the time these considerations were of little comfort. For the next six hours I was assaulted with headaches, nausea, a dizzying weakness, bilateral pain in the groin, and, perhaps most arresting, a sense of anxiety and dread. As the shadows of the afternoon cooled the day below 100° F., I could feel a subtle change for the better. Mark could see the signs of it too. The next morning I had essentially revived; I cherished in a whole new way the sight of elephant trees, boojum cactus, and candy-striped sandstones of Lomas las Tetas. Like Ishmael in *Moby Dick*, but unlike poor Pliny the Elder, I live to tell thee.

CHAPTER 3

1991—THE GREAT GOBI CIRCUIT

The combination of characters, some, as the sacral ones, altogether peculiar among Reptiles, others borrowed, as it were, from groups now distinct from each other, and all manifested by creatures far surpassing in size the largest of the existing reptiles, will, it is presumed be deemed sufficient ground for establishing a distinct tribe or suborder of Saurian Reptiles, for which I would propose the name *Dinosauria.*

SIR RICHARD OWEN.
1842. Report on British Fossil Reptiles. Part II *Report of the British Association for the Advancement of Science,* Eleventh Meeting, Plymouth, July 1841.

Our victory at the gates of Eldorado secured a future for us in the Gobi. Our Mongolian colleagues were impressed by the success of the brief 1990 reconnaissance and expressed their support for future collaboration. But the good spirit and earnest work involved in launching something as elaborate as a joint dinosaur expedition to the Gobi would come to naught without a critical element. There had to be an international agreement, one sanctioned by the appropriate people. When we returned to Ulaan Baatar we negotiated and drafted a protocol over several days, which finally was signed by Dr. Rinchen Barsbold, director of the Geological Institute, Mongolian Academy of Sciences, and me, wearing my hat as vice-president and dean of science of the American Museum. Our memorandum of understanding allowed for full collaboration for three years. Malcolm McKenna, Mark Norell, and I would be the principals from the American side. I would be leader of the overall planning, funding, and administration of the expedition and the international program necessary to keep it going. Mark would be leader of field operations and strategies, and Malcolm would direct logistics, navigational preparation, and communication (maps, GPS, and radios, etc.), as well as the activities necessary to supply the expedition. As it turned out, we all overlapped broadly in our responsibilities. Demberelyin Dashzeveg continued as the principal collaborator and leader from the Mongolian side. He would be complemented by the participation of Dr. Altangerel Perle, a dinosaur expert, as well as possibly Barsbold himself. The American side would provide for basic expenses and would support the travel of Mongolian scientists and administrators to the United States. The specimens would be property of the Mongolian people but could be taken to the United States for preparation, illustration, and extended study. Casts of the material could be made, and there was provision for exchange or purchase of material at some future date. For their part, the Mongolians promised to help us in preparations each season and to work with us in both the field and the lab. A later extended agreement for 1994 and 1995 had much the same terms, as did our agreement for 1996 and 1997.

The agreement made in 1990, however important, would be worthless parchment without money and equipment appropriate to such an ambitious enterprise. The central problem was transportation. Our experience with Max and the GAZ in 1990 taught us that the most important equipment needed were reliable four-wheelers, for which we would serve as drivers. This is not to denigrate our Mongolian drivers. In subsequent years I have seen spectacular feats of steering those huge hulks through the Gobi. But there is nothing so fulfilling as piloting your own craft and "setting it down" wherever you choose. Soon after our return to New York we set about the business of obtaining money for cars and equipment.

The vehicles we used in the 1991 season turned out to be three four-wheel-drive Mitsubishi Monteros. The Mitsus were appealing because they were not only roadworthy, but they had already invaded Mongolia. A Japanese archaeological expedition of 1990 in search of Genghis Khan's tomb had used a fleet of these models. We figured we could, if necessary, find some extra parts in a pinch and draw on the expertise of some of the Japanese mechanics. The vehicles were not bought in Japan, though—the jeeps had to be customized for rugged desert travel. The best place to do this was the great automotive empire of Southern California. Here, stock four-wheelers were often bizarrely altered into sand and gravel monsters ready for the infamous "Baja 500." The trio of vehicles—red for Mark, beige for Malcolm, and green for me—were appointed with sturdy steel top racks. Two of the jeeps were "front-ended" with a very heavy winch—a motor-driven spool of heavy cable for extracting other vehicles or dragging sleds with dinosaur casts. Our jeeps were hardly Spartan tough though. The only available Mitsus on short order were deluxe models, which came with air conditioning, tape decks, and multispeaker sound systems. We could entertain our Mongolian comrades with the soaring sounds of Neil Young or the Pixies (and later Nirvana, Pearl Jam, and Nine Inch Nails) or, as Malcolm preferred, Mozart and Beethoven.

The customized Mitsus were taken to a warehouse in Compton, an industrial appendage of L.A., where they were loaded into containers with our other equipment, which included plaster, lumber for boxes and crates, shovels and other field equipment, a good and comprehensive auto toolbox, an elaborate medical kit, various spare vehicle parts and fluids, a gas-powered generator, a portable (but ultimately inadequate) refrigerator, rope and cable, food (freeze-dried, canned, and other non-perishables), miles of tape for specimens, toilet paper for fossils and conveniences, and camping paraphernalia—tents, drop cloths, beach chairs, etc.

Not all the necessary equipment for an expedition can be put in containers or even air freighted. Some are too precious and sensitive. These we would take ourselves: cameras, laptop computers and walkie-talkies, satellite navigators or GPS (for global positioning systems), miraculous little receivers that fix our positions in latitude and longitude and even elevation by receiving signals from satellites that orbit the earth. These are very useful for nailing down the positions of localities or charting a course through the desert. Malcolm's wife, Priscilla McKenna, who joined the expedition in 1991 (and remained with it since that time), would make a detailed road log based on continual readings from our GPS. Not that these readings always kept us on course. Even if a satellite can give us a fix on where we are, there is no reason to be confident that the route to our destination is correct. The Gobi is not an unbroken sea where a bearing can be faithfully followed. It is still possible to become lost as the result of a bad route, an intricate canyon, or a circuitous mountain pass.

All this electronic gadgetry requires power, usually from batteries. We stockpiled and shipped great quantities of D, AA, and AAA batteries for the purpose. Certain cells, like those for the portable radios and the GPS, required recharging. For this purpose, a cigarette lighter could be fitted with a plug. We could also get voltage with the gas-powered generator, though the loud buzz of that contraption in camp proved annoying. Malcolm quickly coopted new solar technology and brought along two metal suitcases with solar-powered cells. Over the years these have proven very handy and much quieter than the generator. We found ourselves tempted to run our small refrigerator off one of these suitcases, bringing the forces of nature—from solar bombardment to ice making—full circle. But the fridge was feeble and proved unworthy of prolonged charging. In recent years, those miraculous fan-fridges that run again off the plugs in cigarette lighters have kept things relatively cool while vehicles are moving.

Prepackaged food of course was vital, considering our experiences with Mongolian fare in the summer of 1990. Basic items—meat, potatoes, and flour—were available in Mongolia but we needed to supplement that with food from the United States. Priscilla did the heroic job of purchasing and organizing most of the food (and, for that matter, most of the supplies). But all this preparation turned out not to be sufficient. When we returned to Ulaan Baatar in June 1991, we were shocked to find that the Mongolian economy, by this time in free fall, could not provide us even with the basics. Food was nearly impossible to buy. Thereafter we shipped ample food, treating this aspect of field preparation like a trip to Antarctica or a distant planet.

Other supplies were less problematic. Mark and I had particular fun shopping in L.A. for the vehicle accessories. Although our Mitsus were "Baja proven," there was still a need for a great array of paraphernalia. Mark took me to the famous Dick Cepek's in his home "town." of Downey, California, a lavish palace for "off-roaders," where we bought bunge webbing for the roof racks, countless tire repair kits, emergency "sun guns"—blinding lights that were powered off the car battery—and heavy jacks of different design. Our favorite item was something we simply called the "thing": a heavy metal implement known as a "deadman," about two feet long, made from two intersecting triangles of steel, whose sharp point could be driven into the sand to serve as an anchor for pulling out a stuck jeep.

Planning an expedition to a remote desert requires attention to detail. Even in the Andean towns of southern Chile we could find quantities of sundry, like toilet paper, tape, and rope. But that was out of the question in Mongolia. We shipped miles of tape and toilet paper—the crucial items for wrapping fossil specimens or coating a fossil before it receives a plaster jacket—probably two hundred rolls of toilet paper in all, an amount that we supplemented every successive season. In much the same way, I've purchased practically a flat stacked with "TP" in stores in Chile or Wyoming, eliciting giggles from the Chilean checkout ladies, and wisecracks from shoppers at the Rock Springs Safeway. "Food not treat'n ya right?"

Shipping Out

In the spring of 1991 three ocean-vessel containers were loaded with their precious contents—Mitsus, gear, toilet paper, and sundry—a cargo amounting to well over $100,000 worth of supplies and equipment. In addition we had paid several thousand dollars to purchase the containers themselves. In most countries the container can be rented on the assumption that it can be reloaded and returned. In 1991 this was an impossibility for any container on its way to Mongolia. Many shipments were going in, but nothing was coming out. The West was moving its wares to Mongolia, but there was no reciprocation with furs, meat, minerals, or vodka. Landlocked by China and Russia, Mongolia continues to have problems sending its exports overseas. This was not a problem for us. The Mitsus needed a winter home, and the giant containers would make convenient storage bins, general stores, and tool shops in our compound next to the Natural History Museum in Ulaan Baatar. We could even sleep in them if the hotels were overcrowded. Our shipping company had the containers repainted—in unintentional honor of the expedition—the pleasing beige of a Gobi sand dune.

The containers headed off across the Pacific in March 1991. They eventually made their way to the Chinese coastal city of Tien Jin, where the containers were loaded on flatcars and pulled by a Chinese steam locomotive to the border town of Erlian.

It was at this point where our shipping scheme was most vulnerable. The elaborate ritual of swapping Chinese train wheel sets with Russian wheel sets is not the only reason for delays on the Mongolian-Chinese border. Thievery and black marketeering are rampant at this checkpoint. Erlian, along with its sister town Erdensu on the Mongolian side of the border, is a tough place. Mongolians bring down meat and furs and endangered species to trade for metal tools, tires, gasoline, and Chinese beer. There is even rumor of illegal transfer from Mongolia of "red mercury," or plutonium, across the border, dark tales told by our Mongolian truckers around the campfire. Here, as in many frontier towns, unusual things can happen. Whole cargoes are broken into and dispersed. At the least, cargoes are often sidelined in the railroad yards for needless days or weeks. Luckily for us in 1991, this was not the case; our containers arrived in Ulaan Baatar on schedule, about fifty days later.

Even with this compulsive attention to logistics and supplies there were inevitable oversights. On several occasions while driving back from the field into Ulaan Baatar we have been signaled to pull over by a truculent-looking policeman, who kicks our mud-encrusted tires and draws his white-gloved finger over the dust on the windows, all the while murmuring, *"Moh! Moh!"* Our vehicles look fairly unpresentable after a day of slogging through the muddy roads south of the mountains ringing the capital city. To Mongolians, unaccustomed to the reality of expensive foreign cars, this kind of filth is a disgrace. We have been thus ordered back to the Tuul Gol *(Gol* is the Mongolian word for river) for a thorough washing of the jeeps, and for that matter probably ourselves. In more recent seasons, our last stop on the return road has usually been at the river, in anticipation and fear of the vigilance of the "clean police."

THE EASTERN TREK

By June 19, 1991, our team, as well as the vehicles and equipment, were assembled, and we were ready to enter the Gobi on our first "full-fledged" expedition. But the expedition launch was rather piecemeal. Dashzeveg could not get a GAZ truck from the Academy in time for our scheduled departure. We decided to strike out ahead and let "Dash" intersect with us later, at the oasis of Naran Bulak in early July. Our three Mitsubishis carried ten people, including our Mongolian dinosaur specialist Altangerel Perle and his son Chimbald, as well as the *New York Times* writer John Noble Wilford and photographer Fred Conrad. We loaded up the interior of the vehicles with gear and food, and piled lighter supplies—such as cans of freeze-dried food—on the roof rack. Mark, traveling with Dr. Jim Clark, a tall, sharp-eyed paleontologist who was doing postdoctoral work at the American Museum, had the honor of piloting our "Molotov Cocktail," a red Mitsu carrying a 50-gallon drum full of 93 octane gasoline in its back storage. It was not the safest way to set course for the Gobi, but the economic and logistical chaos in Mongolia offered no other option. Malcolm drove the beige Mitsu with Priscilla by his side. I shared driving honors with Dr. Lowell Dingus, our team geologist.

We embarked on an audacious route, one that followed the "Efremov tour" to the eastern Gobi. This meant a road that traveled over the sweeping steppes parallel to the Beijing–Ulaan Baatar railroad. About two hundred miles south we would reach the Early Cretaceous rocks of Hurrendoch, where our Mongolian colleagues Barsbold and Perle reported remains of important dinosaurs. Depending on our luck, we would then proceed to the railroad town of Saynshand, replenish our gasoline and fresh water, and strike westward into the Gobi proper. We would explore a few fossil-bearing badlands in the eastern Gobi, pushing ever westward toward Dalan Dzadgad. The prospect was hazardous. The route transected a blank on the map, an empty wilderness even by Gobi standards. For a scorching three hundred miles there would be no gasoline supply depots, indeed virtually no towns, until we reached the comparative civilization of Dalan Dzadgad. The route made the journey to Flaming Cliffs from Ulaan Baatar look like an interstate highway. Fortunately, Perle, who had done important work at several of these areas some years before, would be our vital guide. Moreover, we were now armed with GPS satellite navigators, and Priscilla would painstakingly log our positions via satellite signal with every turn of our road.

Less than an hour (roughly ten miles) out of our base camp in Ulaan Baatar we said goodbye to the crater-blemished pavement and set off across the steppes. By the second day of travel, the unbroken rolling hills of rusty grass became oppressively boring. As Paul Theroux wrote, "Featurelessness is the steppes' single attribute, and, having said that and assigned it a shade of brown, there is nothing more to say."

But the nights on the steppes were beautiful. The sweet smell of grass and wild onion was accentuated in the heavy, moist air. When the conversation lowered we could pick up the calls of wolves roaming the distant hills. Our fires, built from the meager pieces of scrap wood found along the road, crackled under one of the clearest, star-studded skies on the face of the planet. The ruler of this night sky was Scorpius; it was one constellation that lived up to its name. There were other brilliant constellations in the Gobi heavens—Hercules, Lyra, Boötes—but these were high overhead, requiring us to bend our heads back to appreciate them. Scorpius, crawling its way just above the southern horizon, was easily in view, a backdrop to any conversation. This constellation also lies near the ecliptic, the pathway of the planets, and we debated whether that bright red object in the middle of its thorax was Mars or the star Antares—indeed Antares is Greek for "rival of Mars." I remember exactly those same debates decades ago on a starlit night above a New Mexico desert.

Perle was the center of attention here in his homeland. His dark rounded face vividly reminded us of the noble countenance of the Navajo warrior. With the pleasures of American company and the wonders from foreign lands—spices for foods, superb Swiss-made binoculars, and car stereos—his face seemed to be one of perpetual happiness and excitement.

Perle had a smile or a laugh or a charming twist of an English phrase for everything. Upon contending with his first freeze-dried dinner of the expedition, Perle mustered the best compliment he could think of for such bland fare. "Quite palatable!" he exclaimed with delight. Around the after-dinner campfire, he sang passionate Mongolian love songs and related the highlights of his country's long history.

To many Mongolians, their history really begins, and in some ways ends, with a phase marked by the birth of Temujin in 1162. This young warrior fought for twenty years to unite the Mongol clans and turn them out on the world. By this time Temujin was called Genghis Khan, "universal king." In later centuries, during years of Chinese and Soviet domination, Mongolians were expected to regard the exploits of Genghis Khan with shame and contrition. But the new democracy holds no such constraints on its people and many citizens look back with pride, and a little wistfulness, on the glory days of the conqueror. Before we left Ulaan Baatar on this summer's expedition, one young Mongolian, whom I encountered at the hotel bar, said to me, "You are lucky, you are from the great empire of the United States, we have not been such an empire since the thirteenth century."

By the morning of June 20, we reached Hurrendoch. It was an unspectacular outcrop, with low brown sand exposed in the shallow green valley, technically still surrounded by drab high prairie or steppe land. But we knew that Barsbold, Perle, and other paleontologists had recovered some magnificent dinosaurs at this locality—bipedal iguanodontids, the parrot-beaked *Psittacosaurus*, and an intriguing theropod, *Harpymimus*. More than anything, we simply wanted to get to work. It had been nearly a year since we romped the red hills of the Gobi, extracting a hefty ankylosaur skull and retrieving that spectacular specimen of "Cretaceous Komodo," *Estesia*. Now, after months of planning and purchasing and hashing out details of logistics, we were anxious to set on those outcrops and practice our profession.

TOOLS OF THE TRADE

My professional training for the Gobi did not really begin in the vacant lots of my youth in Los Angeles but in the scorching, red-walled canyons of southern New Mexico. Toward the end of my undergraduate stint at U.C.L.A., I was asked if I'd like to "come along" on a paleontological expedition. What better way to bypass the ennui of another summer in Los Angeles, I thought. On the cusp of my twenties, I had far less sense of direction than I had as a dinosaur enthusiast at age seven. I was a biology major without much resolve to be a biologist. Outside of school I continued writing and playing music, making hardly enough money to contribute rent to a chaotic household of musicians. It was a predicament that, in later years, helped me understand some of my own students.

Whatever my poor motivations and even poorer qualifications, I headed out in a truck full of field hands just as a more prominent team of explorers prepared for the first manned space flight to the moon. Unlike the lunar venture, our equipment was not unusually costly or high-tech. Indeed, aside from a stake-bed truck and a jeep, it hardly differed from the tools of the trade employed by the bearded bone hunters one hundred years before.

Field paleontology has, like any other science, coopted technology effectively today. Fossil sites are now plotted with GPS to extraordinary precision. Remote sensing is used to pick up a signature that will identify a particular rock unit already known and produce a magnificent geologic map. As I described earlier, rocks that entomb the fossils are now sampled for dates based on radioactive decay or magnetic properties of selected minerals. Yet the essential field items differ little from the contents of the packs carried by Andrews and company. The paraphernalia of a paleontologist striking out for the hinterlands on a day of prospecting is unprepossessing—a rock hammer, some sharp awls, whisk brooms and paintbrushes, a thin glue for hardener, some plaster and burlap strips, and, of course, a liter or two of water. These essential items are complemented by the implements of more extensive excavation as well as the comforts of a home in the field: a pile of pickmatics and shovels, wooden crates full of cooking utensils, sleeping bags, canvas tents, a tarp to shield the sunlight, and extra cans for gas and water.

The Excalibur of the paleontologist is the rock hammer. This unimpressive instrument is not appreciably larger than a typical clawed hammer, but it is heavier, with a squared, blunt striking surface and, opposite, a sharp pick not unlike the form of a pterosaur crest. A good rock hammer, like those made by the Estwing Company in Rockford, Illinois, is a solid piece of forged steel with a handle wrapped in tape or fast-grip polyurethane. The size, weight, and feel of a rock hammer is a very personal matter. A hammer of choice, just like a favorite tennis racket, should be cherished. To lose one, as I have done too often, is a tragic event.

The most versatile field item is probably the burlap sack. It can be used to carry gear if necessary, or a small sackful of rocks. It can be ripped into wide strips and smeared with plaster to form a jacket or a cast for large and fragile bones (new technology is here represented by plaster medical bandages, an application limited to smaller specimens). In a camp, lacking the convenience of refrigeration, burlap can also be soaked with water, wrapped around beer bottles, and buried in the sand for a tepid yet inviting beer at the end of the day. Some years ago I found another surprising use for burlap. Through a comedy of errors, two of us got separated by five hundred miles and three days from the rest of the field crew. After an exhausting drive through the Jackson Hole country of Wyoming, we bivouacked on a frigid plateau in the Bridger Mountains. Much to our horror, we discovered that our sleeping bags were on some unknown route to southern Wyoming in the back of the other vehicle. For an unforgettable five hours I shivered under a pile of dusty burlap sacks.

In addition to assembling the proper equipment there are a series of standard practices, obviously centered on the craft itself. A fossil is not hacked out with dispatch by chipping away the rock close to its edges. This is a felony. "Never prepare a fossil in the field," as my early mentors would say. Indeed, there is no little skill in knowing at what point a fossil is too fragile to remove with a hammer and chisel. In such difficult cases a trench is made around the specimen. The top surface of the rock containing the fossil is slathered with water-drenched toilet paper and then shrouded with burlap soaked in plaster. This plaster cap is then allowed to set in preparation for the critical part of the operation. When the cap is satisfactorily hardened, like the ones that covered our *Protoceratops* and *Estesia* near Eldorado in the 1990 field season, long wedge-shaped chisels are applied to the troughs undercutting the pedestal of plaster. For a heart-stopping moment of truth the cap is flipped. Failure at this point means rock pouring out of the cap and a seriously damaged or destroyed specimen. Victory means you have a block looking like an overturned turtle, with its smooth "bottom shell" awaiting more plaster burlap strips.

Plaster blocks like this come in all sizes. My favorites are ones made for little fossil mammal skulls, that can be picked up with one hand and conveniently stuffed in a day pack. Then there are blocks, like the ones pictured in Andrews' Gobi expeditions, that give a sense of what it must have been like to assemble an Egyptian pyramid. Loading out these monoliths can be an epic. Their weight is usually underestimated as plaster is spread thickly with too much zeal for protecting the specimen. It is a great embarrassment to be compelled to leave one of these monsters behind, and no muscle or implement is spared in transporting them. In the Sangre de Cristo Mountains we pushed such a mighty block out of a gulch with a makeshift sled and the aid of a screaming winch on the front bumper of the jeep. It was a two-day job.

The construction of plaster casts and their removal are matters easy to describe because they represent something of a field technique. But much fieldwork in paleontology is remarkably free-form. True enough, there are great bone quarries, like the Jurassic sandstone slabs of Dinosaur National Monument or the tar pits of La Brea, where field crews swarm all over the bone surface, carefully chiseling out bones of recognizable skeletons, and even more carefully mapping their position and orientation on a quarry map. But such veins of precious bone are rare; fossil discoveries of note can be widely spread over an area of badlands. A great specimen can be camouflaged on the surface as a worthless knob of heavily weathered bone. A little brushing away of the dust or sand from the surface of the rock can reveal to a surprising extent the nature of the find. As Roy Chapman Andrews wrote, this is truly "hunting dinosaurs with a whisk broom." It is a great delight to find that an unassuming fragment is actually a tail of a beauteous beast safely encased in the rock below, free from surface erosion. "Let's go all the way to the head before dinner!" we shout.

Bones, Ruins, and Wastelands

As I set off that first field day with my comrades for the Cretaceous outcrops of Hurrendoch, I carried with me the simple tools of the trade in a lightweight fanny pack—my beloved rock hammer, some thin glue for hardening a fragile fossil, awls, a whisk broom, toilet paper, tape, a multipurpose Swiss Army knife (the *Explorer* model of course!)—as well as a more wicked-looking hunting knife attached to my belt. I also carried a one-liter bottle of water in my pack, and a plastic sandwich bag full of trail mix. I was ready for a full day of prospecting.

But Hurrendoch turned out to be a disappointment. We only came up with the shoulder bones of a crocodile-like champsosaur, a variety of dinosaur fragments, and a lot of petrified wood. We resolved without much discussion that Hurrendoch was only worth that one day. The enormous expanse of the real desert and its fossils to the south drew us on. Indeed, the most passionate conversation the night of the twentieth did not pertain to fossils. There was some consternation about how little food we had. A can of freeze-dried "Chicken Polynesian" had to do for the whole crew.

The next morning we encountered what Perle called a sanitarium, but which seemed more like the Mongolian version of a health spa. A number of smaller gers encircled a huge central ger, whose internal framework was ornately painted. It seemed the spa contained more food than all of Ulaan Baatar. Mongolian democracy had not brought a new order out here; living upcountry was still part of the good life. Perle proceeded to bargain for sheep, fifty pounds of potatoes, rice, and a small chunk of fresh ginger from China. The deal was settled over a bit of morning vodka with the aid of an abacus. We then pushed on to Saynshand.

I've seen Saynshand several times from the MIAT (Mongolian Air Transport) jet at 20,000 feet, and I can assure you that it looks much better from that distance. Its population had that forlorn, confused, and in some cases inebriated aspect we had encountered elsewhere in the country.

Fortunately, Saynshand did have gasoline and water. While Perle met with the local official, showing him our various papers of permission and cooperation, we filled our Mitsus at the fuel depot, strategically mixing 76 octane with 93 octane in a formula strong enough to sustain the cars. Next we filled our many containers with what we assumed to be fresh water (Saynshand does in fact mean "good water"). The location of the well—in the center of town not far from the railroad tracks—raised our suspicions concerning the "sayness" of the water. At any rate, it was cold; and it tasted good.

We left Saynshand after six, heading west. The distant hills were rounded and mauve in the dying light. We could see the black silhouette of jagged ridges; we were back in our wild Gobi.

By the next morning we were lost again. The mauve hills were now green and brown, crowned with chunky rocks; they seemed to block our way to the proper passage. With the help of a lonely trucker and a local herdsman we at last found the right road to our destination—the baking white and pink outcrops of a Cretaceous locality known as Khara Khutuul. We picked what appeared to be the most hospitable spot for a camp—a gully that incised a low rounded hill that faced north, forming a gentle scrub-dotted slope toward the main wash. By 2 P.M. we struck out across the extensive outcrop on the shallow flats formed by the white sandstones of the Khara Khutuul. Before long we were finding chunks of big dinosaur bones—these were the vertebrae and massive limbs of hadrosaurs, or duck-billed dinosaurs.

Hadrosaurs, as I discussed earlier, fit along with ankylosaurs within the dinosaur subdivision Ornithischia. As the common name of this group suggests, hadrosaurs have an elongated, flared ducklike bill. The nasal bone is extended backward into the crest. The tail is long—one and a half times as long as the body and flattened from side to side. The toes of the hind foot are rather broad and spatulate at their extreme ends, giving them a hooflike appearance. Hadrosaurs are known by many genera and species from Central Asia as well as North America (there is also a limited record in South America). They are wholly a Cretaceous group. Central Asia is not really the hotbed of hadrosaur diversity, but it did have some impressively large creatures, like this one at Khara Khutuul.

Nonetheless, three days of prospecting at Khara Khutuul and a nearby set of red beds called Hugin Djavchalant were hardly inspiring. At Hugin Djavchalant we found a few fragmentary specimens of ceratopsian dinosaurs, possibly *Protoceratops* or a close relative. Malcolm came up with some shards of dinosaur eggs and a particularly interesting tiny mammal jaw with one tooth planted in the otherwise empty dental arcade. The slim pickin's, however, did little to instill affection for the broiling sands of Khara Khutuul and Hugin Djavchalant.

Before pushing farther west, we sped back to Saynshand to replenish gas and what supplies we could find. At the town's vodka-bottling factory we caused a riot, when workers left their stations—as dozens of bottles tumbled off the conveyor belt—to pose for my Polaroid camera. Then we continued on our journey through the lonely flat country of the eastern Gobi. From this point the terrain had none of the striking relief of places like the Nemegt Valley in the western reaches of the desert. Instead of mountains there were hills. Instead of vast canyonlands there were isolated outcrops, exposures of gently sloping sands. But the wildlife was glorious to behold—wild ass, eagles, gazelles, and grouse. Perle mentioned that much of the land had been cordoned off as a wildlife preserve and a nascent national park. But we were doubtful that such a huge tract of wasteland could be patrolled and protected as a sanctuary. The eastern Gobi was, unfortunately, a poacher's paradise.

Our tiny caravan, made up of the kind of brightly painted Japanese cars people in the wealthy suburbs of New York City use to pick up groceries at the supermarket, looked a little silly against the backdrop of this vast frontier. This was hardly an expedition that measured up to the drama of Roy Chapman Andrews' winding train of camels. But the Mitsus were sturdy and surprisingly comfortable despite the crowding. The only danger was experienced by the drivers of the red Mitsu, our petrol-loaded incendiary bomb on wheels. The fifty-gallon drum in the back of the car was laid horizontally. Its cap, which opened in the side rather than the top of the barrel, had to point straight up to prevent gasoline spillage. Unfortunately, the drum would slowly rotate in the bouncing car, leaking small streams of gasoline until Mark and Jim could correct the problem and heave the drum around to its proper position. Fortunately, Mark and Jim were not addicted to cigarettes.

We were aiming for a set of badlands called Baishin Tsav that lay many miles west of us in the heart of the Gobi. "Tsav" means badlands, or canyonlands, places with eroded cliffs, dry washes, and little vegetation—places for bone hunters. At Baishin Tsav the Russian and the Mongolian expeditions had extracted some giant Cretaceous dinosaurs—hadrosaurs and snake-necked sauropods. Perhaps the most precious of the finds was the strange theropod *Erlikosaurus*. Perle, however, did not remember with precision the location of this place, and Malcolm and Priscilla could only make an estimate of the coordinates as a means to direct our course. Besides, as is usually the case, the roads were weaving before us, often taking us from our preferred westward tack. The afternoon grew hazy and the sky merged with the dull brown of the low hills to present a featureless diorama—a landscape without a landmark to aim for. The road grew fainter until we were barely on a track at all. The great trans-Gobi route now became an eidolon, a phantom, one that simply vaporized. Our expectation of reaching Baishin Tsav in the early afternoon vaporized as well. We merely held out the desperate hope that the route would eventually become clearer and truer to the westward vector. We followed a curving pathway around a set of rounded hills. The road ended abruptly in an extensive set of ruins, made from the eroded walls of beige mud brick that blended with the surrounding terrain.

Perle flashed me a rare frown. "Sorry, Mike, I don't know the way. This road is dongurous." Perle had a sweet alliance to us, and he took the responsibility of our welfare and success seriously.

"No matter, Perle," I replied. "Let's look around."

Many of the ruins in the Gobi, like the one we encountered, were once places of devotion and harmony, Buddhist monasteries representing the Yellow sect of Tantric or Tibetan Buddhism known as Lamaism. There are also old walls, some so eroded that they can hardly be distinguished from the volcanic dikes that cut through mountain ranges like the Gurvan Saichan. On certain maps, zigzag symbols indicate the presence of such ancient walls. Less than fifty miles south of where we were stopped, several walls are indicated on these maps. These continue southward, like a series of waves running west to east below the border of China. They culminate in the 1,500-mile complex of walls running from the heart of Asia through Kalgan (where Andrews and his colleagues set out for the Gobi) and other towns in the mountains north of Beijing, to the Gulf of Chihli on the Yellow Sea. This is the fortification people generally call the "Great Wall," the largest construction project in human history. The strength and magnitude of a fortification are proportional to its degree of need. Great attacking armies require great walls of defense. Walls were strung across the Gobi and the plains of northern China because this terrain is the battleground of centuries. And the ruins of lost cities and ancient walls are remnants of a violent history.

Mark was already prospecting over the ruins, grabbing up the most enticing shards of the past. We had a hunch that this ruin had been rarely visited; after all, we had stumbled into this site after tracking the wrong road to its dead end. Moreover, there were some interesting objects that clearly would have been harvested by artifact hunters—decent pieces of pottery and a beautiful bronze mask rendered all the more magnificent from the slight distortion and weathering of the metal. But the most striking aspect of the site was the number of bullet holes and rifle shells around. Perle frowned.

"I think this was a place of massacre," he said.

In Ulaan Baatar in the 1930s, the new tyrant in town, Choibalsan, decided that even Buddhists were a serious threat. The martyrs were hunted down in the streets of the city and were tracked to hidden enclaves in the deserts and mountains. The militia spared only four of more than seven hundred monasteries from total destruction. This purge coincided with the extermination of "rightist elements." By 1939, 27,000 people, including 17,000 monks, had disappeared and were most likely imprisoned and executed, over three percent of the country's population at that time. A massive graveyard from this holocaust, recently disinterred in northern Mongolia, was plowed under with thousands of bodies.

Was Perle correct in attributing this lost ruin to Mongolia's brutal past? It was unlikely this was simply the work of an exuberant bunch of hunters. As paleontologists we thrive on carnage—on massive death assemblages of dinosaurs represented by fossil remains. But the fascination of combing this site for bits of history did not mitigate the horror of contemplating the tragedy of these ruins and the unfortunate people who once lived there.

Perle found another faint but apparently more deliberate track in the hills nearby. After several twists and turns this road snaked its way westward, and we rounded a line of higher craggy volcanic hills. There before us was another destroyed city—another remnant of the purge—backed against the hollow cavity of a black mountain. But instead of tragedy this site displayed hope and rejuvenation. In defiance of this history of persecution, the town was repopulated with Buddhists and their lama. Brilliant white gers were nestled in among the dull mud-brown ruins. Bright orange, yellow, and red banners flapped from various points of the new town. The townspeople engulfed us, gawking at our clothes, vehicles, and equipment.

Some of us paid a visit to the local and exalted lama. The Buddhists told us of a good well nearby and bade us on our way. Soon Perle began to recognize a particular profile of the hills. We kept heading west—a good heading—one that would let us stop and admire the dinosaur sites at Baishin Tsav worked by the Mongolians and the Russians. If the roads were passable and the weather forgiving, this heading would also take us across the eastern Gobi, intersect us with our cherished Flaming Cliffs, and eventually return us to the Nemegt Valley, the terrain that most fed our hopes and dreams.

BACK TO ELDORADO

Our tortuous trek brought us back to Khulsan and Eldorado on July 3, 1991, slightly over a year since we had last danced among the dinosaur bones on our "two red hills" near Eldorado. This time we came in directly from the south side of the Nemegt Valley, traveling a decent road that led from the local district center, the Sumon of Gurvan Tes. Far across the green depression the Nemegt and the Khulsan canyonlands were on display for us. The orange-red sands of the lower rock sequences—the Barun Goyot Formation—formed myriad buttes, cliffs, and canyons. Emblazoned by the low sun, they looked like the largest city in the Gobi suddenly set to the torch by an invading army. Looming above these, like an Acropolis on a plateau, were the multicolored pinks, reds, purples, and whites of the upper, dinosaur-rich sequence of rocks, known as the Nemegt Formation. It was, as a whole, Efremov's "illusionary town full of mysteries."

Our plan was to first concentrate again on the lower red beds that had given us our Khulsan dragon the year before. Then, if time permitted, we would begin prospecting the candy-striped Nemegt sequence above. There was ample terrain for exploration and discovery.

It was about time. During our crossing of the eastern Gobi, we had seen some potentials for good dinosaur skeletons at Baishin Tsav, but these were nothing new or extraordinary; a summer would be wasted removing them. Our brief stop at Flaming Cliffs was rather successful; Mark found an attractive skeleton of the mammal *Zalambdalestes*—a long-snouted, long-legged, shrewlike form first discovered at the Cliffs by the 1920s expeditions—near a spot the Poles had named "The Ruins." We also worked for some days on a hefty ankylosaur skeleton, and we found a couple of very nice dinosaur egg nests. But we were after more interesting game—*new species* of theropod dinosaurs, birds, mammals, and who knows what else?

This hunger had led us back to the Nemegt Valley, where we had our best fortune of the 1990 season. Here the country was big, well dissected, and still open for prospecting. It was difficult to resist stopping at the red cliffs when we at last crossed the valley. But it was late in the day; the washes made for extremely difficult sand crossings—two of the Mitsus got stuck and had to be winched out. In the process we ripped out a couple of sets of fog lights and overwound the screaming transmissions. At last, we pitched camp in the general locality of Khulsan, at the entrance to the most impressive maze of canyons called Eldorado. A golden sand dune and the cliffs above would shade us from the morning sun.

But the wind blew unrelentingly for the next two days, and prospecting was miserable. Every time one of us bent down to examine an exposed fleck of bone, a mini-huij sprayed us with sand. The sand was everywhere—in our tents, our cars, our pots and pans. It penetrated under our thin layers of clothes and covered our bodies, lodging itself in crotches and armpits. Even "glacier glasses"—mountaineering sunglasses with side protectors—did not save our eyes from sandblasts. Invariably, one is compelled to take off the glasses for a look at a specimen in full light. On one of these occasions, a whirlwind of sand caught me off guard and blasted my right eye. For a few days my eye was painfully swollen and essentially useless.

The discomforts of the wind and sand hardly helped us get through another problem. We were virtually starving. Our shipment of food—mainly freeze-dried dinners, some cans of tuna and sardines, and a few condiments—was meant to be a supplement to the food we had planned to purchase in Mongolia. This seemed a rational strategy. The shops of Ulaan Baatar were no food emporia, but there was in 1990 an ample supply of bread, pasta, potatoes, onions, and various Russian cans of eggplant, other vegetables, and stews. In a pinch, there was even the rather unattractive prospect of dining on tins of horse meat. Fresh meat—sheep and goats—could be procured at gers, much the same way Andrews and his party availed themselves of the main course. As noted earlier, this reliance on Mongolian fare turned out to be a grave mistake. There was nothing to buy. The cities and towns were experiencing massive food shortages. It seemed an absurd predicament for a country with only two and a half million people and over twenty-five million domesticated animals. But the infrastructure of the young democracy had collapsed. No one had any gas, so it was impossible to get food from the country to the city. Conversely, the country villages and towns could not get basic stores of flour, sugar, salt, or canned goods.

This situation required some careful rationing of what meager food we had. We soon became accustomed to the deprivation; the freeze-dried food made for bland and numbingly repetitious dinners. Besides, I'm not a great fan of packaged oatmeal, canned tuna, or sardines. Unfortunately, fresh meat turned out to be less available than anticipated. It took a major part of the day to acquire a sheep from a local family of nomads. During that hot windswept summer, gers were few and far between, and our limited gas supply did not allow extended expeditions for mutton. When we did get a sheep or a goat, most of the carcass was given to our Mongolian teammates, who had even less tolerance than we did for the freeze-dried stuff. Our Mongolians remarked that the situation was tough. "We eat five meals a day," Perle reminded me. The bare-bones menu, in combination with miles of daily hiking, was the most effective weight-loss program imaginable. I didn't mind losing weight, but the lack of food soon took its toll on our energy stores. We were running on half power.

Despite the sand and the starvation diet, we forged on. On a sandblasted July 4 we snaked our way through the canyons of Eldorado. After a meager lunch that consisted of $1^5/_8$ sardines apiece and a foil-wrapped, freeze-dried fruit bar, we set upon the orange-red sands again. I found a couple of lizard skulls, as did Mark and Jim. Back at camp, we learned that Malcolm had recovered a lizard skull as well as a partial skull of a mammal. Our Mongolian comrade Perle had come up with a fine skeleton of a dromaeosaur, perhaps not *Velociraptor* but something rare, and new. Some of us celebrated American independence that night with a bottle of cognac, crouched behind a pile of wooden field boxes to avoid the whirling blasts of sand.

Much the same weather carried over the next day. Roaming the red hills, Mark found a beautiful dromaeosaur skeleton, a fairly large creature, possibly even something new. But my right eye was now swollen and barely half opened. To stay out of the wind, I volunteered to go with Perle in a Mitsu in search of a sheep at a ger. This was a fruitless expedition; the center of the valley was devoid of a single dwelling. All we found was a hell of swirling sand and powerful winds.

The following day emerged clear and cloudless. Mark was able to quarry his dromaeosaur without being buried in sand. But the respite was temporary. By 4 P.M. we were running back to camp just ahead of a huge dust and thunderstorm. The repeated sandstorms and the lack of good victuals was beginning to wear us down. July 7 was blessedly calm. Unfortunately, Mark's dromaeosaur had shown itself to be critically flawed. At the site, the beautiful skeleton stretched out before us, from the tail to a series of exquisitely detailed neck vertebrae. But there the skeleton gave out. It lacked a precious head. Although Mark was understandably upset, it was still an important find. Informally called *"Ichabodcraniosaurus,"* the skeleton remains under investigation in New York to this day.

One of the members of the team, Lowell Dingus, our geologist, was not preoccupied with extracting fossils. At U.C. Berkeley, he had completed an important Ph.D. thesis on the dinosaur-bearing Cretaceous rocks of the Hell Creek badlands of Montana. Lowell's quiet and easygoing nature combined with a steely professionalism that explains his success at various enterprises. His "day job" in New York involved directing the renovation of our vast halls of dinosaurs and other prehistoric animals—the biggest project in the American Museum's history since it was first established a hundred and twenty-five years ago. He had been focusing his attention on the sandy cliffs of Khulsan, noting the dominance of massive sand sections and the presence of distinctive streaks or curves in the cliff surface. These latter were cross-beds, which indicated an ancient system of migrating dunes. Lowell took measurements of the thickness of various rock sections and, by using a GPS receiver, plotted our localities on a map which he sketched with admirable precision in his field notebook. His job was to put our prized fossils in geologic context. Lowell recorded whether some of the sites were higher or lower in the rock section than others, or whether they came from ancient sand dunes or stream beds. Without this information, the skeletons we collected were merely a jumble of bones without a proper address.

After four days of prospecting and "geologizing" in the wind-battered red beds of Khulsan, we had to face up to a new problem. We were running low on fresh water. For most of our last day, alone or in small teams, we carefully removed slabs of rock with bone or plaster jackets prepared over the previous days. By late afternoon we prepared to retreat to the spring at Naran Bulak, some forty miles west of Khulsan. Dumping the sand out of our sleeping bags and tents, and kicking away the camel ticks and any hidden scorpions that swarmed under our ground cloths, we said goodbye without much regret to the swirling sand pit below the cliffs of Eldorado.

The Land of Tyrants and Dragons

After a day of recuperation at Naran Bulak, we shifted our sights upward to the rock sequence above the orange-red sands, the candy-striped Cretaceous badlands that the Russians and Poles had named the Nemegt Formation. Only a few miles north of our camp lay the Nemegt exposures of Altan Ula I, II, III, and IV, the famous sites of the Russian and Polish-

Mongolian expeditions. These four intricate canyons of rainbow-colored rocks hung like entrails from the underbelly of the great hulk of Altan Mountain. On July 9 we struck out for the westernmost of the canyons, Altan Ula IV. The big obstacle in our route was a spectacular dune field which guarded entry to the base of the mountain. The Polish-Mongolian expedition had reported that these dunes were often impassable, but we had little trouble crossing them in the Mitsubishis. In some respects, the Mitsus were clearly not built to navigate the Gobi—the frame cradled the transmission, gas tank, and other vital parts too low to the ground; in addition, the gas tanks had an annoying penchant for elongate cracks in one weak spot which led to many leaks and repairs through the years. Nonetheless, the Mitsus with their fat tires and high-torque engines were much better in the sand than the Russian jeeps.

By midmorning we were scampering down the steep canyon walls in search of fossils. The effect was shockingly different than our experience in Eldorado. The beds were highly variable in sweep and texture—stream channels, thin shales of lake beds, large sandstone concretions sticking out of hard cliff faces, streamers of rocks and pebbles called conglomerates that indicated the high energy of ancient floods. And the colors were lavishly diverse—blinding white sandstones, purple or mauve shales, stripes of crimson reds, pinks, and yellows that indicated ancient soil surfaces, blue-green where ancient plants had accumulated, decayed and fossilized. The bones we discovered were big bones, bits of leg bones from long-necked sauropods, a complex vertebra of a hadrosaur that would take two hands to hold, a glinting tooth of a *Tarbosaurus* four inches in length. There was the petrified wood of giant trees, some of whose trunks were more than two feet in diameter. Sadly, most of these fossil plants and bones were either shattered and scattered as "float" (the paleontologist's term for the scraps of bone or other fossils lying about on the surface of the ground) at the bottom of the canyon or encased in hard rock on a nearly vertical cliff face. It was by no means easy pickings. Certainly these beds did not provide the delightful fossil opportunities of the lower orange-red beds of Bayn Dzak, Eldorado, and Khulsan, where skeletons could be removed from the soft sands in a matter of hours or a couple of days. I thought of the huge skeletons that the Mongolians, Russians, and Poles had extracted from these canyons. What a stupendous job! Those "tarbosaurs" and hadrosaurs that proudly stand in the Natural History Museum in Ulaan Baatar represent a Promethean quarrying effort.

At Altan Ula, Efremov's party found skeletons of seven giant hadrosaurs embedded in close association in a very hard red sandstone, seemingly the product of a mass death and burial. The site was called the "Dragons' Tomb." The skeletons turned out to be rather undragonly herbivores, the duckbill *Saurolophus angustirostris,* named for the solid crest on the back of the top of the head. *Saurolophus* was a big hadrosaur, with adults reaching lengths of forty feet and some individuals looming over twenty-five feet high.

Hadrosaurs of the Cretaceous Gobi, as well as those in North America, were built to get around quite adeptly on land, but they also may have entered the water to feed on plants. The abundance of lacustrine and fluvial beds near the "Dragons' Tomb" probably provided a rich *Saurolophus* habitat. It is noteworthy that this locality contained not only skeletons but also remarkable fossil evidence for soft parts. Bubbly or pebbly impressions of skin were found in association with the skeletons. The fossilized "chunks of skin" are actually casts made from reverse-image molds pressed into the substrate by these ponderous hulks as they lay in the mud to rest, feed, or die. Skin impressions of hadrosaurs are known from other places, and in some cases they are truly spectacular. The world's premier specimen resides in the recently renovated dinosaur halls at the American Museum of Natural History, essentially a cast of the skin of the entire body—a complete mud "mummy"—of a North American *Edmontosaurus,* a distant relative of *Saurolophus.*

All hadrosaurs have a distinctive battery of closely packed teeth, whose crown surfaces combine to form a mill. Ridges acted together like a rasping file that efficiently broke down tough vegetation. There are actually hundreds of teeth in a hadrosaur jaw but they work together as two sets of grinding mills, one set in each side of the upper and lower jaws. The skull roof and the back of the jaw show prominent struts and ledges of bone. These served as attachments for muscles whose contraction moved the jaws back and forth for effective milling. The skeleton itself is appointed with a rather short but very flexible and sinuous neck, small fore-

Saurolophus (Ed Heck, reprinted with permission, AMNH)

limbs, but large elongated hind limbs. The vertebrae of the massive and lengthy tail are equipped with extended splints or chevron bones, which project downward from the body of each tail (or caudal) vertebra. Covered with flesh, these would give the hadrosaur tail its notable flattened appearance. One of the most remarkable features of the vertebral column in hadrosaurs, as well as in certain other dinosaurs, is the intricate weaving of ossified tendons. These crisscross among the spines, extending above each vertebra in the back (lumbar) and the tail region. They seem to be there for support of the back; to prevent the trusswork of the vertebral column from sagging against that great mass of muscle, fat, and internal organs.

Something, of course, had to feed on big herbivores like hadrosaurs, and the best candidate in the Gobi is the tyrannosaurid *Tarbosaurus*. During the expeditions of the 1940s the Russian team found three skeletons of *Tarbosaurus* in the Nemegt Formation at the Dragons' Tomb site, where several *Saurolophus* lay. *Tarbosaurus* (the name means alarming reptile) is very like the more familiar *Tyrannosaurus* and doubtless a close relative. *Tarbosaurus* is big. Some skeletons are up to forty-six feet long, longer than *Tyrannosaurus* by more than six feet (although apparently it stood a bit shorter than the tyrant king). Like sauropods, these animals required enormous amounts of food, in this case, meat. They were well built for the purpose. The skull of *Tarbosaurus* is more than four feet long. Its jaws are studded with recurved razor-edged teeth, some nearly six inches long. The three toes of the hind feet are appointed with viciously sharpened and curved claws. The lengthy hindlimb bones show scars for attachment of enormous muscles. Everything about it suggests power and agility, an animal capable of lunging its massive body at a hapless hadrosaur and disemboweling it in an instant.

The complex turns and branches of the canyons of Altan Ula IV soon separated our search party. On occasion, I would take a strenuous climb to the top of a promontory to check my bearings, from where I could see an antlike companion up another drainage. Lowell was on a route parallel to mine as he prospected for bones and paused to sketch a rock diagram in his notebook. He sometimes emerged on an absurdly steep cliff face, his attention totally focused on his work—on the nature of the rock, its thickness and texture and relation to other layers—rather than on his precarious position on a slope where pebbles slipped under one's boots like ball bearings.

This up-and-down ballet over gullies and ridges was tiring. On a windless day such as this the heat of the canyon was oppressive, close to 100° F. I nursed my one liter of water and kept an eye on my distance from the cars. By midafternoon I decided to make my way back over the easier pediment surface above the canyons.

Perle had found a partial skeleton of a big dinosaur, probably *Tarbosaurus,* far south, toward the mouth of the canyon. I had found a battered

Tarbosaurus (Ed Heck, reprinted with permission, AMNH)

skeleton too, and Mark had some intriguing remains of what looked like a *Tarbosaurus.* But we did not deem these high on our list of fossil treasures. Nor were we in the mood to spend the rest of the summer in the white-hot crypt of Altan Ula IV, like an army of Egyptians chiseling a granite obelisk out of the ancient Nile quarry at Aswan.

The Gobi is by no means the hottest desert in the world. And as in many deserts of high elevation, the nights can be crisp and chilly. But the fact that the Gobi sits at over forty degrees north latitude has a major drawback. The daily arc of the sun seems an eternal journey. High noon is merely a prelude to the oven-hot afternoon. It soon became apparent to us that the hottest part of the day was often quite late, around 5 P.M. Thereafter, the light for collecting fossils slowly degraded, as every bush, rock, and pebble cast a shadow, muting the reflection of fossil bone on the ground surface. We worked long days, though; the time in Mongolia for a summer season was too precious and too short to work a normal eight- or nine-hour "Wyoming day."

Resting in the meager shade of the Mitsu, Mark said, "I'm not in the mood to chunk and drill a huge hadrosaur out of this rock all summer. Let's give it one or two more days, then move on to Kheerman Tsav."

"That's doable if Dashzeveg shows up in camp with a GAZ as planned," I said. "We need him to take us into Kheerman Tsav."

Our conversation drifted back to jokes about digestive conditions—our paltry, freeze-dried fare had wreaked havoc on our empty intestines—and lost days of our youth in sunny California. We spend little time "talking shop" in the field. A "killer" fossil (our slang in the field for a great specimen) is worth a thousand words.

Although it was clear and windless where we rested, staring toward the southern orb of the Gobi, I saw huge clouds of dust that seemed to reach the stratosphere, torn from the desert by a ferocious vortex, far away, somewhere over the Gobi of northern China, assaulting towns with complicated names like Ts'un-ching-ch'a-nu-tau-erh-tifang.

I could not help thinking about the lost world of Altan Ula IV, a world full of long-necked and scimitar-toothed monsters, swamps, marshes, huge rivers, enormous trees, and mirrorlike lakes. A world so different than the one around me.

HELL IN THE GOBI

It turned out that Dashzeveg, Bayer, and a small crew arrived at our camp in Naran Bulak that day. Malcolm and Priscilla had spent some very successful days collecting mammal skulls. Jim Clark had found the first of these beautiful skulls encased in a nodule during a leisurely afternoon walk. These skulls were from younger Tertiary sandstones—beds some forty-five million years in age—that surrounded our camp. The fossil-loaded rocks put on record the glorious and burgeoning age of mammals that postdated the great Cretaceous extinction event. Dashzeveg was passionate about this terrain too, and he set up a crew to shovel out some of his favorite quarry sites. The loose sand and silt was loaded in bags and "washed" with water from the spring in boxes floored with a fine-meshed metal screen. This had the effect of breaking down much of the finer-grained rock and leaving a residue of tiny pebbles and chunks along with tiny mammal teeth, jaws, and bones on top of the screen. This screen-washing technique was pioneered by Malcolm and others in the 1950s, for the purpose of retrieving precious microscopic fossils from Tertiary and Cretaceous beds in Wyoming. But here, Malcolm and Priscilla preferred to comb the slopes for those spectacular skull-filled nodules.

Our platoon from Altan Ula, attracted by the diversion of riches from the age of mammals, joined in the nodule hunt. But soon we were back on the Cretaceous trail. On July 11 the whole team struck out for a set of magnificent canyonlands called Kheerman Tsav about thirty miles southwest. "Kheerman" derives roughly from "Kremlin," the great walled city in Moscow central. It is a perfect word for the place—the imposing red ramparts of the outer walls of Kheerman Tsav surround a crypt of great pinnacles and spires, the Gobi's Monument Valley, that is virtually unparalleled.

Unfortunately, Kheerman Tsav was also an unparalleled hell. Not only was it scorchingly hot, but its swarms of flies marauded us at every step.

"Maybe there'll be fewer flies at the main locality," I said hopefully.

It was not to be. By early evening we reached the outer westward gates of Kheerman Tsav; these magnificent pinnacles burned like red torches, catching the long rays of the sun. But the flies still hung over us in great clouds.

To make matters worse, the route into the canyonland was horrendous. Steep hills of sand entrenched our trucks. We had to work constantly to extract our vehicles, one hand on the lever of the Mitsu's winch, the other swatting away the flies. It took three long hours of this labor to cover the half mile or so to our campsite in the heart of the canyon. At 10 P.M., with the sunken sun still issuing soft orange streamers across the Gobi sky, we pitched camp. The flies were diminishing in numbers, trailing off to regroup for tomorrow's epic battle. Too tired to cook a meal with the stove, we settled for some salami and some old, rather decrepit potatoes, roasted in a tin can in the fire. There was little talk.

As we retired to our tents, Mark whispered to me, "This is going to be hard, but it's our only chance out here. We gotta find some good stuff."

I nodded and retired to the sanctuary of our tent, intent on enjoying the last night before our death march through Kheerman Tsav.

Flies, rather than fossils, were the chief topic of conversation at breakfast. Someone remarked on the incredible fierceness of the rising sun. Kheerman Tsav is a thousand feet lower in elevation than the Altan Ula badlands, and the blistering temperatures bore witness to this topographic change. We expected the temperature to easily break the 100° F. mark. I don't remember the conversation much. I was obsessed with sweeping the hordes of flies off my face. To them, we were the oasis in the dunes—the life-giving, perspiring source of water. There was no other water for miles.

Mark and I staggered off at nine o'clock, crossing a deep wash that marked the main drainage of the canyon and prospecting the slopes on the other side. There Mark discovered a fragmentary skull and skeleton of the bird-beaked theropod *Oviraptor.* The specimen was battered and weather-beaten, but we were elated—*Oviraptor,* a very rare dinosaur, was one of our primary targets. We marked the specimen for later retrieval. We relished any brief gust of wind that momentarily scattered the flies. Each of us now walked with his own dipteran cloud. Heat often stills the activity of insects, but this species seemed remarkably unaffected by the inferno of the canyons. The afternoon brought darkening clouds and what we hoped was an impending storm. But the storm grazed past us, and the air became sultry and more suffocating.

Our meanderings took us eventually back to the *Oviraptor* site. Nearby I found some bits and pieces of lizard skeletons and a nice jaw and partial skull of a protoceratopsid. These were a cut above much of the bone we had found; most of the discoveries of the day were scraps of limbs and jaws and isolated "proto" teeth. Where were those brilliant skeletons of the sort retrieved by the Russians, Poles, and Mongolians? With some disappointment, Mark and I removed our one prize, the *Oviraptor* skeleton, near the end of a long workday. By 7 P.M., in the stubborn 100° F. evening temperatures, we headed back to camp with our specimens cradled in our arms. The flies gathered around our faces and arms, attacking with an unprecedented vengeance. I can't remember a more miserable desert mile.

Back at camp, we discovered Dashzeveg was still absent and had not returned even at midday to replenish his water.

"He is not young man," Bayer said. "I am very worried."

It was now 8 P.M. To my horror, I realized that Dashzeveg had been out in the blazing canyons of Kheerman Tsav for over eleven hours with only a liter of water. The situation was extremely serious, and could be tragic. Mark and I volunteered to search for Dash. We knew that he had taken the twisting corridors to the north of camp that eventually intersected with Kheerman Tsav II, a place where years ago Dash had found his precious Cretaceous mammal skulls. We headed out in the main drainage, carefully taking note of landmarks that would guide our return. From time to time we could see the telltale sole prints of his desert boots, only to have them frustratingly disappear on the steep talus slopes that bordered the wash. Later, we would pick up the tracks reemerging in the sand at the canyon bottom. Dashzeveg seemed to be stalking fossils like a restless *Velociraptor* stalking its prey. Over an hour of this desperate tracking passed before our hand radio crackled.

To our great relief, we heard Lowell saying, "Come on back, we see Dash headin' in."

We turned off our radio to conserve the battery and headed back, dreaming of dinner and sleep. But our decision to shut down radio contact turned out to be dangerously presumptive, as was our camp's decision to call for our return. The man taken for Dashzeveg out on the northern horizon turned out to be one of the drivers, walking slowly back from a latrine spot. We got back to camp only to discover Dashzeveg was still missing. After wolfing down some freeze-dried chili, Mark and I headed out again. It was now after 9:30 P.M., and the light had receded. We shifted our search to the high ridges where some light survived. Jim and Bayer headed down the main canyon, their radios on, waiting for any sighting from us. Mark and I desperately scanned the impossible shadow-filled maze of gullies below us. Suddenly we saw a figure on a high ridge only half a mile away. Dashzeveg! But it turned out to be Perle instead, who had also joined in the search. The light withered away and I began to face the possibility of a disaster. Dashzeveg was tough, but no one could hold up for an extended time out in that desert inferno without water.

"Mike, what's that?" Mark said, squinting through my binoculars toward a canyon flanking the main drainage searched by Bayer and Jim.

I grabbed the binoculars and saw a lanky man in a desert storm cap, walking slowly but deliberately up the wash. It was him!

"Jim, we see Dash. He's one, maybe two canyons over from you on your right," we called on the radio.

But Dash soon disappeared in the shadow of a canyon ridge, and we saw no more sign of either him or the search party. A few uncomfortable minutes passed before we heard Jim's voice over the radio.

"We got him. He's all right—I think."

Back at camp, I found Dash as exhausted as I had ever seen him. In the heat of the day he had developed a painful migraine and "slept" for a few hours in the afternoon heat. He awoke tired and thirsty and realized he was a great and confusing distance from camp. The return trip had been endless and excruciating.

"Ho, ho! Kheerman Tsav too hot now!" He managed a husky laugh before retreating to his sleeping quarters.

Mark and I were feeling pretty bad ourselves, from both the heat and the sheer fatigue. But the excitement of the rescue precluded sleep. We stayed up long into the night socializing around the fire with our Mongolian comrades, long after our American colleagues had gone to bed. The flies had at last retreated and the air was comparatively cool and bracing.

The constellations mapped the black vault above the region's jagged black outlines. Kheerman Tsav was transformed into a wonderland. It was a glorious desert night, like one of those nights in the Sahara described by Paul Bowles in *The Sheltering Sky*, when the moon "had slipped behind the earth's sharp edge," a landscape that lacked "comforting terrestrial curves."

The next day, as expected, was one of unabated misery. The heat had driven us all into the shade by the late afternoon. When the heat settled in, all the flies of the Gobi Desert seemed to assault us in waves, converging on Kheerman Tsav as if part of an allied invasion. We downed some freeze-dried awfulness called chicken stew, waving off the flies in a mad, Saint Vitus dance. Even the night offered little relief. The heat from the embers of the fire mixed with a changing weather front, a cloud bank drifting in from the south. The inside of the tent was suffocating even with the flaps pulled back. In the middle of the night I was startled by the flapping of the tent in a stiffer breeze. I opened the netting of the door, scampered out, and stood barefoot in the sand. The campfire that we had carelessly deserted was now raging in the strong wind, catapulting embers toward the tents like streams of orange flak. I quickly put it out. I remained outside in the lukewarm sand for a few moments of comparative relief. Toward the south and west, I could see the celestial vault nearly to the horizon, revealed by the only gap in the rock cathedral around our camp. The familiar sky of midnight was now obliterated. Scorpius, our summer constellation, had slipped below the horizon. In its place at the meridian were a grouping of ragged, esoteric constellations, of which I recognized only the ellipse of Piscis Austrinus. The paradise of night was now only a span of three hours before dawn.

July 14 began as had the last previous two mornings—with the incessant buzzing of flies colliding on the outer shell of our tent. A Kheerman Tsav wake-up call. We could see their myriad shadows against the yellow nylon. It was to be our last day at Kheerman Tsav. The unexpected heat of the place had drained off our water supplies; some of the Mongolian drivers were even less accustomed to these conditions than we were, and it was hard to impose a strict water ration on them just to prolong our agony.

Our prospecting of the last two days had not brought much reward for all our discomfort. Doubtless our intensity and concentration were dulled by the inhospitality of the place. Both Dashzeveg and Perle claimed that Kheerman Tsav was usually a much more tolerable locale (as indeed we were to find out in future seasons). But for the moment our concern was an orderly retreat. I glanced with annoyance at a pot of water boiling for instant coffee. Our water supplies were at hazardously low levels, and we still had to travel fifty burning miles over difficult terrain before we could replenish our water at Naran Bulak. A windstorm or a vehicle breakdown could add to our risk. I remember reading that the only difference between an epic and a tragedy is that someone doesn't die in the former.

Fortunately we escaped Kheerman Tsav and made Naran Bulak without further mishap.

THE WELCOMING PARTY

By late July we had made the great Gobi circuit: Ulaan Baatar to Saynshand and the nearby fossil beds, across the vast eastern Gobi to the Flaming Cliffs, on to the Nemegt Valley, eventually reaching the hellish outpost of Kheerman Tsav, and finally back to the Flaming Cliffs. But there was one spot we had left to visit—Tugrugeen Shireh, sometimes known as Tugrik—where various triumphs of the past spurred our desire for a more victorious summer. Tugrugeen Shireh is a sand escarpment near an alkaline lake. It lies only thirty miles west of the more famous and more monumental Flaming Cliffs. But Tugrugeen has its own arresting profile. Its white linear southern exposure, from a distance, looks like a bony finger in the dull green of the surrounding hills dotted with desert scrub. The finger points directly toward Arts Bogd, a massive flat-topped mountain over 8,000 feet high. Even in the midst of summer the crests of this peak, and certainly its more western sibling Baga Bogd (11,775 feet), are dusted with ephemeral snows. The butte of Tugrugeen itself runs only about one mile from east to west. Its rim is not utterly flat; it tilts almost imperceptibly to the north, where some lower escarpments and gullies and promontories are exposed on its northern margin. At the southeastern base of the butte is a graceful beige sand dune. The dune—over 80 feet high—is very isolated. Its placement looks rather odd, as if it were added to a half-finished movie set for *Lawrence of Arabia.* Our camp was pitched on the windward side of this dune, high enough to be out of the zak-choked wash that runs toward the alkaline lake. The camp is one of our favorites in the Gobi. The afternoons can be brutal along those white-hot cliffs, but by evening magnificent clouds build out from Arts Bogd and veil the sun.

Fossils are found in all the outcrops along the margin of the Tugrugeen Butte, but the prime fossil-hunting grounds are the steep southern and western exposures. Here, long banks of sandstones flatten out, leaving an array of nodules ripe for picking. The cliffs are also cut by white sand canyons that offer a lot of surface area for prospecting. Although the rocks of Tugrugeen are not the "right" color, the animals preserved there are very typical of the red Djadokhta Formation—protos, theropods, mammals, and lizards well known from the Flaming Cliffs—including the famous "fighting dinosaurs" mentioned in Chapter 2. The preservation of the skeletons is particularly fine here, in many ways even better than at the Flaming Cliffs.

We reached the site on a very humid and hot afternoon at the end of July. After unloading a few crates and equipment we struck out for the slope opposite our camp. Not more than eight hundred feet away, we found a perfect skull of a large adult *Protoceratops,* its beak precisely sculpted and pointed in the air, and its tooth battery pristine and freshly brushed. It was spectacular even to us jaded proto-finders. Everybody scattered, hoping to find their own riches. I wandered in sweeping circles on the concretion-covered pediment, hoping to find a "fighting dinosaur" in the next gully.

We finished early that first day at Tugrugeen, drawn by the luxury of a warm shower from our portable "sun shower" bags. The area around Tugrugeen was well populated with livestock, and the vegetable gardens in the nearby town of Bulgan were at their peak. Our feast that night was spectacular—sheep leg stuffed with onions, scallions, and herbs wrapped in aluminum foil and thrown in a pile of glowing zak coals. The main course was complemented by foil-wrapped baked potatoes and a crisp salad of fresh cucumbers and oil and vinegar dressing.

Our celebrations drew a local crowd. A fat man on a motorcycle I'd seen earlier returned in traditional Mongolian wrestling garb, his great brown belly proudly overhanging his wrestling briefs. With him were his entourage, about ten young wrestlers, all dressed in wrestlers' uniform.

Some were his sons; all looked a bit trimmer and more muscle-toned than their leader. They brought along quantities of irak, fermented mare's milk. It was party time.

That evening we sat and passed the irak cup, and watched demonstrations of wrestling moves. The gringos, Ulaan Baatarians, and Tugrugeenians laughed and joked and had a jolly time.

THE PROPHECY

Our last day at Tugrugeen and my last day in the field in 1991, Jim Clark turned up the skull of a sharp-nosed mammal. Returning to camp, he extracted from his pocket a small collecting bag containing a nodule carefully wrapped in toilet paper. When he unraveled the paper we could see the snout region of a small placental mammal skull. Months later, laboratory preparation confirmed our hunch in the field that this nearly perfect skull belonged to *Zalambdalestes,* a species whose relationships to more modern mammals greatly intrigued us. *Zalambdalestes* has elongate front incisors, a gap between the incisors and the anterior premolars, and lengthy hind limbs. This combination of features is somewhat reminiscent of rabbits. Indeed, my colleague Malcolm McKenna had a hunch that *Zalambdalestes* could be a granddaddy rabbit, a rather dramatic connection since the first undoubted members of the Lagomorpha—the order to which rabbits and pikas belong—occur some twenty million years later. I was skeptical of Malcolm's idea and we had a running debate. Jim Clark's Tugrugeen skull, I realized, might settle the matter; it is certainly the finest yet discovered for *Zalambdalestes,* indeed was one of the best ever found for any Mesozoic mammal. But our best hunches would have to await detailed study of the skull back at the lab in New York.

Sauntering up a long broad canyon near the west side of the butte, Jim next found a smattering of theropod bone exposed in the sand gully. It turned out to be parts of a decent skeleton of *Velociraptor,* as well as a good bit of the roof of the skull, parts of the snout with teeth, and the bony braincase. Our last day at Tugrugeen was turning out to be our best one.

Meanwhile, Mark and I had spent most of the day organizing specimens, repairing plaster jackets, and checking the field catalogue—the numbered list of specimens with brief descriptions that had been collected throughout the summer. Since we were departing for Ulaan Baatar with Lowell, Perle, and Chimbald the next day, we planned to haul out some of the most precious fossils. Tomorrow, Malcolm, Priscilla, Jim Clark, Dashzeveg, and Bayer would head northwest to the older Cretaceous outcrops of Khoobor for a few more weeks of work. They wouldn't want to be encumbered by the task of securing those delicate and already fragile fossils over more bumpy roads. It was four in the afternoon, about an hour before the height of the afternoon heat of the day, before we had a chance to escape the ennui of this work and head for the hills. Our field of operations was a white sand slope below the main section of the butte. In less than an hour Mark picked up a concretion with a beautiful little lizard skull. A moment later I bent down to a concretion nearby and found a complete skull of a mammal about the size of a chipmunk, its long, gnawing incisors extended along the surface of the rock.

Two "killers" in one hour—not bad at all. As we went our own way wandering over the slopes, I reflected on our weeks in the Gobi. Tugrugeen was rich no doubt, but what had it been like when the Polish-Mongolian crew first came upon it? How glorious it must have been to find a score of small vertebrate skulls and a pair of fighting dinosaurs among a battlefield of *Protoceratops* skeletons.

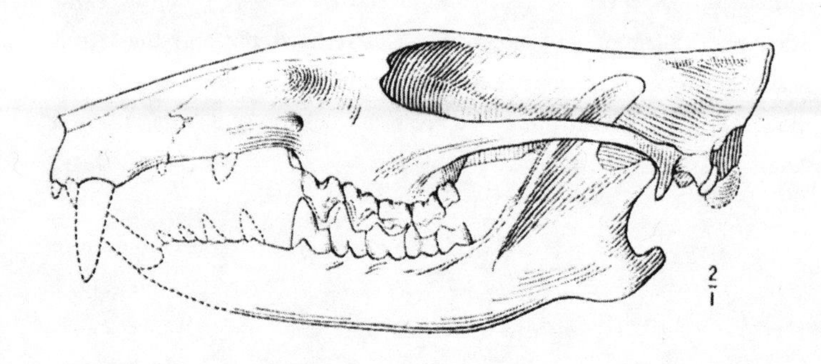

Skull of the mammal *Zalambdalestes,* about twice natural size *(Reprinted with permission, AMNH)*

These thoughts absorbed me for a longer time than I realized. It was now 7 P.M. and the shadows were obscuring the glint of bone on the desert surface. I headed back to camp, where I saw an extra GAZ parked next to ours. Barsbold, the Mongolian dinosaur specialist, who was conveying a small group of French paleontologists and an Italian philanthropist on a Gobi tour, had stopped by for a visit.

Cultured, personally impressive, and congenial, Barsbold clapped my hands in a fond greeting and smiled broadly. His hair, despite its whiteness, was thick, and his round face and open smile gave him a good deal of boyish appeal. Barsbold asked us about our long summer, interested in our reports of mammals, *Velociraptor* skulls, and headless dromaeosaurs. I expressed a muted sense of accomplishment to all this.

"It's been great, but not yet like Andrews in '23, Efremov in '46, or you and the Poles in '71. No fighting dinosaurs, no new localities."

Barsbold laughed dismissively and squinted intensely into my eyes, speaking slowly in articulate English.

"But I think you Americans will be very successful. We have heard you work very hard. And it is only just that you will make in the next years the truly great discoveries. For your work must be the closure, the realization of what the American Museum team began seventy years ago."

His assurances were golden to my ears, and very moving, but there was a certain unintended edge to them that I found disquieting. We had already spent two summers in the Gobi and, despite our successes, had yet to measure up to our predecessors. Would the next few summers prove otherwise? A triumphant discovery was, I realized, not only the expectation of our Mongolian comrades; it drove much of the interest, and frankly the funding, back home, too. The pressure I felt we were under was real.

CHAPTER 4

THE TERRAIN OF EONS

I pace upon the battlements and stare
On the foundations of a house, or where
Tree, like a sooty finger, starts to the earth;
And send imagination forth
Under the day's declining beam, and call
Images and memories
From ruin or from ancient trees,
For I would ask a question of them all.

W. B. YEATS.
1928. "The Tower."

Paleontologists have only certain battlements upon which to pace. Although fossiliferous rocks take them on a global hunt, they are found in only particular places in these widely scattered destinations. Tropical jungles are not good places for such rocks. Their lush vegetation conceals the texture of the land. (I have a paleontological friend who says he cannot stand the beauties of Costa Rica because "there are too many trees.") Their torrential rains sweep away sediments; huge masses of decaying vegetation and microbes keep recycling the soils which serve as the burial sites for future fossils. Many high mountain regions are also less than optimal places for bone hunting. These regions are heaped with volcanics and other igneous rocks and are tortured, thrust up, and reheated. The rocks in high mountain ranges as well as the poles are also, of course, typically buried in ice and snow. This general prescription does, however, have its exceptions. Dinosaurs and other fossils are known from the limited collar of exposed rocks on the mainland and islands of Antarctica. Many paleontologists, including teams in which I have worked, have found impressive suites of fossils in rocks sticking out of the crests of mountain ranges. There are even instances where some lost plateau of sandstone or limestone brimming with fossils is left standing in the midst of the tropical forests, like the famous Santana site in Brazil, or the fossil outcrops in the thick jungles on Cuba. But these are on the whole unlikely spots for such discoveries. I do not anticipate traveling to the neotropics in the very near future. More importantly, it is not likely that such a folly would be supported by reviewers and research foundations. Paleontologists place their bets on the best terrain.

Such terrain can itself vary considerably in its texture and profile, but it shares a number of characteristics. It should be cropped of annoying, if aesthetic, vegetation. There should be some relief—some cliffs, gullies, and hills—but not so steep as to frustrate ambulatory searching. Of course, the terrain should be overwhelmingly dominated by sedimentary rocks. Moreover, these rocks should be relatively horizontal or gently tilting (in geological parlance, gently dipping). Their exposure should not be befuddled by crisscrossing dikes or offsetting faults. Lastly, the matter of degree of exposure is critical. There should be a high surface-to-volume ratio in fossiliferous sediments. That is, the rocks should be exposed in a way that increases the chances for revealing their contents. Thus places of aridity, where erosion sculptures the surface rock into great infolding sutures represented by canyons, gullies, and badlands, are good prospecting areas. Valleys footing great mountain ranges like the Nemegt are also ideal, as long as they are not so old that their rock surfaces are buried by accumulations of desert pavement or sand. Tablelands, like the mighty Colorado plateau of western North America, provide expansive terrain for such fossils.

The Gobi of Central Asia combines all these qualities in an extraordinary, perhaps unmatched way. Remember that a place like the Nemegt

Valley actually exposes Cretaceous rock facies representing two basically different environments. The upper beds, those of the Nemegt Formation, are sandstones and shales representing lakes, streams, and flood plains, a series of habitats very close to that preserved in the famous dinosaur beds of North America. Altan Ula—where our team scampered up and down steep cliffs and narrow washes in 1991 and subsequent seasons—is a canyon carved out of a very rich series of sedimentary rocks of the Nemegt Formation. This rock unit has a far reach. Its type section is stacked on top of the red beds forty miles to the east of Altan Ula, at the Nemegt locality itself. The unit is also lavishly displayed in the great canyonlands of Bugin Tsav and Kheerman Tsav, in the blazing windswept frontier west of the Nemegt Valley.

In contrast to the ancient sand dunes exposed at the Flaming Cliffs or Khulsan and Eldorado, the Nemegt Formation represents a "lush and soggy world." Numerous rock facies indicate Cretaceous stream and river channels, mudflats, and shallow lakes. Fossilized remains of fruit show that *Nyssoidea,* an extinct relative of the sour gum tree *(Nyssa sylvatica),* was a successful shrub along the rivers and seashores. Members of this group today are cultivated for their dazzling autumn foliage. Today seeds of *Nyssa* are dispersed in the feces of small birds and mammals, and it is likely that such a mode of transport and seed dispersal pertained to the Cretaceous Nemegt as well. The *Nyssoidea* gallery was also populated with *Bothrocaryum,* early members of the dogwood group, also today a popular ornamental. Petrified wood, some trunks reaching sixteen to twenty-six feet long, indicate impressively large and mature trees in dense forests. Simply cruising the steep slopes of places like Altan Ula IV, we have found generous stacks of this mineralized wood. These appear to be remains of Araucariaceae, a conifer group that persists today as thirty-two species in the Southern Hemisphere, attracting fanciful names like the "monkey puzzle," "bunya-bunya," and "Leopold Astrid" as well as the more prosaic "Norfolk Island pine" and "Paraná pine." Some araucarians today tower over 250 feet high. Thus it is likely that the Cretaceous forests of the Nemegt were expansive and thick, with tall conifers forming the high canopy.

Yet it is also probable that during this time the habitat had a variable character—lakes, marshes, river valleys, forests, and open, boulder-sprinkled parkland—that maximized its carrying capacity for big and diverse dinosaurs. The rocks of the Nemegt Formation indicate that this enriched habitat was a mosaic offering abundant and diverse food and good living space. It was in some sense a biological Garden of Eden.

The great sinks of the world's diversity today are the tropical rain forests. There, the accumulated biomass represented by hordes of insects, invertebrates, fungi, leaves, branches, logs, and moss per square acreage of tropical jungle is astounding. Nonetheless, the highest biomass—a measure of both the size and abundance—as well as species diversity for the large vertebrate animals today exist in savannah regions of East Africa. Biomass figures range from 100,000 pounds per square mile in Serengeti savannahs to only about 400 pounds per square mile in tropical rain forests. The Cretaceous world, too, might have experienced its greatest biomass in areas like the Nemegt, where vegetation and terrain produced a rainbow of habitats.

NAMES FOR STRATA

The Gobi dinosaur skeletons from the Nemegt Formation are remarkably complete, rivaling or surpassing that quality in many other regions of the world. But the more delicate and very critical skeletons of smaller dinosaurs and other vertebrates are not preserved in the Nemegt Formation, a situation very like the sequence in North America. It is in this respect that the lower and older strata, the red-orange sands of the Barun Goyot Formation (and in the region of the Flaming Cliffs, the Djadokhta Formation), are so remarkable. The contrast between these rock units and their fossils gives us an opportunity to consider the science involved in recognizing and naming strata.

Originally the lower red beds of the Nemegt Valley were deemed by the Russian 1940s expeditions as the boneless "blank series." But, as noted earlier, the Polish-Mongolian team returned to Efremov's Nemegt wonderland and filled in the "blank series." These rocks have abundant material but of a different quality than the upper Nemegt series. Unlike the latter, they were not lavishly appointed with the most massive dinosaurs, hadrosaurs, sauropods, and tyrannosaurids. The bigger stuff of the "blank" zone was instead represented primarily by skeletons of armored ankylosaurs. They also contrasted with the upper beds in having a significant sample of protoceratopsids. And, unlike the upper Nemegt beds, they were loaded with mammals and lizards.

With further collecting and further descriptions of rock types comes further refinement of the geologic sequence. Such refinement requires names. The geologists with the Andrews expeditions, Berkey and Morris, named the red sandstones at the Flaming Cliffs the Djadokhta Formation. The name is a variant of translations of the Chinese phrase for Bayn Dzak, where "Djad" is actually equivalent to "dzak" or "zak."

Thus rock formations, like time intervals, have formal names. But time units and rock units are not equivalent. A formation is a unit of rock of particular quality that helps identify it; the criteria are color, the type of constituent rocks (sandstones, mudstones, silts, etc.), and related features. A reference locality or type section is used as a basis for recognizing a particular formation elsewhere. In the case of the Djadokhta, the type section is the Flaming Cliffs of Shabarakh Usu. Does this mean that all localities where the Djadokhta Formation is exposed are of the same age? Not necessarily. A formation can in a sense cut through time because its qualities are due to deposition under environmental conditions that might have shifted from place to place over time.

One can visualize this by considering changing coastlines. Let us say for instance that in a particular region the sea level begins to rise. The sea was earlier extensive in the west, now it begins to invade eastward and flood the coastline. A characteristic rock unit, say a limestone, laid down in the sea progressively forms farther east, though at a later time. Thus the bottom of the limestone formation in the east is actually younger than the bottom of the similar-looking limestone formation in the west. Since the strata both east and west are so similar in rock type, they can be thought of as part of the same formation. However, the eastern and western exposures of formation do not represent the same time, only the same environmental conditions. The age of the formation at different places must be established by independent clues, the kinds of fossils preserved or dates derived from radioactive minerals or paleomagnetic signals. Rock formations then can be *time transgressive*. It is not correct to refer to the lower part of a formation as earlier, it is only below the other part of the formation—we call it "lower" or "upper" Djadokhta, not "earlier" or "later" Djadokhta.

Having said all that, it is merely honest to point out that formations often do have an assemblage of particular fossils that represent a particular time. It's just that we can't assume this from the outset. In the Nemegt Valley the Polish-Mongolian team was therefore confronted with an interesting problem. The upper beds were definitely fluvial (meaning stream-deposited) and lacustrine (meaning lake-deposited). The orange-reds below this sequence were definitely deposited in a contrasting arid environment, one possibly dominated by sand dunes and cliffs as well as small streams and ponds. Wherever the two types of strata were exposed, the orange-reds are always below the fluvial or lake deposits. The former not only represented a different environment but also an earlier time. To support this notion, the fossils in the two units were quite different in their composition and diversity. The upper dinosaur-rich but mammal-poor sequence was named the Nemegt Formation.

What about the lower orange-reds? Here there were two choices. Since these looked so much like the Djadokhta, that name might apply. Alternatively, the Nemegt lower sequence deserved a new name, certainly one better than the informal and incorrect "blank series." But in a sense the problem here is not really fair; it has the character of a shell game. This is because the Djadokhta Formation at Flaming Cliffs is in no place capped by the Nemegt Formation; the latter simply doesn't exist in the northern central area of the Gobi. Struggling with this conundrum, the Polish-Mongolian researchers noted slight differences in the rock types between the Djadokhta and the lower red-orange sands of the Nemegt. They also argued that the vertebrates in the latter seemed to indicate a "somewhat more advanced" stage of evolution than the denizens represented by the Djadokhta at Flaming Cliffs and nearby locales (although this wouldn't justify a new formation name). They distinguished the lower Nemegt sequence as its own rock unit—the Barun Goyot Formation.

Naming the Barun Goyot Formation seemed a logical way to solve the geologic problem of the Cretaceous Nemegt. It is, however, difficult to win a shell game. Our own work in the Nemegt, Kheerman Tsav, as well as the Flaming Cliffs, does not in our minds establish such a clear distinction between the Barun Goyot and Djadokhta Formations. By any name, however, the Barun Goyot is an important and dazzling sequence of rock. It is also still loaded with fossils. Indeed, it is puzzling that the very competent Soviet team missed these treasures. Perhaps they were thinking too big. Aside from the pachycephalosaurs and ankylosaurs, the Barun Goyot Formation contains smallish, delicate skeletons of theropods, and minuscule mammals and lizards. The Polish-Mongolian team seemed to have expanded their search image to include these smaller targets.

Other Cretaceous beds preserve delicate creatures like those found in the Barun Goyot Formation at Eldorado as well. In most other places, however, these more fragile skeletons are broken up and dispersed in small bits over the surface of the outcrop. By contrast, there is no place in the world that so exquisitely and completely preserves the range of Cretaceous vertebrate life—from the huge to the most minuscule—as the red beds of the Gobi Desert. To appreciate this distinction, consider a simple fact. By the conclusion of the Polish-Mongolian expeditions, Cretaceous mammals in the Gobi were known from over one hundred skulls, several with partial or complete skeletons. The scientific literature on Cretaceous mammals from North America is based on nary a skull. The hundred and twenty-five years of work on that continent by paleontologists has yet to turn up one mammal specimen that remotely approaches the information content of the Gobi mammal skeletons.

THE ODDS AGAINST IMMORTALITY

It does seem remarkable, and lucky for us, that the remains of fossil vertebrates, like the extraordinary assemblage from the Nemegt and Barun Goyot Formations of the Gobi Desert, are so well represented. A lot of the original animal, namely the soft organs and muscles, has of course been lost through the ravages of decay or scavenging after death. But that bony skeleton, once impregnated with minerals, endures the eons. This is more than a matter of lucky coincidence. For land creatures to attain large size requires reinforcement, in this case an internal skeleton made of hard bone. Some terrestrial creatures solve this problem another way; the insects are made up of a hard exoskeleton derived from chitin. There seems to be a size limit to a chitinous external skeleton. There are a few land giants among the arthropods; coconut crabs come in astounding and rather threatening sizes of three feet from the tip of one jointed leg to another, for example. But no members of this group match the dimensions of the larger vertebrates. An external support system simply isn't suited for stabilizing and moving a lot of weight.

A chitinous exoskeleton is also not the best material for preservation. Of course, certain extraordinary sites with ancient tree sap, or amber, may preserve both hard and soft body parts, and even DNA, to a degree that no other fossil can match. But these amber caches are famously rare and insects for the most part are correspondingly rare or ill preserved. Most of the best-known fossil sites for terrestrial arthropods, in addition to the amber localities, are preserved in special rocks under special conditions—for example, paper-thin shales with abundant plant debris that represent the decaying bottom of a shallow swamp. Here the small invertebrates sink to the bottom and their exoskeletons remain as carbonized films that trace the body plan.

Vertebrate bone, by contrast, is preserved under a great variety of conditions, in sediments that represent ancient sea bottoms, swamps, streams, rivers, lakes, mud flats, channels, flood plains, sand dune fields, and forests. Bone is opportunistically durable, and is available from many ages and places.

Despite the fortitude of vertebrate skeletons, many forces work against their preservation in a state of magnificent entombment. Think of the improbable events between death of an organism and discovery of its fossil remains. An animal dies of disease, old age, starvation or is killed by a predator. Its carcass may be ripped apart and picked over by scavengers. Then the rains come; floods carry the shredded carcass into a drainage system, first a small stream, then a river. The already tenuous framework of individual bones is further broken up; the skull pops off and comes to rest in a gorge on a spit of sand in the middle of the stream. A limb breaks away and is lodged in the left bank; another limb is stuck in the right bank. The rib cage explodes and is scattered. These remains may be uncovered again as the flood recedes, only to be violated and scattered with more rains and flooding. Finally the shards of the beast are enveloped in mud, eventually completely buried and compressed by tons of mud that solidifies to rock. Someday, hundreds or thousands or millions of years later, this layer of rock pokes through the land surface, exposed by buckling of the earth's crust. The nose of the rock is sculpted into a cliff by erosion of the wind and the water, and bits of bone are exposed in the process. A paleontologist happens upon these bones but must be content with only a fragment of the former skeleton, pieces of ancient evidence like bits of broken glass in a recycling pile.

And maybe not even that much. There is no reason to expect that a skeleton will endure the ravages of any step involving its fossilization. It could be shattered and crushed to unrecognizable powder by a suite of scavengers, like those bone-crushing, marrow-sucking hyenas on the Serengeti plain. The moving streams might feed rivers of such power that the bones are broken, beveled, or rounded into indistinguishable pebbles and carried to the sea. The forces of sedimentary compression might squeeze bones like toothpaste into ghostly smears of their former sharp-edged solidity. And why indeed should even the best of buried bones be revealed at all? Many fossiliferous sediments are overburdened by great heaps of other rocks many miles deep. Eons of erosion and uplift are required to shave the encrusting layers of rocks away from these bone beds. On the other hand, the bone-laden sediments may be altogether too naked to the forces of nature. A sedimentary facies may winnow or wash away soon, at least geologically speaking, after they are deposited. Finally, the shudders and upheavals of the earth's crust are not kind to fossiliferous beds. Faults slice up such strata, rising mountains buckle them, volcanic eruptions or masses of underground molten rocks (igneous intrusives) cook them. All these forces violate the conditions most suitable for fossil preservation.

Clearly the creatures who populated the Cretaceous Gobi eighty million years ago beat these odds against immortality. But why are they so exquisitely preserved, even as fossils go? The question will be considered later, in the wake of a remarkable discovery. At this point we can reflect on our theater of operations for the Gobi expeditions. The Nemegt Formation and the lower Barun Goyot and Djadokhta Formations reveal two ancient Gobi worlds, one wet, one dry, both richly populated with strange creatures. Prehistoric beasts enjoyed the verdant valleys represented by the Nemegt Formation. At an older time, in which the Barun Goyot and Djadokhta strata were formed, the Gobi may have looked more or less as it does today—in some places an oasis replete with flowers, trees, and beasts of the fields, in other places rather more like Mars.

Griffins in the Sand

Given the plentiful fossils underfoot and the centuries of invading armies, migrating caravans, and wandering nomads in the Gobi, one wonders if alert, keen-eyed warriors and travelers stopped to ponder the origin of skulls with frills and parrotlike beaks, or skeletons several times larger than the bleached bones of a camel. After all, some of these specimens are not hard to find. We have even seen some large dinosaur skeletons eroded out on the crest of a hill while gazing out our car window.

It is hard to imagine that humans in this region were completely oblivious to these fossils before Andrews and his team explored the area in the 1920s. There is a tradition for use of fossil bones, the "dragon bones," for medicinal purposes in China that goes back centuries. The locations of such fossils were obviously well known to collectors whose motivations were in no way paleontological. Were such things known to the wanderers of the Gobi? Direct historical records are of no help here, but the possibility of recognition has not escaped some authors. Adrienne Mayor, a *cryptozoologist (crypto,* hidden; *zoos,* animal), claims that *Protoceratops* and other such fossils may be the actual inspirations for the griffins. These mythical creatures had wings, a hooked beak, and the powerful body of a lion. According to legend, they fiercely guarded the entry to mines of glittering gold. The author notes the abundance of ancient and modern gold mines in the western Gobi, the Altai Mountains (altai means gold), and other areas of Central Asia. This territory encompasses the ancient realm of Scythians, where the legend of the griffins allegedly originated.

This is not the first time fossils have been linked to well-known legends. Empedocles, writing in 400 B.C., noted that the fossil elephant skulls common in the Mediterranean region could be associated with the Homeric legend of the Cyclops. Indeed the enlarged nasal cavity rather high in the skull of elephants—the place where the long trunk takes root—looks like the socket for an enormous single eye. It is hard to blame Homer in all this; he never saw a living elephant. A giant fossil skull unearthed near the town of Klagenfurt, Austria, was eventually (centuries later actually) used as the basis for a reconstructed head and body (the latter not known from a skeleton). The resultant "dragon" was planted in the town square, where it reminded the citizens of the beast that once terrorized their homes. The original fossil, the inspiration for this art and legend, has since been identified as the skull of an ice-age woolly rhinoceros.

Mayor extends this connection to protos and griffins. Referring to the red sandstone exposures of the Flaming Cliffs and elsewhere, the author surmises that ". . . According to the Issedonian-Scythian folklore, griffins 'guarded' gold; the proximity of the fossil exposures in the badlands and the gold deposits in the gullies accounts for the ancient association of gold and the *gryphs.* Dinosaur remains observed by the prospectors may have inspired the notion of monsters 'guarding' the approaches to gold and chance finds of wind-blown gold dust near or actually in 'griffin nests' would have reinforced the idea. Knowledge about real birds which collect shiny objects (Pliny 37:54.146, 37:39.149) might have been extrapolated to the *gryphs*/griffin, which had a bird's beak and laid eggs in nests. Wings may have been added to enhance the creature's bird like attributes, to indicate 'divine' qualities, or to account for the fossils' odd neck frills."

It's a clever idea, but one strong on anecdotal connections and weak on evidence. No ancient Mongolian writing, which is enriched with myth and legend, explicitly refers to griffins or their bleached bones. Moreover, gold is not found in gullies near the places where the fossils are exposed. Many of the mountain ranges near the Gobi badlands do contain gold, but they are miles from the fossil sites. The possibility of gold dust blowing into a dinosaur nest—and staying there—is nil. And, if the ancient travelers did connect the remains of *Protoceratops* with the griffin, they must have wondered why these creatures unstrategically stationed themselves so far from the actual mines.

Still, one cannot completely exclude the possibility that such legends emerged from the heart of Asia.

Grand Central Asia

Before the 1920s the dinosaurs and other fossil vertebrates of the Gobi were no more real to paleontologists than griffins. But there was a strong hunch that Central Asia held many secrets awaiting disclosure. So, as in other cases, theory and speculation fanned the enthusiasm for exploration. Explorers are driven not only by the desire to see "what's on the other side" but to prove they are right about their particular theory of the earth. Roy Chapman Andrews went searching for fossils in the Gobi to prove that Central Asia was the fertile crescent for evolution, especially the evolution of humans.

Central Asia as a wellspring for mammalian life was not an idea originating with Andrews. It grew from the work of several paleontologists. One was the imperious and wealthy Henry Fairfield Osborn, who from 1908 to 1933 also reigned as president of the American Museum of Natural History. Osborn noted that fossil dinosaurs and mammals occurred in sites as distant as New Jersey, England, the Rocky Mountain region, and western Europe. Assuming these animals had migrated across the Bering Strait, he envisioned a global route whose endpoints were on either side of the Atlantic, in both eastern North America and western Europe. The likely point of departure—the center of origin—for these two great diverging migrations, Osborn claimed, was the Central Asian plateau.

The brilliant William Diller Matthew, also a member of the scientific staff at the American Museum, refined these geographic theories. Matthew gave us a better appreciation of fossils as evidence for evolution not only in time but in space. He took into account other lines of evidence—plant remains and geological reports—that indicated major changes in climate and configurations of the land. In his book *Climate and Evolution,* Matthew laid out several basic proposals. He noted that the northern continents tended to be interconnected from time to time over many millions of years by various land bridges that emerged when the sea level fell. By contrast, the southern continents were more or less continuously isolated like gigantic islands. The northern continents were, according to Matthew and his supporters, the main stage of evolution, where new species emerged, found room to roam, and tested their mettle in competition with other widely ranging species. This required large expanses of land. And there was no land bigger than Eurasia.

Elaborating on these views, Osborn was also consumed by the belief that the roots of humankind were to be found not in Africa but in the vast emptiness of Central Asia. Entombed in the rocks of that region, he felt, should be the evidence of the fountainhead of not only vertebrate life but of our own legacy. Given enough personnel, equipment, time, and money, explorers could bring back the bones of our ancestors.

This chain of events casts a curious historical light on the Central Asiatic Expeditions led by Andrews, who was sent forth by Osborn to find the crucial evidence for the origins of humans. Yet among the tonnage of fossils retrieved from the Gobi Desert is not a single chip of bone of an early primitive human.

BIG PLATES

The failure to find the remains of the earliest humans in Central Asia is not the reason why we have abandoned some of the older ideas cherished by earlier scientists. These early theories about the migrations of species had a fatal weakness. They were projected on an earth that was assumed to be essentially the same over the eons. Continents were thought to remain in their present position and form throughout the evolutionary history, give or take a few sea-level changes, episodes of mountain building, and emergence of land bridges. Species of animals and plants, whether dinosaurs or toadstools, dispersed about this static earth surface. But while Andrews was charting the wastelands of the Gobi, a revolution in geological sciences was taking root. In 1915 and the early 1920s, Alfred Lothar Wegener, a German meteorologist, explorer, geologist—jack of all trades—issued a series of conjectures based on some simple observations. First, he noted, as others had before, that some continents have shorelines that nicely match up, as if at some time in the past they were zipped open and torn apart. Second, very similar ancient fossils—those representing needle-nosed marine reptiles like *Mesosaurus* and the fan-shaped leaves of *Glossopteris*—were found on widely scattered sites on different continents. Wegener drew these facts together to offer a bold theory—continents were not static pieces of crust; they had drifted over the eons. In some cases, former megacontinents were split asunder and major chunks, like Africa and South America, floated apart.

Wegener actually got it a little wrong. He could not effectively answer the question, how could continents drift around? His answer was ultimately off track because he posited that the continents plowed their way through the immobile ocean crust, like blocks of wood being pushed through molasses. Indeed this mixed result caused Wegener much misery. He was roundly attacked by many scientists. As late as the early 1950s symposium volumes proclaimed that the distribution of dinosaurs and other fossils was compatible with the idea that the earth's crust was stable. Wegenerian theories of continental drift and the like were regarded as an element of the lunatic fringe.

Wegener was not the first scientist to suffer ridicule and eventual vindication, but his is one of the more dramatic of such examples in the history of science. Eventually, the "stablists" were swept under the tide of a new geology, one that emerged in the 1960s. It was clear that the continents did move, but for reasons different than those suggested by Wegener. The key was the role of the ocean floor in the process. Some classic theoretical papers conjectured that the oceanic crust was being generated in one place and consumed at another. These creative suggestions seemed increasingly to fit with the rapidly accumulating geological data. Oceanic crust was thinner but more dense than continental crust; its sinking potential was therefore greater. Certain places where oceanic crust and continental crust met, such as the western coasts of North and South America, were regions of particular geologic tumult—volcanoes, earthquakes, and obvious signs of prodigious mountain building. One of the most curious observations had to do with the age of the oceanic rocks themselves. The youngest rocks on the ocean floor appeared to be either in or near the mid-oceanic rifts, like the longest mountain range on earth, the Mid-Atlantic

Ridge. As one moved away from the ridges toward the edges of the ocean basins, the rocks got younger and younger. Remarkably it was found that even the oldest oceanic crust was in no class of antiquity with continental crust. The latter showed rocks billions of years old while no oceanic crust was much older than 200 million years. Where did the rest of that oceanic crust go?

The answer to this question took the form of a general theory, now called plate tectonics, that revolutionized natural sciences. One more piece of evidence was needed to finish the puzzle. This evidence was revealed through a new technique called paleomagnetics. Many rocks contain small minerals that are sensitive to magnetic forces. We can think of these as iron filings that are attracted by a magnet; indeed, iron-bearing rocks are particularly good for paleomagnetics. When a rock cools or is exposed on the surface, these magnetic-sensitive minerals are oriented in a particular direction, parallel to a gigantic magnetic field that girdles the planet. The opposite ends of this field are just like the positive and negative ends of a magnet. Magnetic forces in the field arise from the iron-rich core of the planet. The ends of the magnetic field are the magnetic North Pole and the magnetic South Pole. When rocks cool, their magnetic-sensitive minerals—like iron or magnesium—will point toward or away from the magnetic North Pole. In a sense, an ancient rock gives us an alignment frozen in time—namely the alignment of the minerals at the time the rock cooled.

Apparently the earth's magnetic field from time to time flipflops—positive goes to negative and vice versa. In geology this is called geomagnetic reversal. During a reversal, sediments show that the earth's magnetic field slowly weakens to zero and then builds back up in the opposite direction. We have never experienced one of these events, at least during modern human history, but the magnetic field last reversed less than 100,000 years ago, when it changed to its current orientation. Rocks on the ocean floor also preserve these magnetic alignments—whether the magnetic-sensitive minerals are oriented in a "normal" pattern (rather chauvinistically applied to the present situation) or a "reversed" pattern (there have indeed been about a hundred and fifty reversals distilled to twenty-seven major events since the Late Cretaceous). These patterns can be directly measured by research vessels cruising over wide expanses of the ocean. From these data, a remarkable map of the ocean floor came into focus. Rocks on either side of the mid-oceanic ridges showed symmetrical patterns of alignment—a thick band of "normal" rock was flanked by a thin band of "reversed," flanked by thin "normal," thick "reversed," thick "normal," and so forth. The details on the width of the bands do not matter here. What does matter is that *the patterns were exactly the same on either side of the ridges.* From this, it was easy to conclude that rocks comprising the ocean floor were being pushed aside by new rocks arising from the mid-oceanic ridges. Moreover, paleomagnetic "striping" suggests a spreading of the oceanic crust at similar rates and in similar ways on either side of the ridge. The ocean floor moves. This, in combination with the observations of crustal thickness, the distribution of earthquakes, and other items noted above, at last provided an explanation for Wegener's continental drift. The lighter continental crust could be carried along with the various movements of the denser oceanic crust. The theory of plate tectonics was born.

The Eurasian Plateful

The name "plate tectonics," like *Tyrannosaurus rex,* is economic and expressive. The earth's crust is made up of a system of gigantic plates whose boundaries are marked by major physiographic features. The second word, "tectonics" (from the Greek *tecton,* builder), refers to the crustal forms that arise from plate interaction, such as the great mountain ranges of the earth. Mid-oceanic ridges are one kind of plate boundary, as they indicate places where crust is emerging and spreading. Other plate boundaries are located at deep oceanic trenches, places where oceanic crust is being consumed. The reason for these different boundaries has to do with the varying rates of seafloor spreading in different sectors of the crust. South America, for example, has a westward drift because of the high rates of spreading in the South Atlantic. The South American Plate is impinging on the oceanic plates of the South and East Pacific. The oceanic crust of the East Pacific is gobbled up in the process. This material slides into the ocean floor trenches, down a slope called a subduction zone, which in cross section looks like an escalator.

But that's not the end of the consumed crust. The piling up of the material in areas of subduction creates heat and pressure in the deep boundary between the mantle and the crust. This eventually has to escape in the form of molten rock. Volcanoes are vents for releasing this material. Other igneous bodies are formed and cooled before they reach the surface, like the great granite masses that make up places like Yosemite Valley and many mountains in the Andes. The tension and compression of moving crust deep under these plate boundaries are also expressed as earthquakes, which lead to faulting and folding. Volcanic islands form offshore, mountains rise up, faults offset huge volumes of rock. One can see why subduction zones and other plate boundaries are where the action is.

Plate boundaries can be very complex. They may involve continental plates overriding a combination of other plates, even a spreading ridge or a junction of ridges. The ridges as well are offset by tensions in the crust which form long *transcurrent* faults. The most famous of these is the San Andreas fault system which extends from the Gulf of California to the Pacific Ocean, northwest of Mendicino, California. This system of tortured rock is 1600 miles long! Other plate boundaries represent dramatic collisions, where pieces of light continental crust came crashing together. For millions of years, India made its way northward through the spreading Indian Ocean from the high latitudes near southern Africa and Madagascar. It plowed into the underbelly of Asia about thirty-five million years ago, and the stupendous compressional forces uplifted the Himalayas and the Central Asian plateau, the highest mass of land on earth.

Many complex features on the earth's surface are nicely explained by plate tectonics. That's the power and the beauty of the theory. Not all such features are, however, so easily dealt with, even in light of this breakthrough. Why are the Sierra Nevada and the western Rockies such wide belts of mountains? (The subduction zone necessary to create these features would have to be awfully low in angle.) Why are places like the Colorado Plateau and, for that matter, the Gobi Desert, so inert, so uninterrupted by earthquakes and volcanism, yet at the same time so high in elevation? They average a surprising 4,000 feet above sea level. These are difficult ones. Finally, the basic driving mechanisms for plate tectonics are elusive. These are thought to be currents of convection (heat transfer) in the mantle of the earth, the layer below the crust, but they are not understood with any precision. Just the same, plate tectonics, like the theory of evolution or the unifying approaches in systematics, have taken us to a new level of understanding nature. In science, generalizations matter. We seek explanations that apply not just to local situations but to our world as a whole. Plate tectonics epitomizes the triumph of generalizations.

With this appreciation for the earth's vitality we can look at continents in a whole new way. The Eurasian continent is part of the Eurasian Plate. At one time the plate didn't look anything like it does today; it was lacking the appendage of India, for one thing. At an earlier time, the continent was divided by a vast inland sea running north and south in the area of the present-day Ural Mountains. The *Eur* of Eurasia was severed from its eastern part. In the Cretaceous, Eurasia had broad connections with North America, and its drifted position put it closer to the rotational North Pole. At the time of the dinosaurs, there was no Himalaya, at least in the form we know today. During the Triassic, at the birth of the dinosaur era, Eurasia wasn't very distinctive at all. It was part of the megacontinent of Europe, North America, and the southern continents, that Wegener appropriately called Pangea. In light of these complex histories, the notion of a center of origin of dinosaurs and other vertebrates from Central Asia becomes rather simplistic. On the other hand, the knowledge of drifting continents allows us to contemplate some truly exciting possibilities for the evolution of life.

THE AGE OF FRAGMENTATION

So successful is the new geology that we can actually calculate with fair accuracy the average rates of seafloor spreading in different regions of the world. Some fast plate movement is recorded in the Indian Ocean floor, with a rate of over three inches per year. In the South Atlantic spreading rates are a bit slower; they average less than an inch per year. These movements may not seem very fast, but they are enough to make a big difference over a few million years. Hence, those fast spreading rates in the southern Indian Ocean promoted the northward voyage of the Indian ark to a new hemisphere and a new docking site. Even a few inches a year, accumulated over years or generations, could well have impacts on local environments. Although they are not appreciated in our lifetimes they can be appreciated in an evolutionary lifetime. Years ago, a distinguished professor visiting Berkeley gave us graduate students a lecture on tropical marine paleontology. He expressed befuddlement as to why the marine forms in the present and recent past were so different on either side of the Panamanian isthmus when their physical environments, water temperatures, substrate composition, density, and so on are so comparable. Graduate students are supposed to drill established scholars on such matters. At the time, my fellow student Doug Lawson was already a successful field paleontologist; he had received much acclaim as the discoverer of the world's largest flying creature, the Cretaceous-aged pterosaur *Quetzalcoatlus,* an air machine with a gargantuan wingspan of fifty feet. During this particular seminar, Doug got off a zinger. Of course the environments have been different over the last 3 million years since the isthmus was closed: the seafloor of the Caribbean Plate east of the Isthmus of Panama was spreading at a different rate than the East Pacific Plate just west of the isthmus. Surely that has had some impact on the species evolving in those environments.

This shows how various phenomena of earth history run on different clocks. In our own lifetime, our neighborhood might grow, diminish, get better or worse. Property values change accordingly. In a species lifetime, the situation is comparable. The average duration of a species, as documented from the fossil records of many taxa, can be quite long, although the figure varies with different groups. Mammals, trilobites, ammonites, insects, and birds seem to turn over relatively rapidly, with species durations of one million years or less. Other groups, like marine gastropods (snails and kin), bivalves (clams and oysters and kin), reef corals, and foraminifera (one-celled plankton or bottom-dwellers that have been used for dating marine strata), vary anywhere between ten and twenty-five million years in duration. As we have seen from our brief review of plate tectonics, a lot can happen over that period of time.

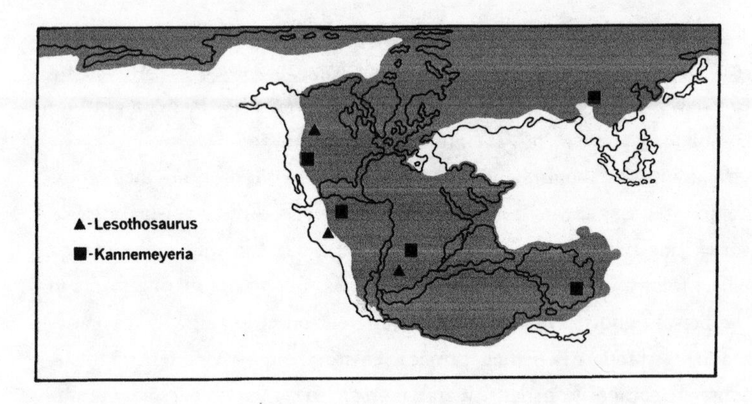

Map, Triassic continents with vertebrate distributions. Shaded areas indicate the Triassic boundaries of land above water. *(Ed Heck)*

When we move up rank to the evolutionary lifetimes of higher groups, not just their component species, we see a vast number of possibilities for changing the stage of action. Dinosaurs are no exception. In the Permian, on the eve of the dinosaur entrance, the major landmasses were beginning to coalesce in an unprecedented fashion. By the time dinosaurs appeared on the scene, these continents had formed the great megamass—Pangea, which was Wegener's name derived from the Greek words for "all land." This chunk of the earth's crust drew in all the major continents, including Australia and Antarctica. Early ornithischian dinosaurs are known from sites in North America, South America, and Africa. Even more widespread, at least by evidence of their fossil occurrence, was a stocky synapsid, *Kannemeyeria,* whose remains hail from all the above-noted regions as well as Australia and southern China. *Syntarsus* is known from southern Africa and North America, *Plateosaurus* from Europe and Greenland.

Triassic land vertebrates could get around. Their translocations were, at first, aided by favorable weather. The Pangea complex had drifted away from the polar regions; Antarctica was comfortably wedged between southern Africa and Australia, a hospitable superhighway for intercontinental travel. Although a gap existed between Asia and Africa in the form of the tropical Tethys seaway, the two continents were well connected along with North and South America in the region of the present-day North Atlantic. The climate was also apparently accommodating; diverse plant life ranged broadly throughout the supercontinent. But by Late Triassic times the climate had become hotter and drier. For some reason this was coincident with the emergence and diversification of dinosaurs and the first appearance of those important creatures of such inauspicious beginnings—the ancestors of mammals.

Meanwhile the pattern toward continental coalescence was reversing itself. By the end of the Triassic—about 210 million years ago—the first signs of detachment were evident with the opening of the North Atlantic. By the middle of the Jurassic—about 30 million years later—the northern continents, called Laurasia (a coinage combining Laurentia—an old term for Greenland—with Asia), had nearly parted company with the southern block. This southern assembly was called Gondwanaland, for the Indian kingdom where fossils typical of the whole supercontinent had been found. Soon Laurasia and Gondwanaland themselves were showing signs of disintegration. The narrow North Atlantic broadened and a seaway isolated Europe from the heart of Asia. In the Southern Hemisphere, India, while still sutured to Antarctica, was separated from the African continent by the nascent Indian Ocean. These events put in motion a pattern of continental breakup. Millions of years later, some of the continental masses were spliced together—India eventually collided with southern Asia and the Isthmus of Panama formed, linking North America with South America. Yet the massive merger of landmasses seen in the early Triassic has not been repeated. We live on an earth that is largely the product of an era of fragmentation that began in the Mesozoic. One of the most plaintive of all appeals (and in these times perhaps the most hopeless) graces my favorite bumper sticker: REUNITE GONDWANALAND.

Despite this early trend in fragmentation, apparently enough connection remained to promote the widespread distribution of many dinosaur groups. In Late Jurassic times the long-necked diplodocid and the lofty brachiosaurid sauropods as well as the iguanodontids are much the same types in Africa and North America. Perhaps abetting this mix of faunas was a fairly uniform, moist, warm climate in a wide belt of middle and lower latitudes around the world. There is, for example, no evidence of polar ice during the Jurassic. This is in strong contrast to the early Permian, which experienced the greatest ice age of all time. Still, the incipient drifting of the continents created a more distinct pattern of climate zones. Plant floras more distant from the equator show a typical temperate habitat.

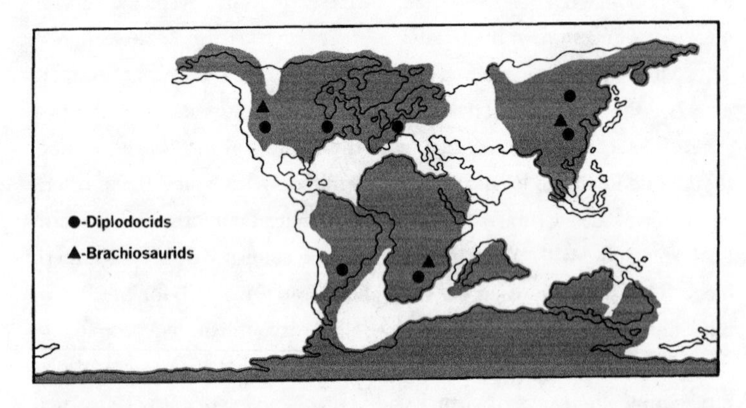

Map, Jurassic continents with dinosaur distributions. Shaded areas are land masses above water during the Late Jurassic. *(Ed Heck)*

As noted earlier, the transition from the Jurassic to the Cretaceous saw a revolution in animal and plant life, marked by the rise of the flowering plants and their insect pollinators. Why this lavish conversion of habitats during the Cretaceous? Perhaps it has something to do with the lay of the land. The Cretaceous was one of the most interesting times for plate tectonics before or since. Continental fragmentation and drift were proceeding full tilt. The South Atlantic had opened. North America was widely cut off from South America. India and its escort ship, Madagascar, had broken free of Gondwanaland and were setting a course northeastward toward Asia. As the Cretaceous rolled on, Africa and South America completely parted company and the Atlantic became a great global body of water confluent with Arctic and Indian oceans.

These basic tectonic changes were accompanied by changes in landforms and seaways. The Cretaceous was a time of major mountain building, when great walls like the North American cordillera—the nascent Rockies—and the South American Andes were violently upthrust. At the same time, the sea-level rise promoted a remarkable increase in seas and coastlines. By the later Cretaceous, the time from whence the dinosaurs of the Gobi are so well known, Asia was further broken up to the west of Central Asia by giant east- and west-trending seaways. Southern Europe had fragmented to the point of an archipelago, like the present-day Philippines or Indonesia. A huge seaway bisected North America, running from the Arctic seas to the present-day Gulf of Mexico in what are now the high plains east of the Rockies. A similar seaway developed in northern Africa, isolating the region that now includes Morocco, Tunisia, Chad, and Nigeria from the great eastern African landmass. The Cretaceous was truly the acme of the age of fragmentation.

DRIFTING DINOSAURS

This enrichment in the texture of landforms, oceans and seas, and certainly the isolation of many regions, fueled the momentum for evolution and divergence. Geographic isolation—by seas, rivers, mountain ranges, or contrasting climate—is a major driving force of speciation because populations that at one time exchanged genes are suddenly cut off from one another. They then evolve along their own pathways. Likewise, this differentiation can be expressed by the divergence among groups containing those species. So Cretaceous dinosaurs can be divided into various regional fiefdoms. To appreciate this, we must abandon a mind set that makes us view continents as they are today. Remember that in the Northern Hemisphere, eastern Asia and western North America were broadly connected in the region of the Bering Sea. On the other hand, eastern North America was cut off from this complex by a great seaway. It was, however, connected, at times, to broken-up parts of Europe in the North Atlantic region. Thus we can think of two continents—let's call them Asiamerica and Euroatlantis—instead of a single northern continent or the customary division of North America and Eurasia. Expectedly, dinosaur divergence mirrors this continental pattern. The armored ankylosaurids, the horned ceratopsians, and the more specialized hadrosaurs, like *Saurolophus* and the crested *Corythosaurus,* were evolving in the Gobi and the Rocky Mountain region and presumably parts in between. In contrast, primarily more generalized hadrosaurs claimed Europe and eastern North America. Although the dromaeosaurids (the group that includes *Velociraptor)* ranged broadly over the northern landmasses, they were virtually excluded from the southern continents. South of the equator there were a number of groups common to the northern continents—iguanodontids, hypsilophodontids, and sauropods—but these were distinctively different genera or even higher groups containing several genera. Also, our denizens of the Gobi, the protoceratopsids, as well as their ceratopsid relatives were notably absent from the southern continents.

We see here, then, a picture of dinosaurs and continents that jives poorly with Osborn's and Andrews' concept of a center of origin for vertebrate evolution in Asia. What the pattern does suggest is a history of dinosaur differentiation that closely tracked the breakup of the continents. Originally, certain groups of dinosaurs ranged broadly over the sutured landmasses. Their descendants were allowed the opportunity of divergence once these landmasses were separated. Central Asia was certainly an im-

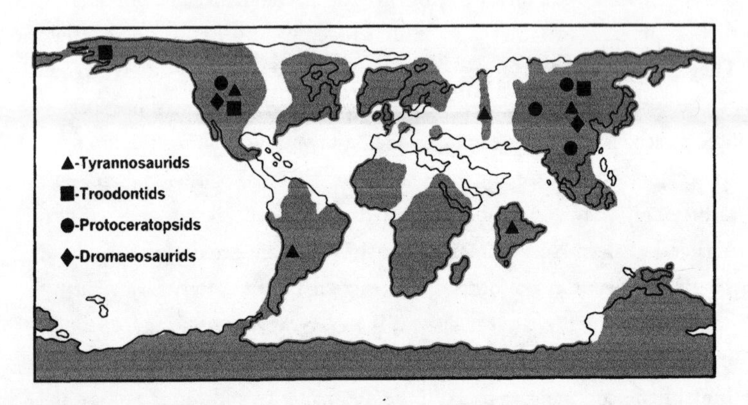

Map, Cretaceous continents with dinosaur distributions. Shaded areas indicate land masses above water during Late Cretaceous times. (*Ed Heck*)

portant region for the emergence of certain dinosaur groups, but it wasn't the only such region.

The Comfortable Corridor

Another point that has much interested paleontologists concerns the affinity of Cretaceous dinosaurs between two major regions of the world—Central Asia and western North America. At closer inspection, the resemblances between these two faunas are even more striking. *Velociraptor* is very much like another somewhat larger dromaeosaur from North America, *Deinonychus.* The hadrosaur *Saurolophus* is known from both regions. The Asian *Tarbosaurus* is a slightly shorter version of the North American *Tyrannosaurus,* but it is otherwise very similar—so much so that some have suggested that they should both be called *Tyrannosaurus. Oviraptor* has a set of North American cousins represented by fragmentary fossils. There was also a closely related couplet of large-brained, big-eyed, slender-necked theropods, the troodontids: *Saurornithoides* from the Gobi and *Troodon* from Alberta, Montana, and Wyoming. (Some alleged troodontids from Romania are represented only by fragments and their assignment is highly dubious.) The ankylosaurs are represented on both landmasses by different genera, but they appear to be a tight-knit group. Finally, although

The troodontid *Saurornithoides* *(Ed Heck, reprinted with permission, AMNH)*

the pachycephalosaurs are known from isolated sites in Madagascar and western Europe, the greatest abundance and diversity of this group are recorded in central and eastern Asia as well as the Rocky Mountain region.

There were, to be sure, some isolated oddities. Certain protos, as well as a number of theropods, like *Erlikosaurus,* are restricted to the Gobi of Central Asia. Groups of organisms unique to a region—that is, found nowhere else—are called *endemics.* There is some degree of endemism in the Gobi dinosaur fauna, but there is also a striking pattern of similarity to the Cretaceous fauna of North America. This similarity is all the more marked when we consider that the richest sites from North America, those with the dinosaurs similar to the Gobi forms, are probably "somewhat younger" (about seventy to sixty-five million years old) than the eighty-million-year-old fossils from Mongolia.

Such a free mixing of faunas suggests a cooperative land connection. Indeed, we know that during the later Cretaceous central and eastern Asia were broadly connected with western North America over what is now the Bering Sea, the gap between Alaska and Siberia. This prompts a question concerning environments and adaptations. Were not climatic conditions too harsh for migrating across this northern route? Indeed, Alaska and eastern Siberia were actually closer to the North Pole during the Cretaceous (because of the northward drift of the continents) than they are today. There are, however, some mitigating circumstances. The abundance of seaways throughout the world—two fifths of the continental landmasses lay under shallow seas—doubtless promoted some insulation of the land and equable climates (conditions where seasonal changes were very mild) over a wide range of latitudes. Just as today, broad coastal regions may have enjoyed the kind of maritime climate we call "Mediterranean" conditions.

We can draw on some supportive data for this picture. There is virtually no evidence of Cretaceous glaciation outside of a few restricted montane glaciers—certainly the polar regions were not ice-choked during this time. Cretaceous coal deposits, indicating relatively high humidity, are extensive in several regions, including higher latitudes. Measurements of isotopes of oxygen preserved in the calcium carbonate shells of various marine organisms (including ammonites and belemnites) show that much of the Cretaceous was rather sultry. A warming trend during the Early Cretaceous reached its acme about 100 million years ago. Thereafter, temperatures declined. In what is now midwestern North America, annual mean temperatures steadily decreased to about 68° F. in the Late Cretaceous. Only in the last few million years of the period did mean annual temperatures take a much steeper decline in this region to about 50° F. Thus temperatures in this region, at least until the very end of the Cretaceous, were rather equable—say a more humid version of Southern California weather compared to the highly seasonal conditions and 48° F. mean temperatures in the Midwest today. One might expect that at least sections of the Bering connection during later Cretaceous times represented a corridor of comfort.

One of the more interesting paleontological developments in recent years is the direct evidence that indeed this northern corridor was eminently livable. For years there had been building data for a very rich flora, plentiful wood, coal seams, and leaf impressions, that suggested a lush forested region some seventy to eighty million years ago in far northern Alaska. These paleobotanical records were soon enhanced by dinosaur discoveries. In the late 1980s a team of scientists from U.C. Berkeley led by Dr. William Clemens found fossil bone in the Cretaceous slopes flanking the Colville River on the north slope of Alaska. These are areas in which rich fossil plant sites are also found. The dinosaur bones were fragmentary compared to the finds of the Gobi, but they were plentiful and readily identifiable. Bits of ribs, limb bones, and vertebrae confirmed the presence of hadrosaurs similar to those known from Cretaceous beds in Saskatchewan. There were even small teeth of *Troodon*, the big-eyed North American theropod closely related to the Gobi's *Saurornithoides*. The team also found remains of juvenile dinosaurs, suggesting that dinosaurs may have bred and grown and lived out their lives in the northern area; they weren't just passing through in great migratory waves. Finally, these dinosaurs lived there despite several months of darkness—remember, northern Alaska at the time was even closer to the Pole than it is today. Although such darkness must have cooled climates considerably, it was probably not down to freezing. Clemens reasoned that, with enough food resources and ingenuity, these dinosaurs could have survived extended periods of darkness. As demonstrated by the occurrence of their fossils, they did just that.

Thus dinosaurs and other organisms apparently lived well in the far north and traveled freely between the two adjoining landmasses. This still leaves unsettled the matter of direction of dispersal. If certain dinosaurs dispersed from one continent to another, which way did they go? If the resultant vector of movement was from Central Asia to North America, were Matthew, Osborn, and Andrews, in effect, partly right? Unfortunately, we have no idea whether such movements went in one direction or the other. The American Museum team in the 1920s cited *Protoceratops* as evidence that the horned dinosaurs emerged in Asia and migrated to North America and there evolved to heavyweights. Subsequently, we have revised the thinking about this plot: *Protoceratops* is indeed a more specialized ceratopsian, and one not necessarily close to the ancestry of the North American forms. Even without such a revision, we still have problems with drawing arrows for migrations. The first occurrence of a fossil does not necessarily indicate its age; it could be much older. It is extremely difficult to use fossil dates to establish that a certain group of dinosaurs appeared first in one place and later in another. The locale with the younger fossils could simply suffer from a dearth of bones in older rocks. A conclusion on direction of dispersal would therefore rely on negative evidence—the absence of data. This is never a very secure scientific position. Given some of the gaps in the Cretaceous faunal sequence, a choice between either a westerly or easterly immigration is hardly justified. All we can say is that there was a great deal of blending of these northern faunas during Late Cretaceous times.

The Cretaceous communities of the Gobi then did not evolve in isolation. They were part of a dynamic worldwide system; they were tied to developments on other broadly connected continents. And they evolved during one of the most dramatic time intervals in the history of the earth. When the Cretaceous began there were only forests of conifers and other gymnosperms, mosses, and ferns. When it ended the modern floras of flowering plants and deciduous trees had taken root. At the dawn of the Cretaceous there were essentially two huge landmasses and a few errant blocks, like India. At the end there were—from a combination of seafloor spreading and the inundation of shallow seas—at least twelve landmasses. The Early Cretaceous also marked the transfer of power, particularly in the northern continents, among the dinosaur rulers, from the waning sauropods and iguanodontids to the diversifying ceratopsians and hadrosaurs. In the end, of course, even these were extinguished, leaving behind only one subgroup of feathery, winged theropods to carry on the dinosaur legacy. The Cenozoic, so to speak, was for the birds.

CHAPTER 5

1992—The Big Expedition

As I rode I became part of the world. I lived in it, I didn't just pass through it. Everything that happened and existed around me had direct effects, moulding my day, my mood and my whole being. Nothing was simply visual. It was all tangible.

CHRISTINA DODWELL.
1979. *Travels With Fortune*

In 1992 the Mongolian Academy-American Museum joint expedition suddenly got more complicated. During the year preceding the expedition we were in long conversations with producers eager to capture the venture on a television special. We finally settled on a small group from BBC Horizon working with the producer Anthony Nahas and Philip Clarke of Diverse/Aperture Productions. The special would air on BBC in the spring of 1993, and a U.S. version would be shown on the public television series "Nova" some months later. Anthony recruited the brilliant British science film maker John Lynch and a small crew for the job. In addition to John we would be joined by a cameraman, Jerry Pass, and a sound man, Tim Watts. The negotiations for the project went amicably but intruded on the frenetic weeks of preparation in late May and early June just before our departure. The lawyers were still hashing out the details of the contract as we boarded the plane for Mongolia.

Further complications came with the larger crew for this season. Among the Americans were the core team—Mark, Malcolm, Lowell, Jim, Priscilla, and myself. We also brought a vital new member, an expert mechanic from Minnesota, Kevin Alexander. The president of the American Museum, George Langdon, was to accompany us for the early part of the journey, assuming we could safely fly him out of the Gobi town of Dalan Dzadgad at the beginning of July. We also knew from our sketchy communications that Dashzeveg and Perle were planning to come out with a larger Mongolian team.

The timing for such a complicated venture couldn't have been worse. Mongolia had reached its nadir of democratic infrastructural chaos. Nothing seemed to be working.

The first bit of bad news hit us soon after we touched down in Beijing on June 25. Jim Clark and Kevin Alexander were already in Ulaan Baatar, sending messages 13,000 miles back to New York which were then relayed to me, waiting in a city only a few hundred miles south of where the messages originated. Barbara Werscheck, my assistant, called me at the Xi Yuan hotel.

"Are you sitting down?" she asked.

Our container of food and equipment (our vehicles had been stored in Mongolia) were stalled at Erlian at the Mongolia-China border. Representatives of our Japanese shipping firm telexed us that they were "heading for Erlin with a suitcase full of money." There was nothing to do but carry on to Ulaan Bataar, hoping that the shipment would eventually arrive. The loss of this cargo had grave implications. The shipment had the two new transmissions, one for the red and one for the green Mitsu. The previous summer the transmissions in these vehicles had been mistakenly drained of transmission fluid and their gears had been ground to metal shavings on the long drive back to Ulaan Baatar. Kevin was forced to do busy work on the vehicles while we waited for the new transmissions. Our mechanic's time was precious and well remunerated; he could only afford a couple of weeks away from his business. Moreover, it was unlikely that we could keep a large crew sustained without the additional food shipped with the container. In 1992 the food situation in Mongolia was little improved.

This crisis brought the expedition to a grinding halt. At our compound near the Natural History Museum, a large number of people hung around in the rain of a dreary June, waiting for the supplies and equipment that might never arrive. The BBC crew worried about their expenses and delays, but were preoccupied by the complicated and less than perfect arrangements with an outfit known as Mongolian Television and Radio, or, as we called them, MTV. Meanwhile, we were squandering our money on fatty, gravy-laden meals in the hotel's dollar restaurant. The only thing attractive about this dining room was its charming menu, wherein entrees took on as many different spellings as Mongolian place names. "Fried Fish" alternated with "Fried Fush" or "Fried Tush." While morale at the Ulaan Baatar hotel where we stayed deflated, our bar tab at the hotel's disco inflated.

The bar in the UB hotel is a real spectacle, and somewhat a place of ill repute, but it was the only safe place for us to hang out that didn't require a walk through the dark and dangerous night streets of Ulaan Baatar city. Mark and I remarked on its dramatic transformation from our first summer there in 1990. At that time we called it the Star Wars Bar because it resembled that seedy den in the movie that seemed to collect every strange and dissolute renegade in the galaxy. There were many Russians and eastern Europeans, who on occasion were removed by a phalanx of Genghis Khan warriors dressed in military uniforms, high boots, and red berets. It turned out that these poor clients were sneaking in their own vodka and getting outrageously drunk without paying the required dollars for the bar fare.

By 1993 the Russians were virtually absent from the Star Wars Bar, and in their place were swarms of Japanese businessmen and a mosaic of other foreigners—Australians, Brits, Belgians, and a few Americans. They were attended to by a brigade of prostitutes, an extremely aggressive lot, who hung with some very surly Mongolian pimps, who conveniently also worked as "cab drivers." Such business could not be accomplished within the hotel, so the prostitutes had to induce their clients to go home with them in a cab.

At last we received the news that the goods had been relocated and put back on the train. We were able to retrieve them from the Ulaan Baatar railroad yard on June 29—two days after our originally scheduled date of departure for the Gobi. The team plunged into repacking and other details with furious dispatch. Under Priscilla's supervision, containers of food were stacked according to various meal plans, a procedure that had become suddenly formalized with an enterprise much larger and more complicated than in previous field seasons. This year Priscilla and Malcolm had acquired, through purchase and company donations, a great mountain of cans of freeze-dried foods. Some of these were army issue. A popular item in 1992 and thereafter turned out to be freeze-dried pork chops, which when resoaked and cooked were delicious. They could be sautéed and stir-fried with noodles, vegetables (freeze-dried of course), onions, and various sauces procured in Beijing. They could also be slathered with barbecue sauce and grilled over a zak fire. We had an extraordinary abundance of cans of pork. They were apparently meant for the Gulf War but were never shipped because of the Islamic restrictions of Saudi Arabia. Our Mongolian friends, however, had no such restrictions and they enjoyed the pork immensely.

Finally we were also able to replace the transmissions in the Mitsus. But there is no such thing as a hoist in Mongolia. The cars were tilted against the edge of a container, and Kevin, with the help of Mark and others, had to lie flat on his back and heave the new transmission into its niche.

There were more logistical problems. Gas was hardly trickling from fuel depots in the city. Worse yet, the few fuel depots in the Gobi—places like Daus, Saynshand, or Dalan Dzadgad where we had refueled in previous seasons—were either unreliable or extinct. We adopted another strategy, renting a rather archaic-looking tanker truck with 1,500 gallons of 93-octane gasoline. But this did not solve all our problems. This year we had purchased our own GAZ with Perle's help; in addition the expedition would require at least two other GAZ and a Russian jeep or WAZ. These vehicles required 76 octane. We filled a few barrels with 76 and loaded them on the GAZ, betting that we would at least be able to get this grade of fuel at some spot in the desert. It turned out to be a bad bet.

Despite all these hassles, we departed on July 4. Our convoy comprised eight vehicles and twenty-five people, including the BBC. The fragmented caravan stretched out over the length of the city like an *Allergorhai horhai*—the giant sand worm of Mongolian folklore—that had been cut into independently wriggling segments.

As in previous seasons, we were limited by the impoverished gas supplies and the sparse occurrence of watering holes. Our plan was to get this complicated caravan to Flaming Cliffs as soon as possible. From there the group could reassemble, spend a couple of days prospecting, and BBC could begin filming. We next planned to shift westward about thirty miles to Tugrugeen Shireh—the home of the Polish-Mongolian expeditions' famous fighting dinosaurs, as well as our own precious *Velociraptor* and mammals like *Zalambdalestes*. Later, we would take the whole enterprise southwest either over the formidable Gurvan Saichan Mountains or, if mud and rain impeded that route, on a jog farther west, skirting the mountains but contending with uncertain and potentially treacherous territory. Our eventual destination of course would be the Nemegt Valley with its fossil wonderlands of Khulsan, Altan Ula, Eldorado, and, we hoped, our own vault of Cretaceous treasures.

THE DESPERATE CARAVAN

A convoy is only as strong as its weakest vehicle. Only a few hours and twenty miles out of Ulaan Baatar, Mangal Jal, Dashzeveg's distinguished and highly skilled driver, brought his truck to a screeching halt near a river crossing.

"*Moh*, busted differential," Perle said.

We unloaded Mangal Jal's GAZ and stuffed most of its contents in the other vehicles. Dashzeveg and the old driver would limp back to Ulaan Baatar, repair the vehicle, and rejoin us in a few days at Flaming Cliffs.

We shuddered onward. In the rearview mirror I could see the dull blue of our decrepit gas tanker—the "benzina machina"—trailing us, spouting dust and exhaust fumes as it headed over a crest of a hill. In a soggy field not many miles from the point of our last breakdown, the tanker had stopped dead, engulfed in a cloud of steam.

"Mike, this tanker is veery baad. I think it is dongerous," said Perle.

"Will it travel at all, Perle?"

"Maybe only at night. Then maybe we can fix it at Bayn Dzak."

"Okay, Benzina should head on out as soon as night falls and his machina cools down," Mark told Perle.

Boyin Tok-Tok, the Mongolian driver we had all named Benzina, after his truck, stared blankly at us, baffled by the frenetic exchanges in a foreign tongue and Perle's elaborate set of instructions.

We filled a couple of barrels in the back of the surviving GAZ with 93 octane from the tanker. Perle and Batsuk, Perle's muscular brother-in-law, would escort the benzina machina with a GAZ, as soon as the tanker cooled off and the sun had sunk low to the horizon. The rest of the fleet would head out the next day.

By this time we were exhausted from the theater of the absurd that represented our first field day of the 1992 season. Marooned in a pleasant field of grass and flowers, we pitched camp at the top of a nearby knoll. As we ate a cold meal, some country folk came by on a couple of motorcycles. They asked for some gas. One of the Mongolians gave them a tin cup full of a small amount of our precious fuel. I was astounded to see our visitors sipping from the cup and passing it around.

"Dadanzurad!" one of our visitors exclaimed, the Mongolian word for "76," referring to the proper level of octane for their Russian motorcycles. These hardy folk distinguished 93 octane from 76 octane by taste. It was a powerful testament to the barter value of gasoline in this rugged, resource-deprived land, and reminded me of some crazy scene from a Mad Max movie.

Expeditions like ours often set up large army tents of canvas, anchored by a plethora of ropes and stakes, for sleeping quarters. Experience had taught us, however, to stick to smaller, more resilient tents used by mountaineers. Most of our tents were North Face VE-25, which had proven reliable not only in the raging winds of the Gobi but in the roaring southern storms that assaulted us during our expeditions in the Patagonian Andes. These tents are domes easily stretched out by threading them with four flexible poles. But on this clear and cloudless evening I gave in to my exhaustion and eschewed the tent-setting ritual. Mark was of the same weary body and lazy mind. Several others followed our example. That night the Gobi decided to christen our expedition with a torrential rainstorm. On future occasions, expedition members followed their own instincts about setting up tents.

By midmorning the next day we were resolutely on our way south. Despite a few problems—a recalcitrant ignition and a stripped bolt in the steering linkage of our GAZ—we made it to Erdenandalay, a hundred and fifty miles from Flaming Cliffs. The big news of the day—encouraging news—was that the tanker and the other GAZ had chugged through town some hours before us. We had expected to see a dead tanker on the far side of every hill we crested, but to our surprise it remained ahead of us.

The morning of July 6 came with less wind, but a disturbing set of clouds were closing in on us from the west. Soon we reached the cobblestone plain at the edge of the dry river—our gateway to the Gobi. From the red Mitsu, Mark Norell grinned through a bristly, embryonic beard and gave me the thumbs-up sign.

At Mandal Obo, the forlorn outpost near the volcanic hills just north of Bayn Dzak, we learned that our smoking gas tanker and rumbling GAZ had passed through town only a few hours before us. We caught up to our "vanguard" at the high point of the road through the volcanic hills. Perle and Batsuk were lying below the tanker in the shade, as if they had passed out. Eventually Perle's body stirred.

"How are you, Perle?"

"Oh, I am miserable. This tanker is no good. The benzine is worth more than the tanker. When we use it up, maybe we will leave the tanker in the desert."

It was easy to see that the tanker would be the bane of the expedition.

But at that point I didn't care. I stretched my legs and took a brief walk up one of the nearby volcanic hills. There to the south was the desert. It looked like the Mojave Desert, or the great basins between the Nevada ranges, or the empty plains of Wyoming's red desert, or the baking flatiron of bare rocks east of New Mexico's Caballo Mountains, places where I grew up, places that I loved. Some fifteen miles away rose the Flaming Cliffs, their battlements hidden among lines of pale sand dunes and the shimmer of rising heat. Before me all was flat and empty, vaulted by an enormous sky brocaded where it met the earth by the jagged fringe of the Gurvan Saichan Mountains.

JOURNEY OF DEATH

It was in the deserts of New Mexico that I became a fledgling paleontologist. This inaugural expedition was guided by Professor Peter Vaughn, who had a passion for 250-million-year-old Permian-aged reptiles. As I mentioned earlier, the Permian takes its name from a rock sequence well exposed near the Russian town of Perm. Rocks of this age are scattered about the world—they are particularly extensive in Russia, western North America, southern Africa, and southern South America. In many places Permian rocks form magnificent cliffs and pinnacles of gritty sandstones that offer a still life of some ancient river system. Their most striking aspect is, however, their redness. As in the Gobi, the Permian canyons of New Mexico, or the temples of Monument Valley, give the color red new spectral meaning. There are the orange-red massive sands of the main channels; the more crimson mudstones that once formed the bottom of an eddy; the root-choked vermilion shades of an ancient pond; and the pink-red swirly cliffs of a 250-million-year-old field of sand dunes.

No one knows exactly why these rocks are so red. Certainly the iron-bearing minerals in rocks can combine with oxygen in the air to give a reddish cast, and this does happen with some predictability in deserts like the Gobi and other dry places. But red beds are also easily formed from the leached and eroded soils of the tropics. The fact is we're not sure why the red rocks of the Permian (and, for that matter, other red rocks of other ages, like those of the Cretaceous Flaming Cliffs) come that way. Nonetheless, the question is vigorously pursued by geologists.

Professor Vaughn (we called him Peter in the informality of the field) was intrigued with red beds, but he was more interested in what they contained. Permian reptiles are curious beasts. They include the famous *Dimetrodon*, a ten-foot-long dragon with a huge head full of teeth and a prominent sail fin supported by long, vertical spines. *Dimetrodon* and kin are often confused with dinosaurs, but they are actually far off the beaten trail of dinosaur evolution. As shown in the cladogram in Chapter 2, they represent the early members of the line leading to, of all things, the mammals. These are members of the great branch of the tetrapods, the synapsids. Certain openings in the back of the skull for the muscles that close the jaw, as well as the pattern in which the mosaic of skull bones fit together, and the ways in which the separate vertebral bones fit to each other and to the ribs, indicate whether lineages are more on the synapsid side leading toward mammals or on the side leading to turtles, lizards (and snakes), crocodiles, dinosaurs, and birds. These matters fascinated the professor and he wrote important papers on the subject.

In my naiveté I expected during my early explorations to find a magnificent *Dimetrodon* skeleton splayed out on a slab of sandstone, like some frieze of a Greek temple. As the air temperature in the desolate canyons increased, so did my appreciation for the most important fact about Permian vertebrate fossils—they are incredibly hard to find. The initial few hours of bone hunting were truly disorienting. When we first arrived at our destination, a couple of twisting canyons called the Abo and Jeso, at the base of the Organ Mountains east of Las Cruces, New Mexico, the professor simply grabbed a rock hammer and started walking.

"Where's the bone bed?" another student asked.

"You're standing on it," Peter said. Then he offered some advice. "Just keep walking, and scan the ground as you move along. The bone is usually a shiny blue-white."

Although we did this, I confess that that first afternoon I was more impressed with the June heat of southern New Mexico than the prospect of finding *Dimetrodon* bones. By two in the afternoon the temperature was easily 100° F. There was no relief from the sun. The spiky mesquite hardly served as shade and the rocks were too low in profile to cast a comforting shadow. I staggered about, trying to pretend interest in finding fossils. Near the end of the afternoon I had a momentary surge of excitement. On a small mauvish knob of rock I spied a chip of blue bone. I recognized the long flattened shaft as *Dimetrodon* from the professor's lab. It was firmly cemented to the rock but could be removed by resolute chipping with hammer and chisel. A nice prize to bring back to camp and secure some early credibility. However, moments later I felt a shiver of disappointment. A chalky circle around the fossil, no doubt scribed with the professor's rock hammer, marked a discovery apparently not worth retrieving.

For a time it seemed that my rediscovery of the *Dimetrodon* spine would mark the acme of my career in field exploration. Indeed, our team collectively had little other than a fairly scrappy assortment of bone fragments to show for several days of efforts. Even the professor eventually succumbed to the tedium and uprooted camp, heading northwest for a sun-bleached spine of rocks called the Caballo Mountains, rising next to the Rio Grande. Here we worked a series of mesquite-choked canyons draining the hotter eastern side of the range. The canyons opened, with broad fans of loose stones, sand, and gravel that geologists call alluvium, onto a broad flat skillet, a valley of blinding white sand with the forbidding name Jornada del Muerto (Journey of Death).

As with the Abo and Jeso, we found few fossils in the Caballos. The only diversion from our scramblings circa Jornada del Muerto were provided by those thorny mesquite trees. Peter warned us to avoid these at all costs when piloting our vehicles through the narrow canyons. The mesquite thorn, he said, is like no other; it mortally wounds all tire rubber. True enough, our trips to the gas stations in the nearby town of Truth or Consequences—known as T or C by the locals—to repair punctured tires were frequent. With haughty urbanity, we initially derided this dusty pocket of civilization whose residents had the questionable taste to change the town's name from Hot Springs to honor a now forgotten TV quiz show. The name, however, soon appeared to us as having more profound implications. It took a certain strength of character to live out one's short life in the vicinity of Jornada del Muerto, hammered by bitter winter chills and the furnace blasts of summer air rising on the flanks of the Caballos. It was not unlike the realities confronted by Mongolians in Nemegt's lonely village of Daus. Accept the truth of your harsh, unforgiving habitat or suffer its consequences. To us, T or C was like the kind of colony that the Apollo lunar expedition might stumble onto—a place where one could launch a new civilization or lie low during a nuclear holocaust. Appropriately, the big news of the summer was the FBI's apprehension of a group of self-appointed militia and their huge cache of arms in some inaccessible gully west of town.

In time we came to like what T or C had to offer: the profane mechanic and the old lady who owned a shop that sold bits of glass and glass bottles, everything from abandoned opium flasks left by Chinese railroad workers to the coveted insulators from vintage telephone poles that are valued for a purple tint that comes with aging in the sun. We hung around the town square, flirting with the sunburned girls, with no particular hurry to get back to camp following repair of our most recent tire puncture.

As the time oscillated between several days of fieldwork and an afternoon in T or C, we found more bits of *Dimetrodon* and other reptiles. Our collections in the Caballo Mountains, though more abundant than what we retrieved from Abo and Jeso Canyons, were not enough, fortunately, to prolong our summer there. The professor, sharing our discomfort in the heat, talked about our eventual trek north to the cooler Permian badlands of Monument Valley. "It'll feel like Alaska up there," he said. His forecast proved to be accurate. The temperature in Utah barely ever climbed above the high nineties.

There are, of course, hot, insect-ridden, and dusty days in American deserts—like those in Kheerman Tsav—that can be miserable. But working in such places for prolonged periods requires a certain feeling for dusty desolation that goes beyond tolerance. In subtle ways, Professor Vaughn showed us how to pay the land its due respect. He was patient with us if we tied an inept half hitch or ran the jeep into quick mud, but he bristled if we complained too much about the heat, or the smell of the cattle tank we used for a bath, or joked sarcastically about the social life of some small town. At the university, he lectured with such precision and speed that two students often teamed up for note taking. But stopped out on some two-track road in Jornada del Muerto, he could chew on a shaft of grass for an hour, languidly exchanging philosophy with a local cowboy. The professor even adopted a slower, lulling speech pattern in the field, and used local phrases liberally.

Time moved slowly in the desert and we were expected to fall into that rhythm. It was not a sin to feel boredom, but it was a sin to take displeasure in it. Besides, I don't believe the professor ever really was bored.

That first summer in the deep, sun-bleached valleys of New Mexico, the juniper-covered canyons of Colorado's Sangre de Cristo Mountains, and the incredible mazes of Utah's Monument Valley was an epiphany for me. I learned that one could look out of a car at a line of cliffs or into a deep river-sculpted basin and visualize a geologic map, with the layers of rock neatly color-coded in discrete bands. I could see places where faults—fracture zones that record the shivering crust of the earth—split a rock unit, offsetting a whole mountainside from the rest of the range. The American West was beyond anything I had imagined. Mostly stripped clean of human effects, even vegetation, it spread its naked bony ridges of rock, its dune fields and alluvial fans over huge unchartable expanses. And more than developing an understanding for this strange place—indeed that understanding is still elusive—I began to love it.

FIELD SAVVY

At Flaming Cliffs a huge storm moved slowly in on us, drowning our campsite. We tried to go out to prospect but were driven back by sheets of rain. The sand at the base of the Cliffs had been transformed to a mud slide. We huddled together, shivering in the chill of one of the coldest, dank days I can remember in the Central Asian desert. Worse, we discovered there was virtually no gas in Dalan Dzadgad, and, because of the lack of fuel countrywide, there were no flights to convey museum president George Langdon back to Ulaan Baatar. "There are no flights scheduled because there is no gas." For several days we bartered and negotiated for gas and inquired about the miraculous arrival of a plane. But these efforts were in vain. The country was indeed falling apart. Mongolians, who have been besieged by centuries of invasion from the great empires on their borders, don't trust the endurance of their current independence. Frequently, we would hear comments about the impending threat, especially if China, with the world's largest population and most dynamic economy, went looking for more real estate. During this July in 1992, I contemplated the serious vulnerability of this chaotic but beautiful country to such ambitions.

George Langdon was greatly agitated. He had to get back. Finally Jim Clark valiantly volunteered to drive George back the three hundred miles to Ulaan Baatar. He would rejoin us in a few days at Tugrugeen.

We reached Tugrugeen on July 11, where we planned to linger for some days, during a span of good weather. Our team found the conditions ideal for finding fossils. It was also the first prolonged period of work for the BBC film crew.

I could tell that our BBC team was a bit disappointed with the lack of fanfare attached to our regular days in the field. What most surprises people about fossil hunting is that its central activity—walking slowly with your eyes glued to the ground—is so undramatic. Movie producers have shared with me their vision of a multimillion-dollar set with a brigade of workers pounding, drilling, and even dynamiting the rock face in search of a mother lode of skeletons lying deep within. They seem so let down when I recount the usual more prosaic activities in the field.

Not that field exploration is always easy. Over the course of a day, absorbed in daydreams and a desire for discovery, one can readily cut a twisted path of more than twenty miles. The distance covered over the enormous surface areas created in gullies and badlands can be very deceptive. In the heat, with water unwisely consumed too early, and a heavy load of hammers and rock specimens, a walk up the wash to a truck well within sight can be a struggle. It is not surprising that the most common problems of the inexperienced field hand are dehydration and exhaustion.

Of course, too much water is not a good thing either. Knowing when and where to drink water and how much is thus no more trivial to the paleontologist than knowing where to look for fossils. Good judgment in these matters is not to be taken for granted. I once dunked into a cold mountain stream, so dehydrated that I consumed what seemed to be gallons of water. My comrade was more fastidious. "There's a lot of sheep shit around," he said. Indeed, a few pieces floating in an eddy near the opposite bank caught my eyes. A day or so later my head and torso exploded, my limbs shaking and painful. The town nearby had a retired doctor of eighty-five who prescribed a bottle of paregoric and opium. Three days of this treatment offered little improvement, and I was loaded into the truck for a desperate hundred-mile drive on Interstate 90 for a modern hospital in Billings, Montana. Progress was slow. At about every mile marker I fell out of the vehicle to relieve myself. My companions later claimed I was delirious, but I remember some desire on my part to maintain a facade of respectability. Pulling into a rest area, I ran for the bathrooms, but realized I would not make it. I settled for the next best thing—much to the disgust of three elderly ladies, I shared the pet exercise area with their dogs.

Almost everyone I know who shares my pursuits has had some experience like the "sheep dip disease." In the field, the state of one's gut soon becomes a topic of obsessive interest.

THE BEASTS OF TUGRUGEEN

One important assignment of our 1992 stop at Tugrugeen was to collect the remaining skeletal parts of the *Velociraptor* that Jim had found the previous year. Studies by Mark and Jim during the intervening months had showed that this specimen, though not complete, was precious and enlightening. On the roof of the skull, a row of parallel holes perfectly matched the size and spacing of teeth in the upper jaw of another *Velociraptor.* The tooth punctures must have been lethal, as they were strong enough to go clear through the skull and into the brain tissue itself. There is little sign of scavenging on the skeleton, so we infer that this animal was unlucky in a brief fight to the death. There is no direct evidence of cannibalism in any *Velociraptor,* although the Triassic theropod *Coelophysis* from New Mexico does show some remarkable evidence of this behavior. A few skeletons of adult *Coelophysis* have small skeletons of young individuals of the same species entrapped within the rib cage. These are too mature to be examples of live birth, and the position of the juvenile bones corresponds to the likely location of the digestive tract. It cannot be demonstrated that *Velociraptor,* like *Coelophysis,* indulged in cannibalism. It was a fierce, nasty creature just the same.

Jim's specimen of *Velociraptor* from Tugrugeen also supported the suspicion that dromaeosaurs show signs of "higher intelligence." It has a more complete braincase than any other known for *Velociraptor.* Details of this specimen not yet described in the scientific literature clearly demonstrate a remarkable series of similarities with the basic braincase architecture of modern birds. Moreover the braincase relative to the overall size of the skull is quite large.

Inspired by Jim's great success, we scattered among the cliffs and gullies of Tugrugeen, searching for our own *Velociraptor.* "No one looks on those eastern slopes because they're covered with scrub," Malcolm said, and went his own way. His strategy paid off, in ways we hadn't imagined that summer. My field notes for that day blandly state, "Malcolm found a delicate little theropod, which he plaster-jacketed." We would discover later in the lab that this casual field identification, though not technically wrong, was only part of the story. The full story behind Malcolm's theropod, which we didn't discover until winter back in New York, turned out to be much more fascinating.

But theropods were not our only targets at Tugrugeen. In the 1992 season the site continued to produce some very interesting mammals. Part of the reason for the intensity of our mammal hunt at Tugrugeen was the knowledge that Jim had found that magnificent mammal *Zalambdalestes* at this very spot in 1991. Within our first days, virtually every member of the team came up with a gem of a mammal skull. Lowell at first had a dry spell, but he too eventually found a beautiful mammal skull that belonged to a multituberculate, an archaic group of Mesozoic and Early Tertiary mammals that left no living relatives. The name "multituberculate" refers to the curious complex of cusps on each of the cheek teeth of these small beasts. The animals gnawed on seeds, nuts, and plants with their elongate incisors and further mashed their food with their multituberculed cheek teeth. They were the "rodents" of the Mesozoic—in habits if not in kinship. A large multituberculate was actually the first mammal skull found by the C.A.E., the Central Asiatic Expeditions, in 1923 at Flaming Cliffs. Their remains were important and much desired. "I'm on the board," Lowell said to us with a smile.

Mammals existed in great numbers and in diverse kinds in the Mesozoic. But struggling to appreciate this fact has been long and difficult. Fossils of Mesozoic mammals—especially good ones—are very hard to come by, much harder to find than dinosaurs. As Andrews, Osborn,

Multituberculate mammal *(Ed Heck)*

Walter Granger, William King Gregory, and George Gaylord Simpson from the American Museum noted many decades ago, the Cretaceous Gobi cast a brilliant new light on the mysterious first two thirds of mammalian history. Here whole skulls and skeletons of a diversity of early mammalian lineages were beautifully preserved—not just bits of teeth and jaws. Simpson had built a good part of his reputation as a paleontologist by carefully studying the paltry clues of Mesozoic mammals from localities in England and Wyoming. Imagine his delight in describing the entire skulls brought back by the C.A.E. from the Flaming Cliffs.

The Flaming Cliffs continue to yield precious fossil mammals, as the Polish-Mongolian teams and our own field parties can attest. But there are other good places for these fossils in the Gobi. The sandstones of Khulsan, including the cryptlike canyons of Eldorado, were very generous in mammals, as were the red rocks of lonely Kheerman Tsav. Tugrugeen Shireh, however, in our experience, is better. At Tugrugeen, the mammal bones are preserved in small concretions of hard dark sandstone and iron-bearing minerals that look like brazil nuts. The concretions are continually shed as the cliffs are battered by high winds and seasonal rainstorms. These nodules are easily eroded from the surrounding soft white sandstones, yet they provide a durable coating over the more friable fossil bone. The combined factors virtually guarantee the discovery of more mammals every season, even on slopes we have crawled many times before.

DESERT SERPENTS

Our productive days at Tugrugeen had lulled us into a pleasant routine—get up, prospect, find superb fossils, catalogue them, eat, relax, sleep, and begin again the next day. We had become residents of Tugrugeen, planting our camp at the base of the beige sand dune. When driving back from a water run to the spring at Bulgan we could sight this sand dune next to our temporary "home" from more than a mile away. But we all knew that the rhythm would soon be broken. Our next stop was the Nemegt Valley. In my early days in the field with Professor Vaughn in New Mexico and Utah, I preferred to keep constantly on the move. A prolonged period at one spot meant hunkering down in some hot quarry, chiseling a difficult fossil bone out of diamond-hard rock, or walking over and over the same ground in a desperate search for a "killer" *Dimetrodon* skeleton that never materialized. Tugrugeen, however, continued to reward us all with abundant fossils. We were reluctant to pull up stakes.

But the days were slipping by. I wanted to leave plenty of time to explore the Nemegt this summer. It was here that we hoped to find our own elusive mother lode of Gobi fossils. On July 18 we evacuated our camp, assiduously burying our trash, and moved on.

The trip to Nemegt proved more difficult than our trek of the previous season. There was no way to challenge the muddy passes over the Gurvan Saichan with our tanker and its motley escort. Instead, we took a new and poorly known route, skirting the Gurvan Saichan to the west and crossing a broad valley on the south flank of Arts Bogd. The tanker overheated every fifteen miles. Eventually it broke down, requiring major engine overhaul. Even its gas tank developed a mortal wound, leaking precious fuel over the desert pavement. Our mood was at low ebb; there was even discussion of abandoning the enterprise. But somehow we managed to carry on. By July 20, nearly two days behind schedule, we crossed a saw-toothed mountain range north of the valley that divided us from the purple furrows of the Nemegt Range to the south. We were still a hundred miles from Naran Bulak.

At that point a new problem confronted us. We could see a number of dark canyons sliced into the great escarpment of the Nemegt, but which one was the pass to Naran Bulak, the elusive Almas Pass? This cleft was the infamous abode of the almas, Mongolia's version of the yeti, the abominable snowman. When we had first entered that canyon in 1990, it was from the south, where the canyon is easily marked from a distance by the broad gap between the Nemegt Range and Altan Ula. From our vantage point on the north we could see a number of entrances to choose from. Dashzeveg opted for one of the narrow clefts directly south of us, indicating that this branch would eventually lead us into the main pass. I preferred a migration farther west toward a low saddle on the summit ridge that could mark the break between the Nemegt Range and Altan Ula. We first tried the Dashzeveg variant, but the mouth of his canyon was choked with boulders. Dashzeveg, though, stubbornly maintained that this was the correct way and the route would improve once we were in the canyon. I refused to take the risk. It was a tense moment. We decided to bivouac near the canyon entry. The Gobi atmosphere was as silent and brooding as Dashzeveg, as a thundercloud began to materialize above us. I could only hope the morning would bring a solution.

We found a broad pass by late morning of the next day, on a route, as I had suggested, farther west that jibed more closely with the coordinates that Priscilla had provided. On the way up to the pass the trucks made frequent stops to cool their engines. At one point the drivers excitedly congregated around a white serpent winding its way in the sand between the trucks. Despite the many evocative place names like Viper Canyon or Viper Hills, poisonous snakes were a rare sighting in the Gobi. Venomous snakes in the American West, however, were just part of field exploration.

My first encounter with poisonous snakes took place during my first paleontological trip out West. It was far more traumatic than our sighting at the mouth of Almas Canyon. One windy afternoon in the Caballo Mountains of New Mexico, I prospected a cliff pockmarked with natural caves cut deep into the ledges by rain and wind. The experienced enter these places gingerly, but I learned my lesson the hard way. After bending over and stumbling into one of these guano-stained chambers, I turned around to a sound that blended the long buzz of a cicada with the shaking of a baby's rattle. A diamondback rattlesnake, with the girth of a coiled rope on the deck of a tugboat, was expressing its annoyance. Opposite, four other smaller snakes were in the midst of uncoiling.

I have colleagues who would have been delighted with this situation. Some might even have attempted to snatch one or two of the snakes for their collections (the risks here are not trivial; herpetologists on occasion lose fingers or limbs because of snakebite). But I was duly terrified. The cave floor leading to the entrance seemed just within the striking range of the flanking snakes. Half closing my eyes, I leapt back through the cave opening in an adrenalin-aided blur. Outside, unbitten, but wildly unnerved, I almost fainted in the sun.

Some miles down the canyon I encountered the professor perched on a cairn of rock resting, his narrow-brimmed Stetson hat slanted over his eyes. We exchanged the usual salutations before I remarked with some excitement that a freshly molted diamondback was curled up in a crevice of the cairn some inches below his rear end. Given my recent encounter, the professor's reaction to this warning seemed unreasonably stoic. He sat slightly more erect, raised his hat for a moment, then, lowering the brim again, he resumed his relaxed position and continued sucking on a long shaft of grass.

The Abominable Canyon

Soon even the Mongolians got tired of poking and taunting the poor viper that sought refuge in the scrub near the trucks. We proceeded up the pass and into the mouth of what we hoped was our "abominable canyon." After an hour of driving in the steep horseshoe bends of the wash it looked like the right canyon—we expected to see an almas waving its shaggy arms at us around the next bend. As it turned out, though, both Dashzeveg and I were right. His choice was indeed a tributary of Almas Canyon, one that eventually joined up with our route in a rocky confluence. But it looked like this eastern cutoff was, if anything, even tougher than the route on which we now struggled.

Many country folk in Mongolia do believe in the existence of the almas, a belief that seems to have a deep history. The cryptozoologist Mayor notes: "A recent linguistic analysis of the Scythian word *arimaspu* (recorded by Herodotus in his discussion of griffins and Arimaspeans) shows that Arimaspeans can be identified with *almases,* elusive, shaggy, monocular creatures of Mongolian folklore."

By 1992 there had been several forays into the Gobi in search of the dreaded almas. Upon my return to New York after the 1992 season, I received, via George Langdon, an excited inquiry from an international team of almas hunters led by the well-known British explorer, Colonel John Blashford-Snell. Their expedition had spotted an extremely large footprint in the gravelly wash of Almas Canyon. The footprint did not have well-defined toes but the outlines were expectedly blurred in the gravel; the proportions, though, were quite large, nearly fifteen inches in length. A photograph of the specimen was sent with a request for information and opinion. I replied that our team had crossed the canyon that summer, probably only a matter of days before almas hunters entered it. Six-foot six-inch Jim Clark—our "giant man," as the Mongolians called him—also has very large feet, and likely the spreading gravel around his impact would have given his footprint even greater dimension. There is no doubt that Jim from time to time, like the rest of us, jumped out of his vehicle to urinate; the washboard road of the canyon necessitated frequent stops. Jim indeed wrote a letter to the museum president in support of such alternative scenarios. It was the best theory we could muster. The almas of the Gobi remains undiscovered.

THE WESTERN FRONTIER

After various frustrations and delays our sprawling caravan threaded through Almas Canyon and at last limped into Naran Bulak late in the afternoon of July 21. Naran Bulak, as in previous years, would be our base camp. Dashzeveg even rented a ger for us, for a few dollars for three weeks of use. From here various parties would strike out to prospect. Dashzeveg, Malcolm, and Priscilla could search the rich Cenozoic rocks around the camp, looking for nodules with primitive rodentlike skulls. Several of us formed "team Khulsan" and took the BBC crew with us out to the canyons of Eldorado. It would be disingenuous, however, to report the next few weeks as ones of drama and triumph. Discoveries were made, and the film team got some choice footage before leaving us a week later. But despite the enormous effort in getting this complex convoy to the fossil wonderland of the Nemegt, our hunting was not unusually glorious.

On July 27, the day after the departure of the BBC, a small team—Mark, Jim, Lowell, Dashzeveg, and I—struck out with two Mitsus and Mangal Jal's GAZ to a wholly new area. This sortie represented our farthest push westward—to a frontier as hot and windy and uninhabited as lonely Kheerman Tsav, a maze of sandstone cliffs and spires known as Bugin Tsav. The Russians and Mongolians had worked this place extensively, but we had never been there and we were anxious to see it. Bugin Tsav, whose baroquely intricate canyons exposed the multihued Nemegt Formation, had been declared a national park. The park lay about twenty miles west and slightly north of Naran Bulak, the lonely windswept complement to Kheerman Tsav directly to its south. Despite the seclusion of Bugin Tsav, we had no intention of removing dinosaur skeletons. We were simply interested in reconnoitering the place for skeletons that might warrant work there—pending permission for major excavation—the next field season.

We did indeed find the myriad canyons, washes, cliffs, and hills of Bugin Tsav full of big bones. Some of these were the skeletons of lumbering sauropods or hadrosaurs. At one spot, I spied a long string of tail vertebrae snaking its way around the edge of a hill. As I walked around the other side of this knob, I was amazed to see more of the skeleton, part of the ribs and the neck vertebrae exposed. The beasts of Bugin Tsav were of mountain-sized proportions. They can be discovered from a distance. One especially hot July in the subsequent field season of 1993 we were driving through these extensive badlands when Mark suddenly told me to stop the car. Fifty yards back, the giant skeleton of a *Saurolophus*, a forty-foot-long duck-billed dinosaur, was exposed on top of a small sand hill. A *Tarbosaurus* foot, with its distinctive tripod of digits and its long claws, rested on top of the duck-billed skeleton, as if the carnosaur were staking a claim to the carcass.

During this first visit to Bugin Tsav, we marked various skeletons for possible pickup in next year's field season. If we seemed nonchalant, remember that most of the big dinosaurs of Bugin Tsav had been described and were relatively well known. The history of dinosaur paleontology in Central Asia, as elsewhere, involved an early passion for the biggest and the easiest to find. These game were not our primary targets, although we had no intention of ignoring them. We decided that Bugin Tsav could be a viable option to explore in 1993, in case we failed to find a new "red-bed" locality of theropods and mammals. Pending permission and enough plaster, we might invest the prolonged time necessary to excavate one of these monsters.

But in that first 1992 foray to Bugin Tsav our most remarkable discovery had nothing to do with dinosaurs. Jim Clark, returning from a long hot hike southwest of camp, took us to a spot near a butte that was filled with small, rounded plates—turtle shells. We counted shells and skeletal parts of over fifty individuals representing two different taxa in this small depression, roughly the size of a wading pool. This "turtle death pond" represented an interesting but puzzling vignette in Central Asia some seventy-five million years old.

The End of an Adventure?

The thrill of Bugin Tsav, sauropods, turtles, and all, wore off after several days and we returned to Naran Bulak. By this time it was August 1; Lowell and I had only six more scheduled days in the field. We had, to this point, avoided the heat and the flies of Kheerman Tsav, but there was no other option but to return to our hell hole of 1991. This time we traveled less encumbered, with two Mitsus carrying our Bugin Tsav team sans Mangal Jal.

Kheerman Tsav was much more livable this time, despite the heat and some ferocious sandstorms. The camel ticks drove us to sleep in our cars at our first night's camp, but the flies were mysteriously absent. This year, we could give Kheerman Tsav our undivided attention.

This concentration did reap rewards. On the northern escarpment, along the red outer wall of Kheerman Tsav, we stopped at a butte that reminded me of Lomas las Tetas in Baja California. Scrambling up and down the slopes of this butte for a couple of days yielded a reasonable sampling—more than a score of nice skulls, mostly lizards, but even a few mammals. I was pleased that we could cap off the field season with some decent results. It was a good camp, near the edge of a small sand dune with an unimpeded view to the vast, unknown, and uninhabited Gobi west of Kheerman Tsav. We extravagantly doused a dead zak tree in gasoline and torched it for a blazing campfire at the end of my last working day in the field season of 1992.

Despite the pleasures of our Kheerman Tsav camp, however, I was feeling a bit low. This summer had been unusually trying. Too many people. Too many vehicle breakdowns. Too much money spent. And results which were . . . well, perhaps no more dramatic than the previous two outings. True, at this time we had a distorted view of our achievements. We had no notion at the time, for example, that Malcolm's theropod from Tugrugeen would turn out to be one of the most fascinating and controversial fossils ever retrieved from the Gobi. We had not yet marveled at the details of our *Zalambdalestes* skull, seen from animated CAT scans. But great discoveries should be enjoyed in the field as well as in the laboratory. Despite the successes of our Gobi expedition, we had yet to savor the sweet victory that comes in finding a new major locality, rather than just a small pocket like our *Estesia* hill found in 1990. A place like the Flaming Cliffs, an Eldorado—our own mecca. Was there still a chance for such a discovery? It seemed so, but our navigation on this great sea of a desert over three summers had taken us to many of its canyons, gullies, and cliffs without giving us the joy of a first strike. Perhaps it was too much to expect such an implausible outcome. Fulfilling scientific careers don't require them. But we had been sustained under the pressures and labor of a major expedition by the hope of such a discovery.

"The problem," I said to Dash, "is the lack of new sites. Where are they?"

"Mitqua"—Dash shrugged his shoulders—"maybe near Daus. Maybe north of Nemegt."

Dash did not look at me but stared into the torch fire of burning zak. He too appeared somewhat beaten down. After spending thirty years in the Gobi, he was not about to humor an impatient American, anxious to cheat the odds in the brief span of three summers.

I felt a growing sense of weariness and depression. I loved this place. But would I ever be back? Perhaps not. We were running out of momentum. The 1993 season at that point seemed a remote prospect. On the way into the site we had played a tape of Lou Reed, who sang at one point, "It's the beginning of a great adventure." Now I thought, maybe it's the end of a great adventure. It was great fun but it was just one of those things. We had achieved good, respectable, interesting findings—but none that secured a place in the paleontological pantheon. None that seemed to justify the commitment of time and resources that a Gobi expedition involved. Perhaps there were more productive enterprises back home. I began to feel that unwanted desire—the desire to give up.

As the fire flickered, however, the light seemed to beat back the shadows of defeat. The glow of the Gobi itself engulfed me. And I knew that night we must come back. There were still too many possibilities to explore, too many hidden canyons and gullies that could take us to hallowed ground. I could not turn my back on the promise of the Gobi. I could not turn my back on the strange fortune that allowed us to explore the richest dinosaur territory in the world after it had been closed off to so many colleagues for so many decades. I remembered laying those crude maps out on the hoods of the cars and peering at topographic lines and stream drainages and mountain ranges, while Dashzeveg glided his index finger over locales, names that conjured up so many experiences: Nemegt Uul, Altan Ula, Kheerman Tsav, Daus, Gilvent Mountains. He would circle his finger over these places as if he were inscribing targets. He would speak in a low hoarse voice, perhaps with the same intonation that Ahab used when showing Starbuck the routes of the great whales on his secret charts:

"Maybe there, good. New badlandees. Let's go!"

CHAPTER 6

DINOSAUR LIVES—FROM EGG TO OLD AGE

It is most difficult always to remember that the increase of every creature is constantly being checked by unperceived hostile agencies; and that these same unperceived agencies are amply sufficient to cause rarity, and finally extinction. So little is the subject understood, that I have heard surprise repeatedly expressed at such great monsters as the Mastodon and the more ancient Dinosaurians having become extinct; as if mere bodily strength gave victory in the battle of life.

CHARLES DARWIN.
1859. *The Origin of Species* (p. 318, Mentor Edition).

Our desperate search for dinosaur bones in a burning Central Asian desert does attract attention beyond some compulsive specialists. Lots of people—not just paleontologists—get excited when they start thinking about monsters of the past. How did huge long-necked sauropods feed? How much did they consume in their daily meal plan? And for that matter what would a gargantuan meat-eater like *Tyrannosaurus* or *Tarbosaurus* consider a healthy meal? Were dinosaurs gregarious or furtive? How did they mate, reproduce, and grow? These questions and related ones are responsible for tons of paper pulp, an effort to satisfy a relentless thirst for the facts about "dinosaurology," an effort to breathe life into the sepulcher of dinosaur bones that represent the prehistoric past. But one of the central questions for a paleontologist is, what can we *really* say about the lifestyles of extinct and famous? Those prone to more romantic re-creations sometimes find the paleontological take on this question frustratingly conservative. How can we stubbornly refuse to commit ourselves to statements about color, breeding habits, or attack strategies when we have all those evocative bones lying around? But we have no choice. Science is a form of knowledge that is drawn from the things we can see, the observable bits of the universe, in our own parlance what we call *empirical observations.* Despite all the skeletons—despite even those incredible mud impressions of the skin of duck-billed hadrosaurs—we have not one shred of evidence as to what colors Mesozoic dinosaurs were. They could have been a motley green or dull brown as the early paintings of dinosaurs would have them, or they could have been more garishly striped or spotted with brilliant cuing colors on their crests, plates, or spikes, like the chromatic plumage and ornamentation of some of the magnificent animals today. They could have been pink with purple polka dots. But we don't know. In all likelihood, we'll never know.

Sorry about that. It would be nice to claim that the fossil record produces a window to the past without any obfuscating smudges, but that record for all its fantastic wealth and detail still leaves much to the imagination. Yet there is enough here to gain some insight, to eliminate some possibilities and favor others about the lives of the dinosaurs and other prehistoric creatures. I made such an attempt in Chapter 2 when suggesting lifestyles for meat-eaters like *Velociraptor* and herbivores like *Protoceratops.* This kind of re-creation only works, however, if our interpretations are firmly planted on the fossil-laden ground. It requires a consuming attention to anatomical detail in the fossils themselves, even down to minute structure that can be seen at thousandfold magnification under an electron microscope. It also relies on the power of analogy—our comprehensive knowledge of the link between anatomical design and the physiology and behavior we can observe first hand in animals living today.

THE CHALLENGE OF GIGANTISM

I have been preaching here that paleontologists should avoid fabricating "just so" stories about dinosaur lifestyles, at least with the false claim that this has something to do with rigorous science. Yet these matters are impossible to ignore.

Take, for example, the skeletons of those giant sauropods from Mongolia, like the one unearthed in the Nemegt Valley by the Polish-Mongolian team in 1965. This hulking skeleton, which unfortunately lacked a skull and a neck, was given the rather cumbersome name *Opisthocoelicaudia.* ("posterior-cavity-tail"). The name refers to the curious anatomy of the tail or caudal vertebrae. In this animal, the socket of the ball-and-socket joint between tail vertebrae is in the posterior end of each vertebra. Typically in sauropods, this socket is on the anterior or front end of each vertebra. In combination with an unusually massive (even for a sauropod!) pelvic girdle and hind limb, *Opisthocoelicaudia* must have had an extremely rigid framework. The Polish paleontologist Borsuk-Bialynicka argued that the massive tail and hind limbs effected a stable "tripod" that allowed the animal to rear up, lifting its forelimbs off the ground, a posture useful for tree feeding, defense, or sex.

A number of sauropods, including the great beast *Nemegtosaurus,* have been discovered in the Gobi. Sauropods were then well established in the Cretaceous of Central Asia as well as North America. But the group had their heyday in the Late Jurassic, when diplodocids, camarasaurids, and brachiosaurids flourished (although the Cretaceous record for sauropods is improving on many continents). The Late Cretaceous occurrence of the diplodocid *Nemegtosaurus* is in fact rather odd; it is separated by a gap in the fossil record (except for a few sporadic occurrences) of nearly seventy million years, from the time when diplodocids were particularly lengthy and abundant. Sauropods were a successful and persistent dinosaur group; titanic body size seems to have been a design that proved enduring.

Indeed, sauropod body design shows remarkable consistency—a

Brachiosaurus (Ed Heck, reprinted with permission, AMNH)

small head with numerous small teeth, a neck that is often so long it looks like the mirror image of the serpentine tail, massive trunk vertebrae, ribs, and shoulder and hip girdles, and legs that look like columns of a Greek temple. Some sauropods show evidence of armor plating, but most do not. It is likely that the tail in forms like the diplodocids was used as a whip when needed, or if these creatures could rear up perhaps they could also stomp a predator to death with their massive front limbs. The momentum of a 77-ton creature like *Brachiosaurus* is something to be reckoned with. When it comes to size, sauropods were the ultimate dinosaurs. Large diplodocids were over 88 feet long, and large brachiosaurids measured in at lengths of 74 feet and stratospheric heights of 39 feet. Some more recently discovered sauropods like *Ultrasaurus, Supersaurus,* and *Seismosaurus* were even larger, surpassing 100 feet in length.

This huge weight problem induced a great variety of modifications to the massive vertebrae and other elements of the skeleton. Weight of these bones is minimized through an ingenious array of hollows, cavities, and special joints. As a result, there is a great diversity of vertebral structure among various species of sauropods—different solutions for remodeling the backbone for minimum weight and maximum strength. A multitude of taxonomic names have been applied to species thought to be represented by these different kinds of vertebrae. Add to this the fact that highly diagnostic skulls of sauropods are often lacking or are disassociated from the skeletons, and one is confronted with perhaps the most complex taxonomic challenge of any dinosaur group. About 150 species have been introduced in the scientific literature. The majority of these are represented by only small parts of the skeleton: fourteen species are based solely on teeth, twenty-five on one or a few vertebrae, and six on a single limb bone. Less than thirty of the sauropod genera are fairly well established by good skeletal evidence. This is not a problem that can be sorted out with any expediency. Fortunately, Jack McIntosh, one of the world's most respected dinosaur paleontologists, has valiantly and persistently wrestled with sauropod classification.

Something so big as a sauropod skeleton is thrilling to see up close. A few years ago we broke out the stored skeleton of *Barosaurus,* an eighty-foot-plus monster. The original bones (the skeleton is over eighty percent complete) were used to cast a lighter fiberglass skeleton for display in the main entrance, the rotunda, of the American Museum. The giant vertebrae were laid out on a worktable along one wall of the dinosaur storeroom. Some of these, especially along the lumbar and thoracic regions of the back, were as big as a space capsule; one could practically crawl into the thoracic cavity and sit comfortably. It took awhile to walk from the head end to the tip of the tail. It was like walking along a destroyer moored to a dock. Today in the museum's newly renovated hall of saurischian dinosaurs, visitors can get an even bigger thrill: they can dance over the original bones on a see-through glass floor that actually serves as the ceiling over the skeleton.

Sauropods, with their massive trunks, lofty necks, and pillarlike legs, seem to defy gravity. These gargantua serve well as examples of the challenge and ambiguities of reconstructing dinosaur behavior. First of all, how can an animal with this much mass walk, feed, and, for that matter, breathe

Barosaurus (Ed Heck, reprinted with permission, AMNH)

on land? The question has long been pondered. Earlier reconstructions simply put sauropods in the water, where their bloated bodies could be more readily supported. This now seems unlikely. Paleontologists like Elmer Riggs and, later, Walter Coombs and Bob Bakker argued that sauropods were well built for a terrestrial, high-browsing lifestyle. Bakker stressed that sauropod equipment—including super-elephantine legs, short compact feet, and a specialized slablike elongated thorax (breastbone) for the attachment of massive muscles—all were tailored for ambling about on land.

The landlubber lifestyle of sauropods is supported by geological data. Bakker, Peter Dodson, and Anna K. Behrensmayer (she is a curator and director of science at the Smithsonian Natural History Museum in Washington, D.C.) have demonstrated that the sauropod-rich Morrison Formation in the Rocky Mountain regions represents a variety of conditions during the Late Jurassic—from lakes and swamps to much drier terrain. In fact, these authors concluded that the Morrison Formation—the rock unit containing the greatest concentration of diverse sauropods in the world—represents environments which were predominantly on the dry side. However, elsewhere in the world sauropods seem to have occupied diverse and in many cases humid habitats. Those from the Tendagaru Formation in Tanzania and the Glen Rose Limestone of Texas lived near a marine coastline. As pointed out, the sauropods of the Gobi come from the more mesic (more humid and vegetated) Nemegt Formation than the more arid Djadokhta or Barun Goyot Formations. Numerous trackways of the huge monsters confirm that sauropods readily walked on stream or lake bottoms. Perhaps the association of sauropod bones in the drier paleo-environments of the Morrison Formation is the exception rather than the rule. Nevertheless, the old idea that these animals were closely tied to and critically dependent on aquatic habitats is no longer persuasive.

Whether sauropods were partial to watercress salads or treetop foliage, the question concerning the manner in which huge, absurdly small-headed beasts ate invariably comes up. The long necks of sauropods remind one of giraffes, suggesting that these dinosaurs raised their serpentine necks into the canopy for browsing among the treetops. Much has been written about the engineering problems this erect feeding would entail. Some paleontologists note that the vertebrae at the base of the neck are not equipped with large neural spines. These bony projections would be necessary for the attachment of epaxial muscles, which work to keep the neck erect. A popular image, then, is one of a sauropod with a neck stretched out horizontally, like an enormous boom, sweeping side to side and occasionally vertically from a fixed position. In this re-creation sauropods did not consistently feed in a giraffelike manner, with the head and neck erect. But there is no reason to expect that all sauropods ascribed to one feeding strategy. In fact, these animals vary markedly in the relative lengths of their necks: camarasaurids are rather short-necked (well, less than twenty feet long) while sauropods like *Diplodocus, Barosaurus,* or the Chinese *Mamenchisaurus* have awesomely stretched necks, some over thirty feet and nearly equal to the tail in length (in *Mamenchisaurus* the head and neck equaled the length of the rest of the tail *and* the body!). This different design in booms and cranes doubtless had implications for different styles in feeding.

The question of sauropod posture has given rise to one of the more notable controversies concerning dinosaur lifestyles. Some claim that to hold the neck erect for prolonged periods of time would be a physiological impossibility. The arrangement of muscles and ligaments as inferred from the geometry of the neck vertebrae, it is asserted, would not allow a lot of neck-raising. The more famous dispute over the matter, however, concerns blood pressure. It has been estimated that the pressure necessary for the perfusion of the brain with blood in an erect sauropod head would place extreme and dangerous loads on the heart pumping system. Some have further calculated that the systolic blood pressure necessary to perfuse such brains raised far above the chest cavity would be extraordinary: anywhere between 367 and 717 mm. Hg. These are apparently pressures not approached by any living vertebrates. Based on comparisons with living mammals, it is estimated that an 80-ton sauropod had to have at the very least an 800-pound heart to effectively pump its blood to its extremities. A 120-ton blue whale may be expected to have a heart that weighs 1,200 pounds. But whales live in the water, and the support provided by this medium helps whales deal with problems of blood pressure, circulation, and cardiovascular pumping. Land-loving sauropods probably faced enormous difficulties in circulation under much greater vascular stress and pressure. It has been suggested that such a heart could weigh between one and three metric tons!

This extraordinary problem for the blood-pumping machinery in sauropods could have had several solutions. Sauropods may have never, or rarely, raised their heads more than several feet above the ground. Alternatively, the cardiovascular system may have been radically redesigned for shunting blood from heart to head. A special net of blood vessels at the base of the brain may have been more flexible to abrupt changes in blood pressure. Perhaps these animals experienced, but tolerated, some degree of hypoxia (oxygen starvation) during short bouts of vertical feeding. Perhaps there were carotid reservoirs or sinuses high in the neck and at the base of the head that supplied extra blood when the carotid arteries collapsed from a drop in systolic blood pressure. There has even been a recent, but far-fetched and implausible, suggestion that these beasts may have had three or four enormous pumping hearts to get the blood up the long gantry tower to the head.

The work of circulation in an animal, whether sauropod or snail, requires energy derived from food. The effectiveness of the digestion and the feeding apparatus in sauropods is also a matter of much discussion. Different kinds of teeth are found in different sauropod species. Diplodocids tend to have slender peglike teeth good for nipping softer plants and the weeds of aquatic soup. Camarasaurids have more spatulate teeth that show notable wear; perhaps they ingested coarser plant material. In places in the Morrison Formation, five or six species representing both dental and feeding types are found together. It is possible that these beasts milled the food much like birds by means of gizzard stones. Although smooth and polished spherical stones have been found among the bones of certain skeletons, the evidence of gizzard stones is not decisive. It is possible that sauropods employed fermentation in a hind gut in a caecum or pouch where microbes broke down and metabolized the massive mowings. Again, there is no clear and direct evidence for such an adaptation, despite the logic of such an idea.

However they processed food, sauropods must have eaten it in copious amounts. This is a curious situation, because sauropods have relatively tiny heads. Those front-end feeders must have been overworked by the need for siphoning huge masses of plant material. Based on models extrapolated for information on elephants, it has been suggested that a sauropod could have consumed about 400 pounds of food a day. One can readily see how this might pose limits on sauropod size, given the huge intake required for animals weighing over 50 tons. It is also likely, given this enormous need for food in bulk, that sauropods were not overly fastidious. Large herbivorous mammals today are generalists when it comes to consumption of food. They cannot afford to carefully pick and select the preferred items when there is such a premium on bulk meals. They simply take in a variety of plant material which is then digested by microbes in their voluminous digestive tracts. Ferns were all over the Jurassic landscape and it has been suggested that ferns, despite their low nutritional value, were probably mowed down constantly by herds of sauropods.

Related to the questions of neck orientation and feeding in sauropods is the issue concerning posture and stance of the whole body. Like most dinosaurs, sauropods were able to keep their serpentine tails off the ground. Large muscles extending from the pelvic region would have ably supported the base of the tail. Moreover, numerous sauropod fossilized trackways do not show tail impressions. Unlike the venerated old dinosaur paintings of Charles Knight and others, these animals did not drag their tails. In some, like *Brachiosaurus,* the forelimbs are very massive and actually higher than the hind limbs. This gives the animal a "high shoulder," with the back sloping down to the hip region. In other sauropods, like *Diplodocus,* the forelimbs are shorter and less massive than the hind limbs, so the animal probably had a back that sloped up to its rump. In either case, however, the limbs fore and aft supported and moved the body. As trackways readily show, these animals were quadrupeds; that is, they moved about on all four limbs. Nonetheless, Henry Fairfield Osborn and many succeeding paleontologists have suggested that sauropods, especially the diplodocids, were perhaps able to elevate their front limbs off the ground, using the massive hind legs and tail as a tripod. Others have objected that this could not have been a likely behavior for an animal that for physiological reasons could hardly raise its neck.

Given these controversies about sauropod posture and behavior, one can easily appreciate why our decision to display an 80-foot-long *Barosaurus* in the main entrance of the American Museum, with its neck fully erect and its forelimbs high off the ground, was greeted from some corners with incredulity and protest. Perhaps this action might seem rash, but at least several of us involved in the decision have no regrets. We believe that all the erudite discussion about what *Barosaurus* could or could not have done does not exclude the possibility that these animals occasionally reared up. Indeed, sauropods must have been able to go bipedal in some instances—mating would have been impossible without this ability. Moreover, one could imagine, in the stress of defending against a predator, these animals might well lift up on their haunches, ready to bring their massive pile-driver forelimbs down on the attacker. Thus our "mamma" *Barosaurus* is depicted protecting its young one from a lunging carnosaur, *Allosaurus.*

When it comes to biological qualities and behaviors, sauropods have been the object of much pondering and speculation for a simple reason. They are, as far as we know, the ultimate heavyweights of all terrestrial animals. It is likely that the characteristic physiological challenges of big dinosaurs were even more accentuated in the sauropods. Thus sauropods have served as models for the thresholds of physiological tolerances or anatomical architecture. Unfortunately, sauropods are extinct, so most interpretations are intriguing but hardly concrete. For example, the problems of blood pressure have been raised forcefully as decisive evidence against the idea that sauropods could hold their necks vertically or raise their front legs off the ground. But all we can say is that there were undoubtedly some physiological problems that came with this behavior in such a big animal. This says nothing about whether or not such problems were overcome. Since there have been many suggested possibilities—carotid sinuses, vascular networks, quadruple hearts—there seem to be many possible solutions to this problem.

Living animals are capable of extraordinary things, and we presume that extinct ones were too. In this regard, one would never have imagined that air-breathing mammals like dolphins and sperm whales could dive for incredibly long periods (sometimes over an hour) to depths of 3,600 feet, where, according to hydrostatic principles, tremendous pressures would normally crush the body. Certainly we would be hesitant to ascribe such abilities to whales if we only knew them from fossils. On the other hand, we cannot *prove* that *Barosaurus* and other sauropods sometimes held their heads high and their front limbs off the ground. That's the very point, though. Arguments for a certain behavior of a fossil creature are often as consistent with scientific evidence as arguments against the behavior. The *Barosaurus* display at the American Museum dramatically illustrates the point.

TAKING TEMPERATURES

Food is digested to keep the body working. In the process, some energy is lost in the form of heat. This simple equation is also related to the metabolism of sauropods and other dinosaurs. Indeed, the question of dinosaur activity as related to their internal body temperature is at the top of the list of all dinosaur controversies. The debate seems all the more engaging because of what we know about the stark differences in metabolism among living vertebrate animals. Lizards, crocodiles, and turtles are *ectotherms* (from the Greek *ektos,* outside; *thermos,* heat). They are evocatively but misleadingly labeled "cold-blooded." Body temperatures vary with external conditions because the body has not evolved the machinery and energy budgets necessary to keep internal heat constant. In contrast, birds and mammals are *endothermic* (Gr. *endothi,* within) or warm-blooded; they maintain a stable body temperature despite heating or cooling of the external habitat. Indeed, the variance in internal body heat in these forms is a sign of malfunction or illness. The traditional "reptilian" reputation of dinosaurs typecast them as cold-blooded, stupid, and sluggish animals. One of the major intuitions in dinosaur studies over the past two decades is that these creatures don't fit the sluggish ectothermic stereotype. But the evidence that dinosaurs were endothermic or "warm-blooded," despite all the fanfare, is not absolutely established. Proponents of this theory stress several points:

1. Dinosaurs have postures with legs set directly below the trunks and powerful, gracile frames that suggest active locomotion. This kind of speed and activity—especially evident in smaller theropods—is what we would expect in endothermic creatures.

2. The larger brains of certain dinosaurs, again notably the theropods like *Velociraptor,* are more characteristic of active and alert creatures. Typically these creatures are endotherms, like birds and mammals.

3. Dinosaurs were widespread on the globe, from the tropics to high latitudes, even, as we saw in Chapter 4, in polar regions. This kind of tolerance of the rigors of climate tend to be more typical of endotherms. Ectothermic turtles, lizards, and crocodiles are generally restricted to more tropical and lower temperate latitudes.

4. Certain fossil localities rich with dinosaurs show a very high proportion of prey items, usually herbivorous or plant-eating forms, to meat-eating predators. This is typical of modern animal communities, like that on the African savannah, where endothermic predators require comparatively large amounts of food for their active metabolism. A lion consumes on average about the same mass of food as ten similar-sized crocodiles.

5. Studies of the fine structure of bone (histology of bone) in di-

nosaurs demonstrate remarkable similarities in detailed architecture, the blood nutrient system, and other traits to the bone of warm-blooded mammals and birds.

6. Since birds are a lineage of theropod dinosaurs, one would expect that many extinct dinosaurs had a metabolic motive similar to that in living birds. Perhaps not coincidentally, birds are warm-blooded.

7. Recent work on isotopes preserved in some dinosaur fossils suggests a warm-blooded metabolism.

These all seem promising points to buttress the theory that dinosaurs were hot-blooded. But most points are rather indirect. Predator/prey ratios are not necessarily locked in to the kind of metabolism found in the eaters and the eaten. Indeed, the sampling of dinosaurs at many sites is fraught with questions concerning collecting bias or problems in preservation. There are sites where the number of carnivorous dinosaurs is in fact unexpectedly high for an agglomeration of warm-blooded creatures. And what if these concentrations were misleading because they indicated unusually high numbers of scavenging meat-eaters at mass death sites? Similarly, the case for bird-dinosaur relations is a strong basis for suggesting that these groups shared similar physiologies. Yet the extrapolation from endothermic birds to inert fossils can't be guaranteed.

The key and only direct evidence here resides in the studies of bone histology and isotopic studies. Unfortunately, the above-noted similarities between dinosaur bone and the bone of living endotherms are oversimplified. Careful analyses by Armand de Ricqlès and Robin Reid show that the amount of blood channeling, or vascularization, in bone and the degree of warm-bloodedness are not perfectly correlated. Some ectothermic reptiles have highly vascularized bone, while some small, albeit endothermic, birds and mammals have poorly vascularized bone. Moreover, there seem to be two kinds of vascularized bone; one kind meant for fast-growing endothermic animals and one that is designed to support a heavy body. Dinosaurs could have vascularized bone for either or both purposes, but it cannot be concluded with comfort that the histology of their bone indicates they were warm-blooded animals. Finally, some late-breaking CAT-scan studies by John Ruben and his colleagues suggest that dinosaurs lacked the detailed internal structure in the nose region—namely, they lacked the complex scroll-like bones called nasal turbinates—that is found in endothermic mammals and birds. This complex nose architecture seems to provide not only an elaborate surface for a refined sense of smell. It also offers a filtration system that seems necessary for the rapid breathing associated with metabolically charged, warm-blooded animals, a system that, as far as we know, dinosaurs don't seem to have.

Thus, the popular infatuation with the theory that dinosaurs were vibrant "warm-blooded" endotherms is not resoundingly supported by the fossil evidence. Just the same, it seems that warm-bloodedness in some dinosaurs, whether or not we can demonstrate it at this time, is a reasonable possibility. Moreover, on two counts—the emerging studies of isotopes in fossil bone and the fact that living warm-blooded birds are a type of dinosaur—the theory of endothermy for some extinct dinosaurs seems compelling.

These thoughts relate to the special metabolic predicament of the huge sauropods. In these animals, microbial fermentation in the gut, as well as the thick layers of skin, muscle, and fat, must have created enormous amounts of body heat. The internal temperatures of these animals may have remained relatively warm despite external fluctuations in temperature. In this sense, sauropods could have been passively warm-blooded, *passive endotherms.* "Warm-bloodedness" is customarily a product of the more stepped-up pace of body functions—breathing, locomotion, muscle contraction, etc.—that make up the metabolism. Animals with high metabolic rates—like mammals and birds—produce a lot of body heat through these metabolic processes and are insulated by hair or feathers to help them retain that heat and keep their internal temperatures constant. They are *actively* endothermic. However, with enough body mass and layers of fat, animals with rather sluggish activity patterns and lower metabolism can *passively* mimic the endothermic mode.

Certainly sauropods seem to have the right profile for such passive endothermy. Insulation in some endothermic mammals can be so effective that the problem becomes one of losing rather than retaining heat. This problem is particularly critical under high ambient (external) temperatures. Sweat glands and large blood-enriched skin surfaces help in this heat loss. Desert antelope, for example, have a complex network of blood vessels called a *rete,* wherein the increased surface area cools the blood and prevents the brain from parboiling. What kind of heat-dissipating engineering did sauropods have? It is not known, and perhaps will never be known. If these beasts, however, maintained high internal body temperatures, they must have also developed some mechanisms for losing body heat when the situation called for it.

High Society

Whatever the pace of their metabolism, and their capacity for regulating body temperatures, we can be confident that dinosaurs were highly active creatures. The upright postures, the long legs planted firmly under the body, and whiplike tails all suggest muscle power, dynamic bursts of movement, and—as in the case of dinosaurs like the small theropods—lithe, bounding grace. But did dinosaurs act alone or with others? In some ways this is an even more difficult question to address than those pondered for dinosaur function, physiology, and behavior. The societies of certain living species—the ants, the wasps and bees, flocks of birds, herds and hierarchies of mammals—are some of the most complex and refined products of evolution. Projecting such subtle social relationships back over sixty-five million years to the age of the dinosaurs is fraught with difficulty.

Sometimes the circumstantial evidence allows us to infer something about the "high society" of the dinosaurs, but that evidence allows us to go only so far. For instance, let us consider those stratospheric sauropods once again. In places, sauropods occur in huge heaps of bones, as if the animals lived and died together. Also trackways often indicate many individuals. One may speculate that these animals traveled in great herds, crashing their way through vegetation as they consumed enormous amounts of biomass. There is no direct fossil evidence of this clear cutting of forests by thundering armies of sauropods, but it seems likely that they had great impact on the available resource. The suggestion that at least some sauropod species were highly gregarious has been taken one step further. Dodson, Behrensmayer, Bakker, and McIntosh noted that sauropods were essentially homogeneous over great distances—as much as six hundred miles. Such a range, coupled with the recognition that the Morrison represents environments with a rainy season and a dry season, reminds one of the Serengeti today. Martin Lockley's ingenious work on fossil trackways certainly supports the vision of great herds of sauropods and other kinds of dinosaurs stomping through wide expanses of territory.

It has been proposed therefore that these sauropods migrated annually in huge herds, following the opportunities for more food at different locales. There are indeed clues that suggest that these dinosaurs were capable of ranging over wide areas. Very similar sauropods like *Brachiosaurus* and *Barosaurus* are found in places as far flung as Tanzania and the Rocky Mountain region. Sauropods, either in herds or in small groupings, must have had impressive capacities for dispersal. Can we then proceed automatically to the theory of annual or some kind of regular migration in these beasts? I'm afraid not. How one can demonstrate sauropod migration, as opposed to a simple tendency to herd, is not at all obvious, at least when one considers the limitations of the fossil record.

Another fascinating but risky probe of dinosaur social behaviors concerns the possibility of a mating hierarchy, a pecking order, within the herd. We see such extreme competition and rank in many living species. Magnificent rams of bighorn sheep battle for a flock of females. African gazelle bucks aggressively keep an impressive herd of does away from other ambitious bucks. In a reversal of roles, the top-ranked female hyena—who is much larger than the males and even develops a penislike structure from the clitoris—dominates her den. But these species hardly measure up to the severe hierarchical ranking in elephant seals, where a snorting, ugly, stinking, three-ton alpha bull may inflict death on challengers, and even crush young pups in the heat of battle, in order to maintain his breeding dominance over a harem of as many as 400 females. These realities of rank and competition extend even into the branches of our own sector of the evolutionary tree. Alpha male gorillas beat their chests in dominating and self-aggrandizing ways, putting would-be challengers in their respectively more submissive places. Gorilla relatives—humans—have systems of myriad diversity and complexity that defy a simple biological or evolutionary description. Nonetheless, in our own species we see all kinds of correlations between attributes and societal rank—attributes that include physique, money, and political power.

One might reasonably expect that dinosaurs—large, active, and in some cases even perhaps congregational creatures that they were—might have exhibited such brutally stratified social organization. Indeed there are dramatic paintings showing dinosaurs of the same species in furious mortal combat over the possession of the herd. But once again we must return to the question—what is the real evidence for such depictions? One thought is that some of the ornamentation of certain dinosaur species—emblems that draw a parallel to plumage, horns, antlers, and other adornments in living mammals—provides a clue to such behavior. In Chapter 2, I briefly described the horns and frills in *Protoceratops* and other rhinolike ceratopsians. All those variations in horns and frills and spikes are impressive. They certainly serve to differentiate the kinds of ceratopsians. But did they have any social significance?

The question has been addressed by research on rich ceratopsid samples from North America. Some excellent work in this vein has been recently completed by Scott Sampson, who studied marvelous bone beds at Landslide Butte, Montana, that contain huge amounts of concentrated bone of horned ceratopsids. Unlike the Gobi situation, the bones are not articulated in skeletons, but they can be sorted into different batches that represent various taxa. The sorting to a great extent relied on the bizarre array of headgear. But it soon became apparent to Sampson and his coworkers, Michael Ryan and Darren Tanke, that these head ornaments were not always helpful in classifying ceratopsids. It seemed curious that so many grades of horn development existed in certain rich samples. The investigative team recognized an important growth pattern that seemed to account for the bewildering variation in an important subgroup of the ceratopsids, the Centrosaurinae. All juvenile centrosaurines were alike; they had simple, unadorned heads. Subadults or adolescent centrosaurines attained adult-sized skulls but not yet the fancy horns and frills of adults. Their horn cores, where present, were simple, small, and sometimes backward pointing. Juvenile and subadult forms all had essentially the same frills, thin flanges with a scalloped margin. But adults went to baroque ex-

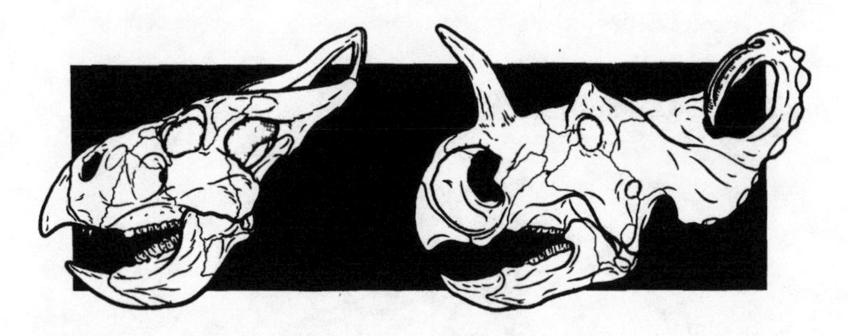

Centrosaurine "headgear" (right) compared with *Protoceratops* *(Ed Heck)*

tremes in developing hooks, horns, and spikes on the frill. In addition, not all adults seemed to show marked development of horns. There was some suggestion that some of the sample of adult skulls had more poorly developed horns because they represented the sex (either male or female) not involved in signal dominance or combat.

Thus the variation in these headpieces, if carefully analyzed, tells us something important about the growth patterns and sexual dimorphism as well as the taxonomy of ceratopsids. Now think of some parallel situations on the African plains. Young gazelles, wildebeest, or kudus are virtually hornless. Horn growth is a function of getting older. Horn size is related to body size and often social dominance over the herd. Moreover, one sex, in this case males, shows a much more pronounced development of horns than the other. The pattern in samples of horned dinosaurs studied by Sampson and others suggests a strikingly similar story for social interactions. Just the same, only the possibility, not the confirmation, of a social hierarchy in ceratopsians can be entertained.

This pattern of head accouterments and their possible bearing on social rank or role is not confined to ceratopsians. The duck-billed hadrosaurs are differentiated mainly by the development (or lack thereof) of their emblematic head crests. Some forms have virtually no crest at all. This may have been a primitive feature of the group, which is found in both early and later hadrosaurs (remember, the fossil record doesn't always perfectly mirror the advancement of steps in evolution). In crested forms,

Hadrosaur "headgear": *Corythosaurus* (above) and *Parasaurolophus* *(Ed Heck, reprinted with permission, AMNH)*

the name of the game is weird elaboration and variation. *Saurolophus,* our beast from the Gobi (also known from North America), has a prominent bony ridge on top of the snout and face that ends behind in a small spike. The Gobi species of this group shows some distinctiveness in its somewhat longer and more fan-shaped head spike. But the really bizarre species are some of the North American forms. An extremely broad platelike crest is known in *Corythosaurus,* and there is a long bony tube that extends backward nearly three and a half feet from the skull of *Parasaurolophus.* Stranger still is the fact that the nasal passages actually extend for some distance into these hollow crests. Their caliber and design vary, just like the differences one sees in the passageways of trombones, saxophones, and tubas. Not all these crests are hollow; for example, hadrosaurines, which include the Gobi beast *Saurolophus,* lacked such passageways.

These crests have excited some paleontologists to a considerable degree and there is no shortage of speculation on their function. As with many reconstructions of fossils, none of these explanations can be decisively verified, nor are any of them necessarily false. To complicate matters, the crests may have taken on different functions in different species. Yet it is possible to determine which of these ideas make more sense. The suggested use of the crests for breathing, air storage, or air trapping while underwater seems on the whole rather unlikely. These beasts were adept in water, and they might have retreated to lakes, rivers, and seas to escape a rapacious *Tarbosaurus* or *Tyrannosaurus.* Nonetheless, their well-supported bodies and their tooth batteries indicate that hadrosaurs probably spent most of their time on land, eating relatively tough branches and leaves of trees and bushes. Fossilized conifer needles, branches, deciduous foliage, and numerous small seeds and fruits have been claimed to be the "stomach contents" of a hadrosaur *Edmontosaurus.* If this identity is correct, one can assume that the duckbills did not have an overpowering need to feed underwater for long periods of time.

For lack of a better notion, the correlation between the hadrosaur crest development and a signaling function endures. The idea was refined starting with some thoughts expressed by the paleontologist James Hopson at the University of Chicago and later elaborated by Peter Dodson at the University of Pennsylvania and Dave Weishampel at Johns Hopkins University. Animals that use such obvious cues today often share a number of qualities. They have a keen sense of eyesight and/or hearing. They are often social and sexually dimorphic (males are larger and more aggressively built and armed than females, or vice versa), with individuals in frequent threatening behavior or combat for competition for mates of the opposite sex. Finally, species living in the same area that rely on such signals, like the antelopes of the Serengeti, often have very distinctive, highly different head ornaments like horns, to cue their own species.

Hadrosaurs in a broad sense fit this picture. They have large eye sockets and intricate ear bones, indicating acute vision and hearing. Weishampel developed some ingenious experiments to suggest that the hollow tubular crests of *Parasaurolophus* and other hadrosaurs were effective sound resonators. Moreover, Dodson's studies have shown that crests are accentuated in adults and, even in the case of adults, both big-crested and smaller-crested individuals are found in samples representing the same species. As in the case of the ceratopsians, this suggests a difference in the sexes pertaining to social behavior. Either the males or the females were establishing mating hierarchies or involved in rituals of signaling threats and combat. Finally, in a few localities in North America more than six different species are found together and most of these are easily discriminated by their varying head crests.

What about our Gobi creature *Saurolophus*? One might conjure up a picture of massive animals feeding in the marshes and deeps of a lake. Crests on some individuals may have indicated their position in the mating hierarchy, a signal backed up by the honking sounds of protective males. Is this vision the reality of seventy-five million years before Efremov and his band came upon hadrosaur skeletons in the Nemegt Valley? We'll never know. Some ideas, like the use of crests for visual cues, seem to match some circumstantial evidence. Elaboration of the scene is not so easy however. The myriad published color schemes for dinosaur crests, shields, trunks, and tails are purely imaginary. At any rate it's fun to think about them.

But of all the examples relating dinosaur head structures to social and competitive behavior, there is perhaps none more bizarre and entrancing than that provided by the pachycephalosaurs (*pachy-ceph;* thick-headed) or "bone-heads." Both the Nemegt and Barun Goyot Formations of the Cretaceous Gobi entomb these "bone-heads," as the Polish-Mongolian team found out. Pachycephalosaurs were originally described around the turn of the century from Cretaceous beds in North America. They vary markedly in size, ranging from five feet to twenty-six feet in length, and are ornithischians with the characteristic pelvic structure. The skeleton shows that these animals were bipedal with strong and long hind limbs and much shorter but well-developed forelimbs. The skeleton also shows several areas of reinforcement, as if these animals were adjusted for

Pachycephalosaurus (above) and *Stegoceras (Ed Heck, reprinted with permission, AMNH)*

a body-shaking shock. Those ossified tendons interlace the vertebrae of the back and tail. The back vertebrae have distinctive interlocking ridges above the body, or centrum, of the vertebrae, the part which houses the vital spinal cord.

These seem to be the parts of a machine that propelled a battering ram. Indeed, as the name indicates, most pachycephalosaurs had extraordinarily thick skulls. More specifically, the bone of the skull roof is dense, extremely thick, and arched upward, like the great cupola of a cathedral. To reinforce this object, the skull roof is studded with many individual plates of bone. Although not all pachycephalosaurs had these accentuated dome heads, in virtually all species—even the "flat-topped" *Homalocephale* from the Gobi Cretaceous—the skull roofing bones are thickened and encrusted with bony studs. At the back of the skull are distinct depressions and scars for the attachment of powerful muscles to the neck vertebrae.

This list of features seems indicative of the armament for an aggressive attacker, ready to take on even the most obstinate ankylosaur, matching armor and mace with like weapons. Pachycephalosaurs, however, were not predators. Their teeth were small, flattened utensils with a weak serration along their edges, teeth not meant for slicing meat but merely for shredding plants. Why, then, the thick studded heads and reinforced skeletons? One is forced to the conclusion that these features account for some kind of head-butting. Although there is no direct evidence, it seems that the battering ram was not just used in defense against predators. Two pachycephalosaurs might have indulged in head-to-head combat, much like that seen in rams of bighorn sheep today. It is not hard to imagine the drama of such a contest. Two animals lowered their heads and charged at each other full speed ahead, propelled by their powerful hind limbs. The combatants met head to head with a mighty blow resounding through the Cretaceous forest. How could these hulks survive such a joust? Since at the point of contact the skull was pointed down, its skull roof was oriented forward and in line with the neck and the horizontally held vertebrae that were meant to buttress the head during impact. Moreover, Ned Colbert and later Dr. Hans Dieter-Sues and Peter Galton have described aspects of

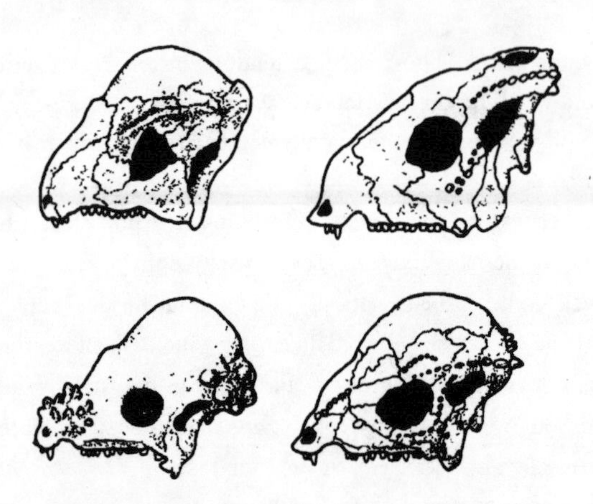

Skulls of the "dome heads" (clockwise from upper left) *Stegoceras, Homalocephale, Prenocephale, Pachycephalosaurus* (Novacek)

the skull roofing bones which seem to suitably disperse the shock waves generated by impact with another solid object.

These features helped absorb the force of the blow in combat. Just the same, it is hard to characterize such behavior as cheerful sport. Unlike battering mountain sheep, there are no horns on pachycephalosaurs to take the first impact; the bony studs on the skull are set low and seem ornamental rather than protective. The blow would be applied directly to the skull roof, in which there are no large air spaces to act as a cushion. The shock waves then passed right through the solid skull roof to the brain. Therein the small brains may have been slightly insulated by air spaces and connective tissue. Some authors have floated the rather ludicrous notion that these small brains were a saving grace, as if the dullness of the beasts, like that of a "punch-drunk" fighter, allowed them to tolerate the violations to the health represented by such skull bashing. Perhaps pachycephalosaurs were not the most intelligent of dinosaurs, but there is no reason to assume that they were dim-witted beasts incapable of feeling pain. Their large eye sockets suggest these animals had keen vision and forward-pointing, perhaps even stereoscopic, eyes. Infillings or endocranial casts of the brain cavity reveal very well developed olfactory lobes, the brain centers for the sense of smell. By these indications, pachycephalosaurs were relatively "sensitive" creatures. Besides, even the most intelligent of species often preoccupy themselves with aggressive and self-destructive behavior.

Primeval Predators

There is of course another kind of aggression in dinosaurs and other creatures, one that has more to do with basic survival, and with it culinary needs, than with social dominance. All those big herbivores—those hadrosaurs, sauropods, and ankylosaurs of the Gobi and elsewhere—were potential food items for predators. Theropods like *Tyrannosaurus* and its Gobi cousin *Tarbosaurus* represent the extreme in this predatory lifestyle. As noted earlier, the skulls in some of these monsters are more than four feet long, with recurved six-inch teeth. The skeletons are framed for attack—from the front of their sinuous line of sturdy neck vertebrae through their powerful hind limbs down to the vicious claws on the three functional toes of their hind feet.

Thus armed, one would imagine these tyrannosaurids as the ultimate terror. Indeed, many have proclaimed the group to be the greatest and most "terror-ific" terrestrial predators of all time. Others, like Jack Horner and Don Lessem, reject this reputation for active aggression and violence in tyrannosaurids. Instead, they suggest that these animals were primarily scavengers, preferring the less difficult challenge of removing great chunks from inert corpses. This notion is inspired, in part, by the absurdly shortened forelimbs of tyrannosaurids. These feeble splints of bone end in two-fingered hands with sharp claws, but they don't extend far enough from the torso for the animal to scratch an itch on its lower neck. Horner and others reason that these wimpy appendages were rather extracurricular. And, without their effective use, the animal had no business attacking hulking hadrosaurs or armor-plated ankylosaurs.

The argument, to me, sounds a bit overwrought. Certainly snakes, crocodiles, and other infamous killers have attained this status without the benefits of long grasping forelimbs. Indeed the scimitar-toothed jaws of tyrannosaurids were sufficient to snatch, crush, and carry away a victim, or at least carry away a good chunk of meat from the prey. Moreover, there is some evidence to suggest violent behavior in tyrannosaurids. The famous skeleton of *Tyrannosaurus rex* at the American Museum (AMNH 5027) shows broken ribs and fused trunk vertebrae that resulted from injury. This is backed up by another skeleton (not at the American Museum) that has teeth of an alien tyrannosaurid embedded in its skull. The broken ribs or fractured humeri in some of these beasts might not indicate such combat at all, but instead a violent struggle with a particularly feisty prey, whether plant-eater or meat-eater. Were tyrannosaurids scavengers or predators? Maybe both, but there is no convincing reason to say they weren't terrific killers.

Many scenarios are offered for the murderous manner of assault likely to have been employed by the great carnosaurs. These are based largely on examples of mayhem in the living fauna. A popular notion is that a beast like *Tyrannosaurus* used a "hit-and-run" or "land shark" approach. The predator jumps the prey and inflicts a mortal wound, a gaping chasm of

Tyrannosaurus (Ed Heck, reprinted with permission, AMNH)

muscle, organs, and blood, with its slashing teeth. It then retires to wait out the bloodletting of the shocked and dying victim, zooming in at the right moment to finish off the game. Although this strategy is largely attributed to species like the great white shark, something similar is seen in a formidable land predator, the *ora* or Komodo dragon (*Varanus komodoensis*). This "dragon" is actually the world's largest lizard, reaching a length of ten feet and a weight of 300 pounds. Though the ora indulges in carrion feeding, it often kills its game, which on rare occasions even includes humans, by surprise attack. It hides in the high grass or the bush and lunges out at the prey, which might have had the poor judgment to follow a well-beaten path that serves as the ora's favorite game trail.

The teeth of the ora are designed to disembowel. When chomping on its victim, the ora clamps its broad muzzle onto the flanks and moves its jaws backward. Walter Auffenberg, a zoologist who has thoroughly studied these nasty creatures, notes that if one drew a line connecting all the tips of the upper teeth it would resemble the profile of a scalpel. The longest teeth are about midway along the tooth row. Thus, with the back-

The scalpel-like profile of the *Tyrannosaurus* tooth row *(Ed Heck)*

ward movement of the jaw the taller teeth incise more profoundly into the flesh than the teeth behind them. As implements for both the first strike and the subsequent feeding these teeth rip through a mass of muscle and organs with astounding and horrific efficiency. To make matters worse for the victim, the Komodo is perhaps endowed with the foulest and most lethal breath in the animal kingdom. Its mouth is an incubation chamber for a virulent kind of bacteria to which the ora has built up an immunity. These bacteria, however, can thrive and grow in the wound of the prey, promoting a slow and painful death. The product is a pestilential carcass, but one safe for ora feeding.

It has not escaped the notice of some paleontologists, like Ralph Molnar, that the width of the muzzle and the profile of the tooth row in the ora are remarkably similar to those in forms like *Tyrannosaurus* and *Tarbosaurus*. The largest teeth are near but not at the front of the upper jaw. Retraction of this row would create the kind of gaping wound produced by the ora; of course the wound would be of much greater dimensions.

Mark Norell prospects at the foot of the Flaming Cliffs, the world's most famous dinosaur site. Jim Clark and the author in background. (Photo by Fred Conrad.)

Roy Chapman Andrews (center) in the Gobi, scanning the horizon. (Photo courtesy of the American Museum of Natural History.)

Camel caravan of Andrews' Central Asiatic Expedition at the base of the Flaming Cliffs. (Photo courtesy of the American Museum of Natural History.)

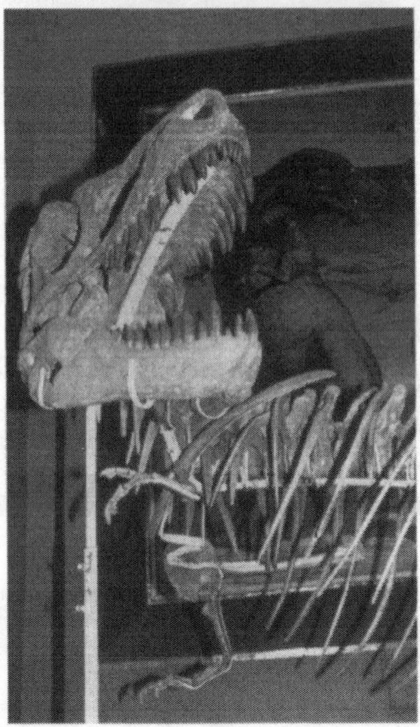

The dinosaur *Protoceratops andrewsi* fi
collected at Flaming Cliffs, next to a
nest of eggs originally thought to belo
to *Protoceratops.* (Photo courtesy of the
American Museum of Natural History.)

Tarbosaurus, a terrifying Asian relative
Tyrannosaurus, in the Natural History
Museum in Ulaan Baatar. (Photo by
Michael Novacek.)

The extraordinary fighting dinosaurs
(Protoceratops and *Velociraptor)* found
Tugrugeen Shireh by the Polish-
Mongolian team. (Photo by
Zofia Kielan-Jaworowska.)

The author (left) with Mark Norell (center) and Mongolian dinosaur expert Altangerel Perle studying fossils. (Photo by Fred Conrad.)

Mongolian scientist Demberelyin Dashzeveg, who knows the Gobi like no other explorer. (Photo by Michael Novacek.)

Where are those fossil sites? Malcolm McKenna (in plaid) and the author scan the map while Mark Norell (background) prepares dinner. (Photo by Fred Conrad.)

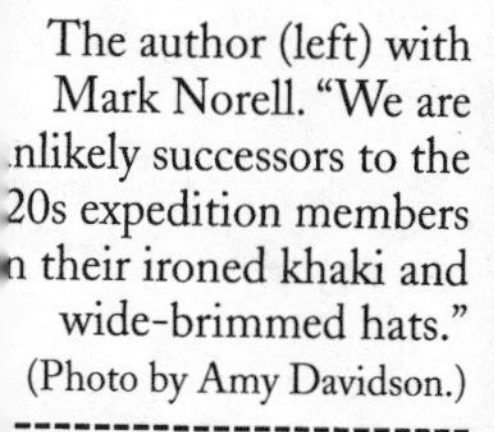

The author (left) with Mark Norell. "We are nlikely successors to the 20s expedition members n their ironed khaki and wide-brimmed hats." (Photo by Amy Davidson.)

Malcolm McKenna's discovery of *Estesia*, a fossil lizard new to science. His ring finger for scale. (Photo by Michael Novacek.)

The Cretaceous lizard *Estesia* emerges from the rock through skilled fossil preparation by Lorraine Meeker. (Photo by Chester Tarka.)

Part of the 1994 Mongolian Academy-American Museum Expedition assembled at Naran Bulak ("Sun Spring"). Back row (from left to right): Batsuk (driver), John Lanns (mechanic), Jim Carpenter (entomologist), Jim Clark (dinosaur and crocodile expert), Amy Davidson (preparator and collector), Lowell Dingus (geologist), Luis Chiappe (fossil bird expert), Sota (driver), the author (expedition leader). Front row (from left to right) Carl Swisher (geologist), Gunbold (field assistant), Demberelyin Dashzeveg (mammal expert, coleader), Mark Norell (dinosaur and lizard expert, coleader). (Photo by Michael Novacek.)

The author with his prized *Velociraptor* skeleton at Flaming Cliffs. (Photo by Amy Davidson.)

Ukhaa Tolgod, perhaps the richest site in the world for Cretaceous dinosaurs and other fossil vertebrates. The "Camel's Humps" are at right center. (Photo by Mark Norell.)

We virtually tripped over skulls of dinosaurs, like this *Protoceratops*, at Ukhaa Tolgod. (Photo by Michael Novacek.)

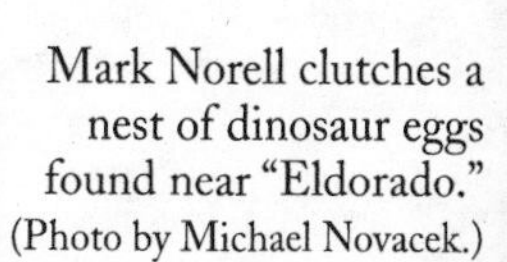

Mark Norell clutches a nest of dinosaur eggs found near "Eldorado." (Photo by Michael Novacek.)

Perle and Mark Nore
(right) remove an armored ankylosaur skull from the red sands of the Flaming Cliffs. (Photo by Micha
Novacek.)

Luis Chiappe (right) and Amy Davidson prepare the nesting *Oviraptor* for removal from the sands of Ukhaa Tolgod. (Photo by Mark Norell.)

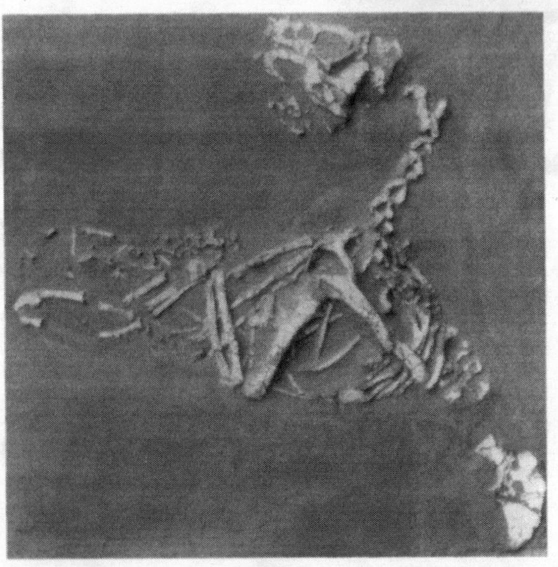

The first skeleton of *Oviraptor* found in the 1920s at the Flaming Cliffs. (Phot
courtesy of the America
Museum of Natural History.)

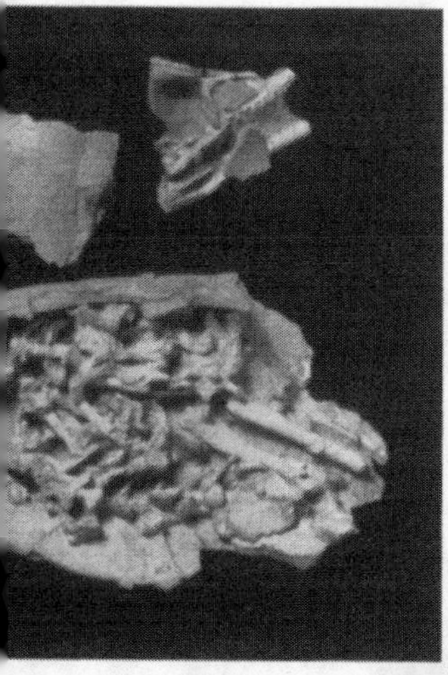

The egg with the skeleton of the oviraptorid embryo (below) found with a small juvenile *Velociraptor*-like form (above right) from Ukhaa Tolgod. The first known embryo of a meat-eating dinosaur. (Photo by Mick Ellison.)

(Below) The nesting *Oviraptor,* "Big Mama," from Ukhaa Tolgod. The forelimbs with their huge claws embrace a clutch of at least twenty eggs. (Photo Dennis Finnin.)

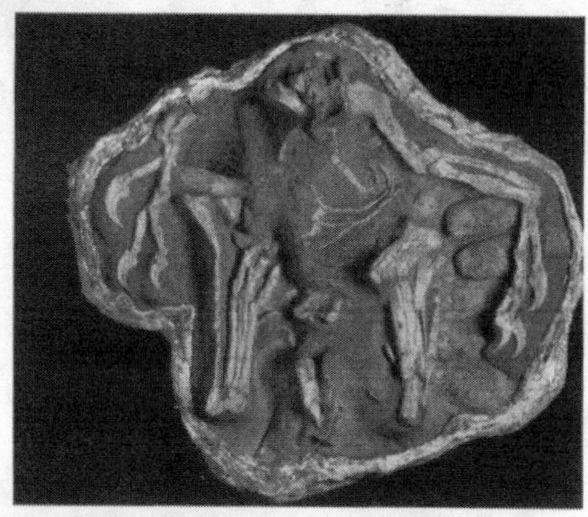

ow) Reconstruction of oviraptorid embryo in gg. (Drawing by Mick Ellison, tesy of the American Museum of ıral History.)

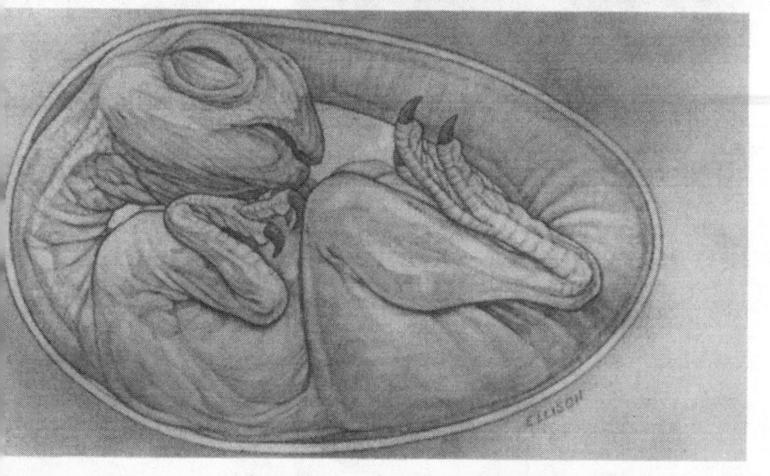

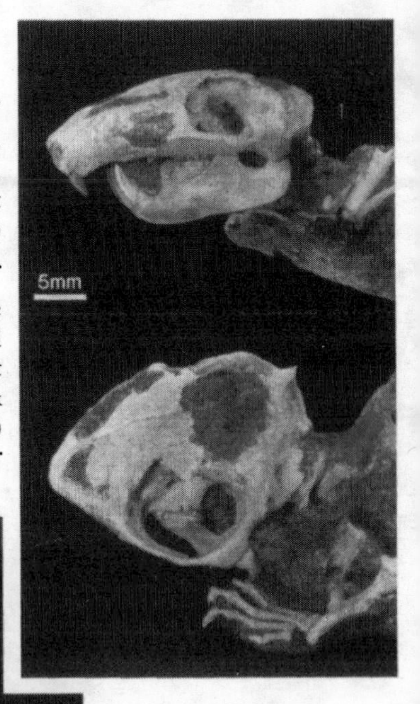

The skull and forelimb of a "gemlike" multituberculate mammal, *Kryptobaatar*, from Ukhaa Tolgod. Note the "chipmunk-like" front incisors. (Photo by Chester Tarka.)

(Below) The stunning skull of the tiny placental mammal *Zalambdalestes*, found by Jim Clark at Tugrugeen Shireh. (Photo by Mick Ellison.)

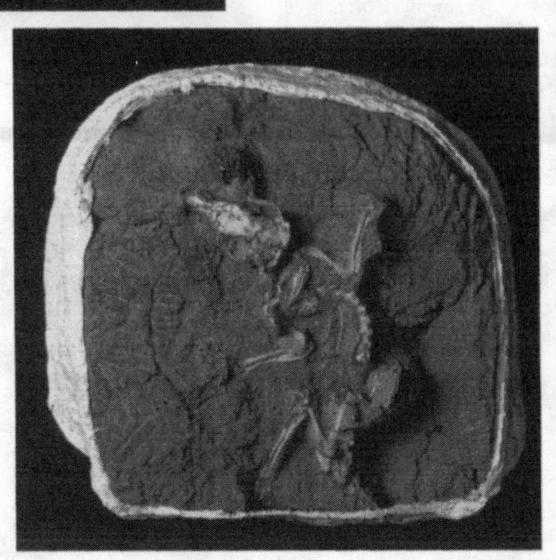

The "Sugar Mountain Mammal," one of five showcase skeletons of placental mammals found at the Sugar Mountain site in Ukhaa Tolgod. (Photo by Mick Ellison.)

Whether tyrannosaurids employed a "hit-and-run" or "land shark" strategy is of course a matter of pure speculation. We have no direct evidence that these prehistoric predators preferred to wait in ambush and then patiently sit out the bloodletting. And we surely do not know whether the tyrant king measured up to the ora in a capacity for mixing the bite with a saliva spiced with virulent bacteria We can at least say that these big theropods seemed to have the right equipment for a horrendous and lethal slashing attack, one that might rival—though on a much bigger scale—the ora's.

I've encountered, first hand, oras in the wild. In February 1991, I led a group of American Museum tourists along one of these "game trails" on Komodo Island. It was a warm and somewhat stifling walk through thick brush in land that time forgot. Komodo Island is a small piece of crust in the steamy Indonesian seas. The island is crowned with ancient, plant-carpeted volcanoes. It is as arresting and primeval as any Hollywood dream for the birthplace of King Kong. We were apparently safe though, escorted by a couple of resident guides armed with rather puny sticks en route to a viewing area some two miles inland. At the observation pit we could watch the oras fight among themselves. The beasts are routinely fed fresh game at this site, a maneuver that is supposed to congregate the lizards and at least allow a better sense of their movement in the bush.

This rather artificial, tourist-oriented procedure doesn't work perfectly. On our way back to the beach, what seemed to be the biggest of all ora—"ora *rex*"—confronted us on the trail. At a length of ten feet and a fighting weight of a few hundred pounds, he was a vicious, snarling monster. He did not back off. Instead he continued to hiss threateningly as our guides furiously tried to beat him off the trail. Meanwhile, an only slightly smaller ora came out on the trail, blocking our retreat and closing in on us from behind. There were some gasps, and a few purses and pairs of sunglasses flew in the air. Much to the alarm of my group, I disappeared into the bush, but I reemerged with my own stick. I helped the guides do battle with our "attacker" at the front of the column. At last we got the beast back to the edge of the trail. A few minutes later a group of excited and grateful tourists came spilling out onto the sanctuary of a white sand beach and a row of zodiac boats waiting to escort them to the afternoon cocktail party on our cruise ship. It was an amusing but not altogether harmless encounter. I couldn't imagine the horror of contending with a dragon of Komodo that managed to spring from the bushes in a surprise attack. A stick wouldn't do much good under the circumstances. Some tourists have been killed and consumed on Komodo Island in this manner.

Dinosaur predators come in all shapes and sizes. While there are things like forty-foot tyrannosaurids, some theropods are smaller than ten-foot oras. *Velociraptor* as I've noted is not awesome in proportions. But its abilities in the hunt can hardly be doubted. Unlike tyrannosaurids, the front limbs of many of these smaller theropods are well developed and armed with vicious, large claws—grappling hooks. And of course there are those killing scimitar-like claws on the hind foot that I described in Chapter 2. Perhaps these animals used such claws rather than teeth to strike the first disemboweling blow.

Perhaps as well they hunted and killed cooperatively in packs. There has been much speculation on the subject of pack hunting in meat-eating theropods. A dinosaur trackway from the Lower Cretaceous Glen Rose Formation in Texas (such trackways are rare in the Gobi strata) show marks of a small group, possibly four theropods, moving in the same direction as a larger congregation of sauropods. But the footprints are ambiguous. They could have been left by a theropod pack on the hunt. Or, as the orientation of some of the prints indicates, they may simply represent the movement of individual theropods in ways that did not track the sauropods. It makes sense that certain predators would team up to bring down bigger game. But, as one might expect, there is no decisive fossil evidence for this strategy in dinosaurs. Just the same, new studies of trackways by Martin Lockley strongly suggest that various and sundry theropods relied on cooperative hunting and killing.

Sometimes the clues left by fossils about the predatory lifestyles of dinosaurs are even more tantalizing and mysterious. Zofia Kielan-Jaworowska discovered one such fossil enigma during the 1965 season in the Gobi, when she spotted a dozen or so bones in soft sandstones at Altan Ula III. She gingerly scraped the surface of the sand to reveal an enormous eleven-inch claw resembling a sickle. When the team returned the next day they uncovered a shoulder girdle and two enormous three-digit

A hunting pack of *Deinonychus*, a dromaeosaur relative of *Velociraptor* *(Ed Heck, reprinted with permission, AMNH)*

forelimbs over eight feet long! It could be ascertained from the proportions of the bones that the beast was big, about *Tarbosaurus* size. But in the latter the forelimbs are stunted and spindly, with didactyl hands and claws no more than two inches in length. This creature was something entirely different—unlike any dinosaur known. The Altan Ula arms were given the name *Deinocheirus* ("terrible hand"), a label any hapless prey on the verge of disembowelment might agree with. Some reconstructions of *Deinocheirus* depict it as a great swarthy serpent with huge clawed forelimbs, suspended from the shoulders like the drooping arms of a giant ground sloth. It has also been suggested that these creatures may actually have been anteaters, who used their great claws to dissect giant ten-foot-high termite nests. But who knows? These depictions are only slightly better than pure fantasy. To this day, all we have of one of the most extraordinary dinosaurs that ever lived are a few bits of skeleton, those amazing forelimbs and claws.

The Great Egg Hunt

Dinosaurs—whether meat-eaters, termite-feeders, or vegetarians; whether solitary or social; whether hot-blooded or cold-blooded—all, of course, had something in common. They all had mortal lives with a beginning and an end. Like other organisms, they were born, they grew up, they mated, they reproduced, and they died. But why state the obvious? How could it be otherwise? I bring the matter up because even the simplest and most straightforward facts of life are not always immediately retrievable in science. Since the time their bones were discovered, scientists could safely assume that dinosaurs, like other organisms, mated and reproduced. The notion that these were indeed "terrible lizards" (from the Greek—*dino*, terrible; and *saurus*, a lizard) prompted the early scholars to conclude that they did not give live birth to their young but laid eggs, as do lizards, snakes, crocodiles, and, for that matter, birds today. But the scientific confirmation of this hunch was long in coming. It came nearly two hundred years after the first described dinosaur fossil was mistaken for the petrified genitals of a giant antediluvian human and more than eighty years after Professor Richard Owen coined the word "dinosaur." The first alleged dinosaur eggs were discovered from the French Pyrenees in 1859 and described by Paul Gervais in 1877. The more dramatic evidence, of course, was found in the late summer of 1922, on the grainy red sands of the Flaming Cliffs. As Andrews somewhat innacurately noted in *The New Conquest* . . . (p. 208), "It is evident that dinosaurs did lay eggs and that we had discovered the first specimens known to science."

With the retrieval of the Gobi fossil eggs by the Andrews team in the 1920s came a new phase of dinosaur research. Eggs were collected and identified in various parts of the world. More large dinosaur eggs from France were reported in the 1920s. The first dinosaur egg shells from North America were identified by Charles Gilmore in 1928 from an obscure butte in Montana near the Canadian border (although he did not immediately publish on these finds). Two years later, egg fragments were reported from the Cretaceous-aged Hell Creek Formation in eastern Montana. And the hunt was on. Eggs started popping up all over the place: China, India, Argentina, South Africa, and Transylvania.

But this chain of findings had an abiding flaw. Although some of the eggs contained bone fragments, they did not contain well-preserved skeletons of baby dinosaurs, either as embryos or as newborns. (However, the C.A.E. may have recovered eggs in association with a baby *Psittacosaurus*, at the Oshi locality. The nest was apparently damaged during extraction of the skeletons.) That all changed in 1978, when paleontologist Jack Horner and his team discovered something incredible. At a rock shop in Bynum, Montana, Horner was shown a coffee can full of tiny bones. He immediately recognized these as the remains of miniaturized hadrosaurs, or duck-billed dinosaurs. Tracking down the source of these remains to a ranch in western Montana, Horner and his crews uncovered a cornucopia of evidence: eggs, hatchlings, juvenile duckbills, adults, even fossil dung, or feces, and plant remains that, Horner claimed, may have been regurgitated by the mother hadrosaurs to feed their young. For this previously unknown duckbill, Horner bestowed the name *Maiasaura*, "the good mother lizard." It was the first slice of dinosaur family life on record.

Horner worked this locality, which acquired fame under the name "Egg Mountain," for several years. His team also found tiny embryonic bones and eggs of hypsilophodontids (medium-sized, sometimes bipedal dinosaurs that were distant relatives of iguanodontids and hadrosaurs), as well as many additional embryos of *Maiasaura*. Some of these were actually "observed" without destroying the egg, by use of CAT scanning. Back in the field, Horner shifted his operations northward, near the Canadian border to the very Landslide Butte where Gilmore had gathered a few dinosaur eggshells fifty-six years before. There he found extensive bone beds and "millions" of bone elements of duckbills and other dinosaurs of all sizes and ages. (This is, naturally enough, the same locale where Scott Sampson and his colleagues later studied horned ceratopsians.) It was an unrivaled modern conquest of North America's dinosaur country.

Horner estimates that his group has accumulated about seven or eight hundred whole or partial dinosaur eggs. Add this to the hundreds of eggs now identified in Mongolia, China, and other regions of the world, and we are confronted with an embarrassment of riches. Because these specimens vary in size, shape, and texture, a classification of dinosaur eggs has emerged. This is a rather risky business, even in cases where bones and eggs are found in association but bones are not actually encased within the egg. For the most part, the categories are based on the kinds of adult dinosaurs most commonly found near the eggs. Sauropod eggs, accordingly, are thought to be usually quite large and spherical. It was for a long time believed that no dinosaur eggs were probably larger than 18 inches in diameter—about the size of the egg of the extinct elephant bird from Madagascar; no eggs found greatly surpassed the size of those laid by ostriches. There are, however, late-breaking reports of 24-inch dinosaur eggs from China. The Andrews team reasoned that the eggs of horned dinosaurs like *Protoceratops* were necessarily elliptical and elongate, because such eggs were found near to numerous proto skeletons. Duckbills, by proven association, also have eggs of this form. Some eggshells are smooth, some textured, but this may vary among species within a certain group of dinosaurs. Through the years, a number of expeditions, including our own to the Gobi, have recovered several different kinds of eggs, some elongate, some with a pebbly surface, or a striated or smooth surface, and some spherical, like those associated with the "sauropod" type. Unfortunately, in the case of most eggs from various regions of the world the chance for assignment to a particular dinosaur group is very poor. Dinosaur embryos—even small fragments of bone—are simply not preserved with most eggs. To concretely identify the species to which an egg belongs requires the kind of Egg Mountain miracle that is all too rare.

Although dinosaur holdings throughout the world now include a large collection of eggs and a few embryos, it must be noted that this sample still does not compare with the myriad skeletal remains of dinosaurs. Moreover, egg sites are highly localized, with the overwhelming numbers of specimens coming from a few choice spots—the Gobi, Montana, Alberta, China, Patagonia, and southern France. Why this comparative scarcity of eggs and embryos? Paleontologist Kenneth Carpenter has speculated that the scarcity may reflect several factors. Dinosaurs, like certain mammals, may have been long-lived and may have invested care in just a few offspring. Perhaps rookeries were remote from adult feeding grounds, or were concentrated in areas of higher elevation where preservation of remains is generally poorer because of faster coursing, steeply descending streams, and harsher seasonality. Juvenile bones are easily subject to damage and disintegration. In addition, juveniles are more commonly swallowed whole by predators, with not even a partial skeleton of the dinosaur juvenile likely to be recovered. Further, tiny juvenile bones and egg fragments have been overlooked by paleontologists over the decades on the prowl for massive skeletons. Small bones have also been, on some occasions, identified as distinct species rather than the juveniles of a previously known species. Some have speculated that eggs with a hard calcified shell (*cleidodic* eggs) may have been a later development in dinosaur evolution; more primitive forms may have laid leathery, non-cleidodic eggs which would not hold up well as fossils (this seems, in fact, highly implausible). Finally, in many dinosaur species, the parents may have eaten the eggshells after the young were hatched, a behavior suspected (but not definitively proven) in living crocodiles. All the above explanations are logical, but they all have exceptions. The causes for this bias may be many and varied.

Bringing Up Baby

Embryos of dinosaurs, when they are found, are of course very small and unimposing. They might be very difficult to pick out in the corner of a hall full of great dinosaur skeletons. But they are just as highly coveted by scientists as any of the best dinosaur mounts known. Embryos and their associations in nest sites give us insights beyond the standard information about dinosaur architecture, form, and possible function. They bring dinosaurs to life not just in terms of movement and eating, but also in terms of what makes life life—reproduction and growth. Unfortunately, some people get carried away with the piecemeal clues available. They start rhapsodizing about mother love, communal care, pair bonding, and other intangibles. My fellow explorer, Mark Norell, in the June 1995 issue of *Natural History,* has considered just how far the evidence can actually take us:

First, that dinosaurs laid eggs—an irrefutable conclusion since we have found dinosaur embryos in eggs. This is the pattern of reproductive behavior most common to vertebrates, including birds. There are some vertebrate groups, including mammals and certain lizards and snakes, that bear their young live. Some have claimed that certain dinosaurs were also live-bearers, but there is absolutely no evidence for this. The possibility cannot be ruled out, but that is not the same as saying we have evidence that some dinosaurs were capable of bearing young live.

There is fossil evidence that dinosaurs not only laid eggs but in some cases built nests around them, although material for the nests is not apparent (it is unlikely it would be preserved—twigs, leaves, and other fragile items are easily destroyed during the phase of burial and preservation). Clusters of eggs in repeatable patterns do, just the same, indicate that some dinosaurs may have constructed nests.

Like living birds, but unlike crocodilians, dinosaurs may have manipulated their eggs after laying them—while many dinosaur egg clusters appear to be random, some, like the Gobi eggs, show a pattern of arrangement that would be unlikely if the parent simply laid them and left them. In these cases, dinosaurs probably moved and arranged their eggs. The Gobi eggs, for instance, are often arranged in a circular pattern of more than a dozen eggs, with the more narrowed end of the egg toward the center of the circle. Often, the eggs are stacked in two or three circular layers. It is hard to imagine that such an intricate and precise arrangement was simply laid without manipulation.

Like living birds, dinosaurs may have guarded and incubated their eggs over extended periods—but awkward hedging here is necessary. It is very hard to extract such subtleties of behavior from even the most remarkable of fossil findings. It nonetheless seems logical to assume that clutches of eggs were attractive to numerous predators and thus required protection. Moreover, sites like Horner's Egg Mountain show a congregation of nests, juveniles, and adults of one species. This rare assemblage suggests some kind of prolonged parental care of the young. Jaws and teeth of the smallest creatures in this nest even show wear, suggesting that the hatchlings were chewing coarse plant material. On the other hand, it has been observed that wear is present on the teeth of young individuals even before hatching. Like human babies, baby dinosaurs inside the egg apparently were grinding their teeth. Thus tooth wear in "nestlings" is not in itself clear evidence of prolonged caring for young in the nest.

Indeed, the subject of dinosaur brooding presents some mysteries, and explanations are often fanciful. But it is important to emphasize that our understanding of *phylogeny*—namely, evolutionary relationships—can help here. The recognition, for example, that certain dinosaurs were closely related to birds allows us to extrapolate from observed bird behavior as some means of predicting the behaviors of such dinosaurs. Birds brood, and build and guard nests; they often migrate and congregate in breeding colonies or rookeries. This establishes a few possibilities for dinosaurs, which might be more or less demonstrated by fossils. As Mark points out in his article, we confidently assume that the earliest African hominids had body hair by virtue of their affinities with later *Homo* and their more distant relatives, the greater apes. Hair is not, however, preserved in any hominid fossil. A knowledge of relationships therefore provides a powerful set of insights to even some of the more intangible elements of dinosaur biology and lifestyles.

GROWING UP

The retrieval of the Gobi fossil eggs in the 1920s also ushered in a new appreciation for the reproduction and growth of dinosaurs. When a dinosaur hatches it comes out into the world; it must survive, grow up, and eventually reproduce its own kind. But what is the manner of this trip to maturity? One might respond that the dinosaur infant simply gets bigger until . . . well, it gets to be the right size, the breeding size or adult size, that is, if it isn't eaten first. The matter of growth is, however, more subtle and indeed much more fascinating than simply a matter of getting bigger or more mature.

When we consider the hierarchy of life we note a fascinating variation in patterns of growth among different groups of organisms. Humans and other mammals and birds grow and change very rapidly but start to slow down, and virtually stop, once they reach sexual maturity. After a brief period of slow size increase at birth, the growth curve is steepened but eventually flattens out at sexual maturity. This pattern is called sigmoid growth because of the S shape of its growth curve. Most other vertebrates show a radically different growth profile. Fish continue to grow throughout their lives, but the rate of growth slows down with each passing year. Most fishes live between five and ten years (we actually don't have firm data on the tens of thousands of fish species identified) and are about four to six inches long even after they reach sexual maturity.

The growth curve of things like lizards and crocodiles is hardly a curve at all. The older the crocodile is, the larger it is. Imagine human patriarchal and matriarchal societies under this growth scheme! Grandparents (and more so great-grandparents) would be the sauropods and the biggest food consumers in the family. These growth curves are important in understanding how species adapt to different situations and how certain changes, or potential changes, in evolution are controlled by development.

Dinosaur embryos are extremely rare and the evolutionary story they tell is incomplete. Nevertheless, the available embryos, along with data on juvenile dinosaurs and egg structure and size, have provided a few insights. Because the largest dinosaur eggs, as noted, are less than 25 inches in diameter, the hatchlings of these sauropod eggs were probably no larger than 35 or 40 inches, but they grew enormously—up to the 70- or even possibly 100-ton giants. Was this growth linear, namely, prolonged through life depending on the amount of food consumed? Or did growth in dinosaurs like these take an early spurt and level off after reproductive age? The truth is we don't have a firm answer to this question. It is, for one thing, very difficult to estimate the age of a fossil dinosaur independent of its body size. For another matter it is quite possible that dinosaurs varied in their growth curves; some may have been linear, others more uneven.

Nonetheless, some biologists, like Ted Case, have tried to estimate growth rates in dinosaurs known from associated eggs and/or good series of juveniles and young adults. Case figured that if *Protoceratops* grew roughly at the same rate as the slowest-growing reptiles they would reach top body size in about 33 years (with a range between 26 and 38 years). Assuming these dinosaurs reached reproductive age before attaining top body size, *Protoceratops* would be part of the breeding population after 8 to 16 years. On the steepest curve for living reptiles *Protoceratops* would achieve maximum size by 12 to 23 years. Case calculated much longer growth spans for much larger dinosaurs associated with eggs, like *Hypselosaurus*, namely 82 to 236 years with an attainment of breeding size after 25 to 72 years. According to Case, some dinosaurs may have lived a long and progressively more ponderous life.

These growth estimates, however, are very shaky. The assumption that dinosaurs grow like reptiles may be invalid. As noted, many dinosaurs may have been "warm-blooded" or endothermic. If so, Case's estimates for their rate of growth could be drastically revised. Studies of living animals show that growth rates for endotherms can be ten times higher than rates in ectotherms. Thus, if protos were warm-blooded they could have attained full body size in 2 years instead of 20 years. Finally, associations of eggs and certain species, even as we shall see in the case of *Protoceratops*, can be suspect. Growth rates based on the size of the supposed eggs and the size of the adult skeleton may be invalid. There is no doubt that dinosaurs, like other biological organisms, grew up, but the nuances of that pattern are not easy to detect.

THE MEANS TO AN END?

A coda is required here for all these observations and inferences about the lives of the dinosaurs. We have marveled at the stupendous size of these creatures, their incredible variety in form, their exquisite design for feeding, and even their possible socialization. But why then did these creatures go extinct? I began this chapter with a quote from Darwin that is the essence of the puzzle. There seem to be many *unperceived* hazards or, in Darwin's words, "hostile agencies" that promote rarity and eventual extermination of species. Dinosaurs, even at the acme of their diversity and wondrous adaptations, were not exempt from this brutal process of extermination. Species, just like individuals and galactic stars, have mortal lives—they are born, they radiate, and they die. In Chapter 4, I noted that, on average, a species seems to last "only" one or a few million years in the geologic record. It then goes extinct and is replaced, sometimes by a related form, sometimes by a completely alien species that takes on its role in the environment. This kind of churning—the loss of species and their replacement by new ones—is known as *taxonomic turnover* in the fossil record.

Sometimes that turnover is much more rapid and profound, as in the case of the mass extinction events at the end of the Cretaceous and the Permian. These are singular events—events somewhat apart from the usual pattern of extinction and replacement. They have particular causes, a subject I will review later. Except for their bird descendants, most dinosaurs were victims of one of these mass extinction events. Does this mean that dinosaurs, as a group, were inferior beings? Some, for example, have postulated that the gargantuan size of many dinosaurs, rather than being a strength that might confer, in Darwin's words, "a victory in the battle of life," was a shattering liability. One might be tempted to embrace this theory, what with all the challenges I described above that come with attaining gigantic body size. Maybe, as habitats radically or catastrophically changed, these oversized creatures required too much food, too much environmental stability, to survive.

These stories don't work. They encounter some well-known obstacles. First off, the mass extinction event at the end of the Cretaceous imploded on dinosaurs big and small, fat and skinny, swift and slow. The same goes for many other non-dinosaur groups affected disastrously by this event. Another stock rejoinder draws on the endurance of the dinosaur empire. Dinosaurs, even if we exclude the surviving birds, had a long and glorious history of more than 160 million years, a span to match many a great group of organisms, and certainly a span that exceeds by millions of years our own brief history on the planet. One can hardly call this history the product of a bizarre evolutionary mistake. Finally, extinction of various kinds of dinosaurs was occurring throughout the Mesozoic. Thousands upon thousands of dinosaur species may have gone extinct and been replaced during that time. A problem forced by environmental changes may have a better solution in the adaptations of a competing species. Dinosaurs, just like other biological groups, experienced the bitter competition and differential survival of its numerous members.

So if some species survive while others fall, does that mean there is an ideal lifestyle? In the early years of evolutionary biology, many paleontologists, like Henry Fairfield Osborn, claimed to document a preferred direction in evolution, one that brought the organisms to an ideal state. We now view such presumptions about the meaning of life as naive and fully unsubstantiated. Who can say a one-horned rhino is *in general* better off then a two-horned rhino? Or that an eighty-foot-long sauropod is better off than a forty-foot-long one? Is a peacock, because of its magnificent plumage and refined social behavior, adaptively superior to a dull sparrow? Or, because of their hugely greater numbers, are sparrows better off than peacocks? Certainly, situations favor particular species, as Darwin's theory of natural selection would have it. There may be a particular set of circumstances—a specific time, clime, and place—where a one-horned rhino is better off than a two-horned rhino. There was a time when oxygen-using, or aerobic, bacteria could not survive; only their anaerobic bacteria relatives thrived. This does not allow us to say that anaerobes or one-horned rhinos will always win out. Nevertheless, we can ascribe to the theory of natural selection without necessarily being able to predict *which* species will win out.

That is why the history of life as read from fossils and inferred from the study of living organisms is so important. It is what actually happened. It disarms naive notions that species evolve toward some ideal adaptive state or attain guaranteed evolutionary superiority. It is a history of great expectations, flourish, turmoil, and extermination. But this does not necessarily mean all is chaos. There is a pattern here to retrieve. What observations we can muster about the nature of birth and growth as well as adulthood in both living and fossil forms prepare us for that greatest of all questions about life, one that relates to all organisms, whether dinosaurs or dung beetles—how did life on earth evolve?

CHAPTER 7

1993—XANADU

DATE: JULY 15, 1993

LOCALITY: ONE HALF MILE NORTH OF THE TOWN OF DAUS, NEMEGT VALLEY, GOBI DESERT

It was a dusty windblown day in Mongolia. We were out of the vehicles, staring at some small pimples of dull red sandstone against the blackened volcanic curtain of the Gilvent Mountains.

"Well, it doesn't look too easy for the gasoline tanker," I said.

We were tired and despondent. The day had not started well.

In the process of heating his morning breakfast, Eungeul set his Russian blowtorch too close to the rear tire of his gas tanker. In an instant, his tire tread melted into a column of fire, producing an enormous cauliflower-like cloud of smoke. Eungeul had to beat the tire with a thick blanket.

After jockeying the trucks through the sand hills we finally reached Daus at 3:30 P.M. We had spent seven hours covering roughly thirty miles. A half-hour struggle up a steep hill brought us to the place where we now stood, stopped dead by an arroyo drowned in sand. The faint tracks we followed seemed to head straight down the hill and plunge suicidally into the very worst and widest part of the wash.

Dashzeveg, always the optimist, said, "I think we go around, on slope."

I shook my head in doubt but we walked the route out with Eungeul and Batsuk. The route was feasible but the tanker would be tilted unacceptably downslope. This might work if the tanker went slowly, but the tanker would have to move very fast to make the crux of the crossing, the narrow neck of the gully that was choked in deep sand. Moreover, the tanker was loaded with something shy of 1,300 gallons of 76 octane gas for the Russian vehicles. It pulled behind it a tank trailer with another 780 gallons of 93 octane for the Mitsus.

Eungeul waited a half hour for his radiator to cool. At last he mounted up, gunned his tanker, and made a sweeping arc around the steep circumference of the hill, like a Formula 1 racer on a banked turn. For a moment he seemed to be heading uphill, reluctant to make the crossing. Then suddenly he plunged down the slope into the gully and stopped with a thud in a mound of sand.

There was a moment of resignation, like a terrible unspoken admission that the expedition, with its buried gas tanker, was over. I could see why Ukhaa Tolgod, the "brown hills" near the Gilvent Mountains that were our destination, had resisted exploration. But Eungeul and Batsuk were not to be defeated. They unleashed the tank trailer and pulled it out with a snorting GAZ. Next Batsuk took on the tanker itself. We heard an awful grinding noise in the bowels of the GAZ. Batsuk persisted. At last the tanker came crawling out of the sand like some *Allergorhai horhai.* On high ground, both men got out and let their monstrous trucks cool. Batsuk laughed nervously. *"Maaike, moh!"* he said.

It was a pyrrhic victory. The tanker pull had left the GAZ with a rough-edged clutch, perhaps even with a few shaved-off teeth on the gears of the transmission.

Unfortunately, this was not our last such crossing. Another tow and more damage to the GAZ was necessary before we at last came to a ridge overlooking our target area. The view offered little to admire, hardly justifying one of the most difficult legs of four summers of Gobi travel. There were a couple of small red hills to the north and east, which blended into a low, flat amphitheater of red sand extending south in our direction. Farther south and closer yet was the main exposure. A pair of pinnacles that looked like shriveled humps of an emaciated camel merged with a bowl of red sandstone. A gap formed by the sheen of desert pavement was broken by another cliff. More cliffs ran down on the west side of the main gully, extending toward the valley and Daus, which was out of view some six nearly impassable miles away. The red hills and cliffs were pretty, but the area was so—well, small. Only a couple of square miles of patchy outcrop lay before us. Besides, we were slightly lost. We had actually aimed for some much more impressive orange badlands to the northwest, but somehow we got stuck in this sand gully and simply followed the path of its drainage.

We squelched our disappointment and remounted. It would be nice to camp directly below the scenic camel's humps, but the sand-filled wash was an impossibility for the tanker. We headed north and a bit east, hoping to outflank the small valley. Alas, the tanker and trailer began drowning in what looked like an innocuous little runnel of sand. This time it was hopelessly mired.

"Oh, oh! Dangeree!" Dash exclaimed. "Camp?" he asked with a laugh.

The next day, we decided, poor Eungeul and Batsuk would struggle with the tanker in the sand while we drove over to the two small red hills on the north side of the valley. If there were fossils, fine. If not—the outcome we most expected—we would move west to the impressive badlands in Khulsan or parts even farther west. It was the west that always drew us.

We bivouacked next to the gas tanker on a plain nearly naked save for a few sagelike bushes and a dusting of dark pebbles that had washed out of the distant Gilvent Mountains. But the evening was still and pleasant. We dined on those delicious "Desert Storm" rejects—rehydrated barbecued pork chops and a healthy serving of unhealthy fried potatoes.

Those of us who bedded down that July 15 evening were part of a much smaller group than the year before. We had decided to retreat from the elaborate expedition plan of 1992. This year there were no journalists, photographers, or film crews, a fact that seemed to relieve Dashzeveg as well as the rest of us. In truth these adjunct members had been delightful companions. But we wanted a summer to ourselves, one that allowed us a last-ditch effort to find a fossil treasure trove. This, Mark and I concluded, meant traveling light with a lean and mean platoon of people used to the work and the rigors of paleontology in the desert. Thus we started the summer with only a few of the regulars and a few new but seasoned recruits. The veterans included Mark, Jim Clark, and me. Malcolm and Priscilla decided not to make the initial foray with us. Malcolm was increasingly focused on the younger rocks of the Tertiary and its rich mammal faunas; he wanted to work Tatal Gol, a set of badlands a couple of hundred miles north of the Nemegt Valley. He and Priscilla would rendezvous with us later in August, as would my wife, Vera, and our daughter, Julie.

To complement the team from the States we added Luis Chiappe, a tall black-bearded Argentinean who was a postdoctoral fellow at the museum. An expert on fossil birds, as well as a highly experienced field hand from the dinosaur badlands of Patagonia and other regions of Argentina, Luis was ideal for the job. We also took on Amy Davidson, whom Mark had the good fortune to recruit as a fossil preparator at the American Museum. She came to us from Harvard, where she trained under the sharp-eyed Bill Ameral, a man whose extraordinary feats with difficult and intricate skeletons were world renowned. Amy, who for many years worked with sculpture, had acquired Bill's astounding finesse in scraping rock matrix from fossil bone, probing a tiny hard steel needle into the internal structure of the skull, or gluing precarious fragments of bone together. These are no mean feats. Neither Mark, Malcolm, Jim, nor I would entrust some of these difficult jobs to our clumsy hands. Amy worked with several other preparators, including Jeanne Kelly, in the museum's Vertebrate Paleontology Laboratory. But the lab effort was largely caught up in the huge amount of exhibition preparation, under Jeanne's supervision, for the museum's renovated fossil halls.

Amy, pretty and petite with a crown of wavy light brown hair and white skin that needed assiduous protection from the searing Gobi sun, had a feminine grace that belied her inner toughness and strength. A key member of the Harvard expedition team that roamed the glaciers and sheer cliffs of Greenland in search of Triassic and Jurassic fossils, she had no qualms about excavating a skeleton under the most horrendous of circumstances—howling winds, bitter cold, and driving ice and snow. The experience inured her well for the Gobi, if one simply substituted hot for cold and sand for snow.

The American contingent was rounded out by John Lanns, a mechanic from Minnesota who worked for Kevin Alexander, our technician from the 1992 crew. John was a young man in his early twenties, a somewhat shy fellow with short blond hair, blue eyes, and a square-jawed, well-chiseled, handsome face. The girls passing through Sukhbaatar Square couldn't help but give a second look as he walked by. "You American pilot?" they would ask with a giggle, adding "Hey, Tom Cruise, Top Gun!" This mixture of Western culture with the traditions of centuries of Mongolian life took John by surprise. He had never traveled out of the States before, let alone to one of the most isolated countries on the planet.

On the Mongolian side, Dashzeveg brought along only Batsuk, Eungeul the tanker driver, and Argil, a handsome, wiry fellow barely out of his teens, youthfully brash, and not yet familiar with the professional activities of field paleontologists. He would prove to be a bright and energetic companion. Perle, our dinosaur expert and comrade on previous outings, would rendezvous with the group later, accompanying Malcolm and Priscilla into the Gobi.

So from the twenty-five strong in 1992 we had whittled ourselves down to ten. Instead of a caravan, we had brought along only five vehicles—the three Mitsus, the GAZ, and a tanker truck in good condition, with a trailer to boot. The strategy here was not only to keep the expedition small and mobile but to be more self-sufficient than on previous forays. This time there would be no dependence on unreliable gas supplies in Dalan Dzadgad or other outposts. We carried all our 2,080 gallons of gas with us. Food would also not be a problem. Mark and Jim had strategically loaded our container in L.A. with a great array of succulent, durable foods.

To the usual fare Mark and Jim added such exotica as canned olives of different varieties, preserved hams, sausages, cheeses packed in oil, and even candied eels, Japanese style. There was a staggering array of chilis and other spices, as well as a stock of fine California wines. The supplement was so easily taken on that we rued not having indulged in this practice before.

Ironically, our determination to be self-reliant was paralleled by a surprisingly better economy in Mongolia. Gas was still scarce but our transactions for acquiring it were not so formidable. Our shipment even arrived on schedule. Packing out of Ulaan Baatar went relatively smoothly and we were on the road by July 5, only a couple of weeks after the lead party (Jim and Louis) had touched down in the capital city, and only six days after our second group joined them.

This year we skipped the Flaming Cliffs. Instead we headed straight for Tugrugeen, arriving there in record time. As in previous years, our white sandstone butte was generous: several skulls of lizards and multituberculate mammal skulls were picked off the nodule slopes. We also got pieces of *Velociraptor* and other kinds of theropods.

Perhaps most interesting of all, we found more evidence of that most unusual avian theropod *Mononykus.* The small theropod that Malcolm had found in the bush-covered slopes at Tugrugeen in 1992 turned out to be identical, except for its smaller size, to a skeleton discovered some years earlier by a Soviet-Mongolian expedition working in the Nemegt Formation at Bugin Tsav, a couple hundred miles to the southwest of Tugrugeen. The animal has a remarkable, gracile frame with long birdlike legs. In addition, like birds, the keel of the sternum or breastbone is very well developed. In modern birds, powerful pectoral muscles attach to this keel as a means of powering the downstroke of the beating wing. However, our Mongolian creature by no means applied its pectoral muscles for the same purpose. Instead of elongate wing bones, this creature had stubby, massive forelimbs somewhat like those of a digging mole. The end of the arm and hand is appointed with a single very large claw; hence the scientific name bestowed on the animal is *Mononykus,* literally, "single claw." (The original official spelling was the more etymologically correct *Mononychus,* but it was disclosed that this spelling had previously been bestowed on a beetle. The

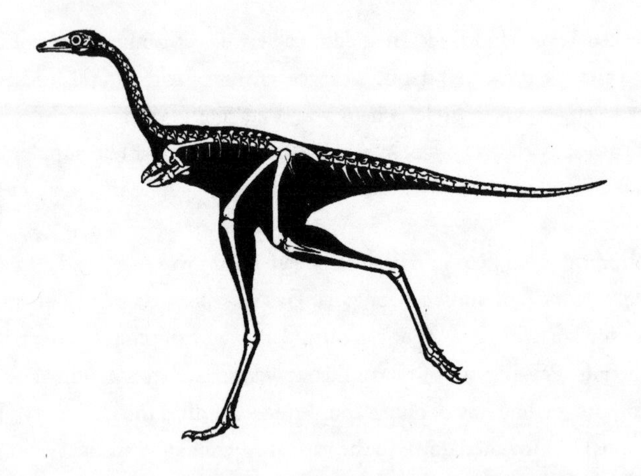

Skeleton of the flightless bird *Mononykus* *(Mick Ellison, reprinted with permission, AMNH)*

earth's most lavishly diverse organisms account for many a name in the scientific literature.) Soon it was clear from studies by Mark, Jim Clark, Perle, and Luis Chiappe that *Mononykus* belongs to that critical branch of the theropods, the Avialae, the one group of dinosaurs, by virtue of including modern birds (Aves), that managed to stick around. Other fascinating implications of this discovery will be described shortly.

Despite these successes, the finds at Tugrugeen predictably began to peter out. By July 12, after only six days at the site, we found ourselves working to diminishing returns. We decided to pack out the following day, confident that we had efficiently availed ourselves of the fossils exposed by this season's spring rains. It would be John Lanns' last day in the field. We planned to get John onto one of the planes that were once again flying in and out of the tourist camp near Dalan Dzadgad before we moved south.

On a previous night, a teapot of boiling water had spilled on top of Luis' naked foot, and he had developed an excruciatingly painful burn that soon oozed with infection. We stuffed Luis with antibiotics and painkillers, contemplating the possible need for his evacuation. Luis was of course violently opposed. But his foot was getting worse, virtually immobilizing him by our last day at Tugrugeen. The predicament gave us pause. Through the last field seasons we had been lucky. Because of our remoteness, serious injury or illness was a grave danger. There were no medical evacuation services out of the Gobi. We were on our own, even more so than those mountaineering expeditions to the Himalayas where helicopter rescue was available. With some trepidation Mark and I decided that Luis could stay under our watchful eyes.

Having said our fond farewells to John at the airstrip next to the tourist camp, Dash and I rejoined our group, who had gone ahead to the hunting camp perched on the high slopes of the Gurvan Saichan Mountains. Now we were only nine. We headed south toward the Nemegt Valley with our small band. Two days and a sand-mired gas tanker later, we found ourselves in a lonely bivouac at the foot of the Gilvent Mountains.

VICTORY DAYS

The morning of July 16 began uneventfully—no portentous accidents, no scorpion stings, no wrenching gut pains, no blinding sandstorms—just a hot, hazy morning with a slight breeze from the west.

The field crew piled into a couple of Mitsus and headed toward the first rise on the horizon. We stopped at a low saddle between the hills. Before I could remove the keys from the ignition, Mark sang out excitedly. I saw several of our group crouching around a splotch of white bone about thirty feet from the car. It turned out to be a lovely lizard skeleton curled in a tight circle, no larger than a pancake, and essentially complete, with a long series of filamentous vertebrae gracefully articulated. Several feet away, near the very apex of the saddle, was a stunning skull and partial skeleton of a proto, a big fellow whose beak and crooked fingers pointed west to our small outcrop, like a griffin pointing the way to a guarded treasure.

Dashzeveg, ever the lone wolf, soon distanced himself from the rest of us. Several others wandered off as well. Three of us, however, remained. The small saddle between two hills seemed too rich with bone to leave so abruptly.

We three started slowly describing circles, intently examining the gently dipping red slopes with our hands behind our backs, like a trio of meditating monks. Mark found a very nice mammal jaw, Jim Clark another lizard, and I a mammal. We continued to pounce on precious specimens with remarkable consistency. Jim started to drift toward the more northern of the small hills, but Mark and I hung back. Mark would sing out, "Skull!" and, almost on cue I would find one too. The surface of the gentle slopes and shallow gullies was splattered with white patches, as if someone had emptied a paint can in a random fashion over the ground. Up close, we could see that each patch was a dinosaur skeleton, either a *Protoceratops* or an ankylosaur, some fairly ravaged by erosion, but clean and articulated and beautifully preserved less than an inch below the surface. We flagged the better specimens with tape, like surveyors plotting a new housing tract. Sometimes we marked a skeleton with a cairn of dark rocks. We continued to trace slow circles over the ground, picking up little concretions or chunks of sandstone with skulls and, often, skeletons of mammals and lizards.

A strange feeling started to take hold in me. At first it was only a twinge, a rush that made me shiver every time I saw yet another mammal skull or dinosaur skeleton sculpted in the rock. But the spark of excitement sustained itself with each new find. It was the feeling of . . . what? Joy? Victory? The rush of discovery that comes after years of pursuit? I had the growing realization that I was having a day like no other in twenty years of field prospecting.

Mark felt the same excitement, though we were both too superstitious to say anything. Like Granger and Andrews when they discovered their first nest of dinosaur eggs at Flaming Cliffs, we couldn't believe that our encounter with this fossil-laden patch of ground was real. Things like this don't happen, even to paleontologists on a major expedition like ours. Two hours ago we had achieved one of the primary goals of the expedition—finding a new locality, one that measured up to some of the well-known dinosaur sites of the Gobi, one with an abundance of fossils. But this was beyond anything we could have imagined. In an area the size of a football field we had found a treasure trove that matched the cumulative riches of all the other famous Gobi localities combined.

By noon Mark and I returned to the green Mitsu for a celebratory lunch. It seemed as if we had managed to walk only a few feet from the car all morning. We emptied our grab bags and opened up tissue-wrapped skulls on the hood. There were multituberculate mammals with long incisors—bigger, squirrel-sized multis, and tiny multis smaller than a fingernail, some with blunt snouts, some with triangular snouts. One of the nuggets we discovered had the delicate jaw of a shrewlike placental mammal. There were also a variety of lizard skulls, some with gently arching jaws and small stubby teeth, several with bumpy teeth, and others with sharp recurved teeth shaped like small needles. I had found twenty-two skulls in all, including thirteen mammals—six more than found by Granger and three other members of his team in seven days during the 1925 season at Bayn Dzak. Mark had a comparable catch. We had marked the location of twenty or so good dinosaur skeletons. Between us, in a period of about three hours, we had found about sixty dinosaurs, mammals, and lizards—a recovery rate of one specimen every three minutes!

"Where are the others?"

In answer Mark pointed in the distance to Jim's spidery figure on a sand hill in the center of the valley. We could see another figure crouched over a hillside far to the left of Jim. "That's Amy," I said, focusing on the small figure with my binoculars.

"Where's Dash?" Mark asked.

I swept the horizon with the binoculars but saw no other sign of human life. Finally I saw a stick figure with a golfing cap far down the wash, near the buttes that looked like saggy camel humps.

Mark volunteered to get Dash. Along the way, he wanted to investigate some small flats just a few hundred yards south of our collecting ground.

The first rush of victory was now gone. As I started to pack out myself, I felt a little melancholy, like a child opening his last present on Christmas morning. It seemed unfair that the most visceral thrills in a lifetime of field exploration had been compressed in three short hours of a July morning.

But a moment later the surge of excitement I had experienced earlier returned.

Mark was walking hurriedly toward me from the direction of the fields which, given my own prospecting, were now only a hundred yards away. Out of breath, he called out, "I found something—and it's the best stuff I've ever seen."

Following Mark to the flats, I saw an even greater concentration of dinosaur skeletons scattered across the surface, as well as an extraordinary abundance of egg shells, egg fragments, whole eggs, and several nests with circles of oblong eggs. Mark led me over to one of these nests. There, in the middle of a circle of large eggs—some complete, some broken—was a broken egg filled with the delicate bones of a tiny dinosaur that looked like an intricate Chinese ivory carving. The embryo in the half shell was accompanied by isolated bones, and not one, but two, skulls of a tiny meat-eating dinosaur with needle-sharp teeth. Mark had discovered the first embryos and possibly hatchlings of a carnivorous dinosaur known to science.

There was more to the story. The eggs themselves were ovoid and about seven inches in length, with a characteristic crinkled surface on the shell—the classic markings of the eggs that had been identified as *Protoceratops* seventy years ago at Flaming Cliffs. Of course, no one had actually found embryos in any of the hundreds of such eggs that had been collected since. But the identity of the eggs was bolstered by the fact that *Protoceratops* and these eggs, so common in some areas, were often found near each other. The nest Mark had discovered threw all that into question. Of course, it was possible that the embryonic bones inside the eggs were *Protoceratops,* while the tiny *Velociraptor*-like skulls were nest raiders. But even in the field the bones in the eggs looked very much like the limbs of a small carnivorous theropod, with their characteristic foot structure. The question would have to be resolved in New York, after Amy had a chance to use her sharp, steel preparator's needle on the soft matrix around the skeletons and skulls of the nest.

We collected on the flats the rest of the day. The site was bristling with not only eggs, nests, *Protoceratops,* and ankylosaurs but other treasures as well. Some of the shattered bone float belonged to theropod skeletons. And the sprinkling of mammal and lizard skulls and skeletons was extraordinary—even richer than the slopes we had prospected in the morning.

By evening, when various members of the team returned to camp, we counted seventy-five mammal and lizard skulls, many with skeletons. We had marked more than forty dinosaur skeletons out in the shallow flats of Ukhaa Tolgod. One particularly rich spot, of course, had a nest with the world's first meat-eating dinosaur embryos and hatchlings. We decided to christen this nesting site, this remarkable field of dreams, Xanadu, after the golden city of the Khan.

July 17, 1993. Day 2. Was I dreaming? No, there were the metal tins—former containers of British cookies and crackers—stuffed with sample bags full of tiny skulls. Mark, satisfied with his stupendous discovery of the previous day, graciously offered to accompany Dashzeveg to the district town of Gurvan Tes for the necessary meeting with the local head of the province. The rest of us, including poor Luis, whose infected foot had confined him to camp on the previous day, struck out for Xanadu and beyond. I slowly migrated toward the Camel's Humps (now a formally named locality), stopping on the way at a small sand hill where I found an exquisite little jaw of a shrewlike placental mammal. The exposures near the Camel's Humps were the most extensive; they formed the outer buttress of a steep bowl about eighty feet high. Before long Jim shouted out that he had found a theropod skeleton high up toward the summit ridge of the bowl. I worked my way over a broad, rock-covered flat to the big set of sand cliffs just west of the humps. There were more white splotches of dinosaur skeletons all around me, mostly large ankylosaurs, the tops of their spiky heads and the coinlike medallions of their body armor barely exposed above the ground's surface. Jaded by the bounty of bones, I continued moving toward a small bump of red sandstone at the base of the western cliffs. Next to these pinnacles was a sand flat about the size of a putting green. A bench of sandstone, no higher than four feet, shaded the flat from the sprinkling of volcanic rocks that usually roll downslope. There, lying belly down, legs splayed out, were seven more ankylosaur skeletons. One of these warriors had a magnificent tail club about four feet long, with prominent delta-shaped spikes exposed at the end of the club. These skeletons demanded more time and interest. I marked the two best specimens, including the one with the tail club, with a small cairn. Then I snapped a roll of film using my rock hammer for a scale.

But I encountered the best specimen a few moments later near the low sandstone bench, a small delicate theropod with a pointed snout and thin splintlike jaws no more than a few inches long. The jaws supported a row of tiny sharp, slightly recurved teeth. Part of a limb stuck out of the rock, and a stream of small delicate bones ran down a mini erosion gully below the skull. This was something I had never seen before, perhaps even a new species.

By the afternoon, Mark and Dashzeveg were back from Gurvan Tes, prowling the slopes above "Ankylosaur Flats," while Luis and I soaked plaster bandages for removing the delicate theropod. The consensus among the group was that this theropod could be one of the most important dinosaurs yet unearthed by the expeditions. It was definitely a rare occurrence of a troodontid, a birdlike dinosaur beauty. Troodontids belong to the Maniraptora, that dramatic radiation of gracile, clawed theropods that actually include the bird lineage. The troodontid *Saurornithoides,* first found by the Andrews team at Flaming Cliffs, has a *Velociraptor*-like skull but the teeth are notably more numerous and are serrated only on their back edges, in contrast to the double-edged serration in *Velociraptor.*

Osborn originally thought that *Saurornithoides* was a small carnivorous dinosaur but one neither as swift nor voracious as its contemporary *Velociraptor.* Nonetheless, the animal does have sharp, hooked dromaeosaur-like claws. Moreover, the second toe is jointed so the claw can be curled up instead of being used to walk on the ground. As John Ostrom at Yale pointed out, this claw in troodontids could be used as an offensive weapon. But Barsbold's study of new specimens of *Saurornithoides,* including a larger beast, *Saurornithoides junior,* showed an unusual configuration of a large swollen braincase with accompanying expansion of the ear region. These special skull features, as well as the differences in teeth and the smaller sizes of the slashing claw of the second toe, prompted separation of troodontids from dromaeosaurs.

It was concluded that these animals, like dromaeosaurs, were predators but that their game was generally less imposing—small lizards and

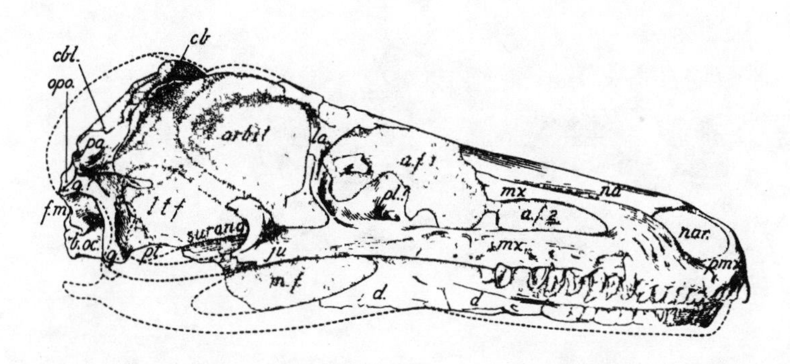

The skull of *Saurornithoides*, a troodontid *(Reprinted with permission, AMNH)*

mammals. The orbits of the known skulls suggest large forward-pointing eyes capable of stereoscopic vision. These orbs, in combination with the large braincase, suggest an alert, intelligent animal, perhaps one of the "smartest" of all dinosaurs.

The animal I encountered at Ukhaa Tolgod was clearly smaller, with more diminutive teeth than *Saurornithoides.* It was likely something new, and would require a new name. Despite the delicate, precious nature of our specimen, it was hard to keep our minds focused on the task of removing it. We kept looking around distractedly, wondering what other fantastic fossils were being located by our colleagues.

The drama continued to unfold throughout the day. Jim nabbed our flightless bird *Mononykus*—although a headless skeleton, it was, nonetheless, an exquisite specimen. Up slope and about a hundred yards to the west of the troodontid pocket, Mark found a very large theropod, a beast perhaps over ten feet long, its wicked claws sticking out of the rock. The big theropod, which Mark thought might be an oviraptorid, unfortunately was also headless. The limbs, feet, and claws of the monster were, however, perfectly preserved. It would take several days to remove this skeleton, and some huffing and puffing to get the 600-pound plaster monolith encasing the dinosaur into the back of a GAZ.

This specimen, however, was worth its weight in gold. Oviraptorids are rare and certainly most unusual theropods. These animals have fasci-

New troodontid from Ukhaa Tolgod *(Mick Ellison)*

nated paleontologists since the time of their discovery in 1923 at the Flaming Cliffs. In 1991 we encountered some enticing bits of *Oviraptor* skeletons in the burning canyons of Kheerman Tsav, and we were on the prowl for more. None of these animals exceed fifteen feet in length. The front limbs and clawed hands are well designed for grasping and the slender but powerful legs are optimally built for speed. In other words, aside from a few specializations in the construction of the tail and other areas, the postcranial skeleton of oviraptorids is what one might expect in a maniraptoran theropod.

By contrast, the oviraptorid skull is a masterpiece of grotesque construction. Basic outlines have been greatly distorted, with the orbits placed high and well forward of the jaw joint. The bones of both the skull and the lower jaw have been extensively hollowed out. As noted, there are no teeth, and the jaws have been completely remodeled into powerful beaks that were probably covered in life by a horny sheath. A very curious toothlike projection extends from the back of the bones that roof the mouth (the palate).

There is, in *Oviraptor,* also an extraordinary development of a longitudinal crest on top of the skull. This structure is reminiscent of the hollow crest of the hornbill, a bird capable of very impressive squawking and horn blasts. The crests differ in construction in the two *Oviraptor* species:

Oviraptor (Ed Heck, reprinted with permission, AMNH)

in *O. philoceratops* it is directed somewhat up and forward; in *O. mongoliensis* the crest is relatively larger and extends directly upward, its convex outline resembling from the side a bony fan. Both species show not only extensive cavities in the skull but also a complex filamentous structure of many of the bone surfaces. This condition, known as *pneumatization*, gives the skull a very spongy appearance.

The development of pneumatized bone occurs in other vertebrates, including birds and mammals. It is often associated with mechanisms for lightening the skull or allowing a complex system of nerves and blood vessels to reach the skin surface. Pneumatized bone, for example, is commonly found in the nasal passageways of the mammalian skull, where it greatly increases the surface area for nerves and blood vessels associated with our sense of smell. The other Gobi oviraptorids, *Conchoraptor* and *Ingenia,* have skulls of a more conservative stamp. Neither has a distinct crest, and pneumatization is not nearly so extreme. These likely represent designs that are more like the primitive skull conditions for oviraptorids, although Barsbold, Maryańská, and Osmólska have noted that the postcranial skeleton of *Ingenia* may be distinctly specialized in its more numerous hip (sacral) vertebrae and shortened fingers of the hand.

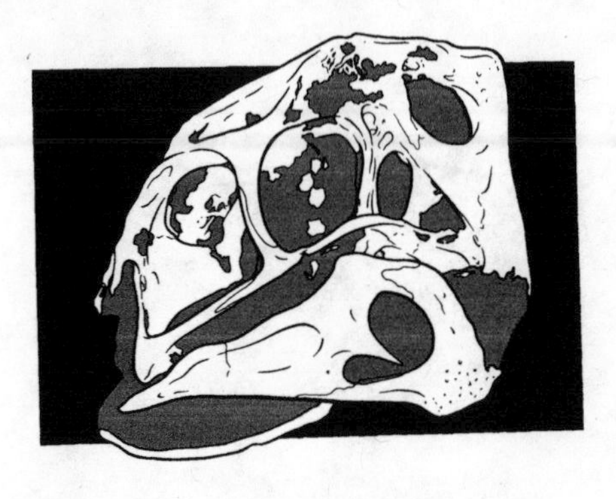

The skull of *Oviraptor* *(Ed Heck)*

The possible habits of these strange creatures have been a matter of fascination since their discovery in 1923. There has been much speculation about what *Oviraptor* liked to eat; and the suggested cuisine includes not only eggs (as the name implies), but plants, small animals, and fresh-water mollusks. A lizard skeleton was actually found in association with the famous specimen first uncovered by Andrews' team, suggesting that lizards, especially abundant at the Flaming Cliffs and other Gobi Cretaceous localities, were part of the diet of *Oviraptor.*

Toward the end of the day our team gathered around as Luis and I finished up the plaster jacket of the troodontid. We quickly made an estimate of the riches thus far retrieved in Ukhaa Tolgod. We had recorded about sixty dinosaur skeletons, preserved in various states of excellence, in the sand flats and cliffs between First Strike (the saddle between the two low hills that was our first collecting spot) and Ankylosaur Flats, a remarkably short distance of about a half mile, and we had continued to pick up small skulls and skeletons of mammals and lizards. When I asked how many each person had found, Dashzeveg held up ten fingers; Jim laconically claimed "maybe fifteen," including a pile of mammal skulls at the base of Camel's Humps.

By now we had well over one hundred skulls, and nearly half of these belonged to mammals. It was, as Dashzeveg said, "A great gift."

A few days brought more surprises. Mark, Jim, and I were working a long set of cliffs extending southwest from the ankylosaur, troodontid, and oviraptorid sites. On one late afternoon we returned to the oviraptorid site where Luis and Amy were industriously excavating. Luis came up to us with a wild look in his eyes. "We think there are . . . eggs under this specimen," he whispered, as though he was afraid that someone would intercept our secret. We went into shock. True, an oviraptorid juxtaposed with supposed *Protoceratops* eggs was found by the Andrews expedition at the Flaming Cliffs. But another such association had not turned up in seventy years. Was this association merely another odd circumstance? We resisted further speculation and further probing around the skeleton for eggs. Instead, we wrapped the specimen in its swaddling of plaster and hoisted it—aided greatly by the brute force of Batsuk and Eungeul—into the GAZ.

We worked Ukhaa Tolgod between July 16 and July 20. By then we had close to 100 dinosaur skeletons and 137 small vertebrate skulls—76 lizard skulls and 61 mammal skulls. But even the richest of places yield diminishing returns as the days go on. We had scoured the surface of the small amphitheater and had excavated some of the real prizes, the embryos and the nest, a couple of oviraptorids, including the big headless monster, a smaller skeleton with a nice skull that looked like *Ingenia,* a few *Mononykus,* the new troodontid, and so forth. It would have been perfectly justifiable to spend the rest of the summer on this hallowed ground, but Dashzeveg for some mysterious reason was anxious to leave. With the pressures of international politics bearing down on me, I agreed to pack out. Later, in August, Malcolm and Priscilla joined the team, allowing a second pass over the site and a chance for them to share in the victory. The tally for dinosaur skeletons and mammal and lizard skulls rose accordingly.

In all we had spent only about three weeks at various intervals at our new locality. But in that breathlessly short time we had reaped rewards that stunningly exceeded by orders of magnitude anything we had retrieved through several years of work in the Gobi. We had no delusions that our scouring of Ukhaa Tolgod was exhaustive. Nonetheless, we felt confident that we had scooped up most of the spectacular remains before another expedition—the Japanese and the Russians in separate parties were roaming the Gobi that year—should stumble into them. It would be a triumphant return to New York.

But our encounter with this pockmarked piece of Gobi was far beyond any triumph we could imagine. We began to uncover facts that were astounding and extraordinary—a realization that would come only after months in our fluorescent-lighted laboratories back at the museum.

CHAPTER 8

FLYING DINOSAURS AND HOPEFUL MONSTERS

I believe there were no flowers then,
In the world where the humming-bird flashed ahead of creation.
I believe he pierced the slow vegetable veins with his long beak.
Probably he was big
As mosses and little lizards, they say, were once big,
Probably he was a jabbing terrifying monster.

We look at him through the wrong end of the long telescope of Time,
Luckily for us.

D. H. LAWRENCE.
1929. "Humming-Bird"

Ukhaa Tolgod gave us dinosaur eggs and embryos, theropod skeletons, and mammals galore. Back at the museum's laboratories, the treasures from the locality fired us up. Research on this material would provide new answers—and doubtless new questions—concerning important evolutionary problems. One of these certainly concerned the pathway between dinosaurs and full-fledged birds. Mark, Jim, and Luis made up the investigative trio that would tackle the problem, with a careful review of anatomy in different theropods, including the archaic bird lineages. Now the cornucopia of evidence from Ukhaa Tolgod provided a new reference system for making these comparisons. Even the newly discovered embryo could be drawn in, as it represented remains of some kind of theropod. Besides, any data on juvenile, newborn, or embryonic dinosaurs were keys to tracing the evolutionary changes through the dinosaur genealogy.

It is important to dispel a common misconception about flying dinosaurs. Pterosaurs, those strange winged creatures, many with large crests and long beaks, that lived at the time of dinosaur dominance were not dinosaurs at all. As shown in the cladogram in Chapter 2, dinosaurs are clearly a member of a major radiation of terrestrial vertebrates, the Archosauria, which includes living crocodiles, fossil pterosaurs, as well as a number of other extinct groups. Pterosaurs are, then, a side branch of archosaurs outside the dinosaurs. The only flying dinosaurs are the birds.

For the question of bird origins we turn to the theropods. Within the theropods are several important branches, including the Coelurosauria and the giant stalkers like *Tarbosaurus* and *Tyrannosaurus.* One member of the coelurosaur radiation, the ostrich-like ornithomimids, is represented well by the gracile *Gallimimus* from the Nemegt. The fossils found in the Barun Goyot and Djadokhta Formations, however, belong to the other complex branch of coelurosaurs, the Maniraptora. Of the smallish, agile maniraptorans, dromaeosaurs like *Velociraptor* are the best known; they were the epitome of the lean and mean predator. In the sense of evolutionary endurance, dromaeosaurs, however, are overshadowed by another clade of maniraptorans, the Avialae. These are the specialized forms that take flight in the form of birds. Despite their extraordinary diversity and the radical alterations, maniraptorans share an unusual design in the wrist. A small wristbone or carpal has a distinctive semilunate, or crescent, shape that affects the way the wrist bends and articulates with the hand. In addition, the three fingers of the hand have a characteristic profile. Although these features are highly modified in living birds, they are found in the oldest bird, the Jurassic *Archaeopteryx.*

To get our bearings, let us reflect for a moment on the evolutionary organization of the maniraptorans. This limb of the dinosaur tree gives rise to at least five major branches: (1) the enigmatic Therizinosaurids; (2) the toothless, high-crested Oviraptoridae; (3) the highly specialized Troodontidae, with their elongate serpentlike skulls and rows of tiny, sharp, recurved teeth; (4) the predaceous Dromaeosauridae, including *Velociraptor* and kin, characterized by the retractable sickle claw on the hind foot and very sharp and formidable claws on the front hands (also in troodontids), as well as a stiff tail; and (5) the Avialae, the most specialized of the theropods, characterized by highly modified front appendages, shoulder and pelvic girdles that indicate the basic blueprint for flying birds.

A formal classification scheme designates modern birds as members of the group Aves within the group Avialae. As noted throughout this book, there is decisive evidence that, technically speaking, dinosaurs are not extinct. At least one of its lineages stays with us in the form of birds, a baroquely elaborate and diverse worldwide group of perchers, migrators, hoverers, and even flightless forms. There are a few detractors from this idea, scientists who still argue that birds and other dinosaurs do not share a close kinship. But these workers are clearly in the stubborn minority, and, as discussed below, their arguments are unconvincing. They remind me of a group of geologists who for years rejected the notion of plate tectonics because of a few problematic observations. No theory in science can be irrefutably proven, but the theory of bird-dinosaur affinity is one of the strongest connections in vertebrate evolution.

Aves is one of the most successful of all the vertebrates, with at least 8,600 living species. Doubtless much of the staging for the evolution of birds took place in the Mesozoic. In the Gobi, birds are not extravagantly represented by fossils. A smallish delicate form, *Gobipteryx*, has been collected in several sites by Mongolians, Russians, and our own field parties. Small ovoid eggs of this bird are particularly abundant in Kheerman Tsav, the brilliant chromatic and multicolored canyon in the frontier west of the Nemegt Valley. Many of these eggs have embryonic remains.

Those small eggs and skeletons of *Gobipteryx* are not much evidence of an avian renaissance during the Mesozoic. There are other regions that do better in providing these clues. The Cretaceous of Argentina, Spain, and China produces the short-winged enantiornithines, which vary from the sparrow-sized *Sinornis* to the turkey-vulture-sized *Enantiornis*. The Cretaceous badlands of Patagonian Argentina also gave us, in the mid

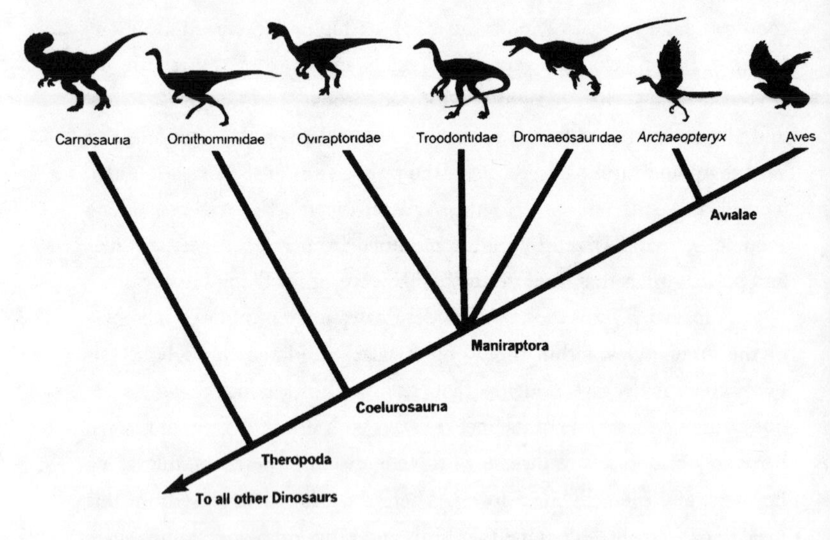

Cladogram for theropods *(Ed Heck)*

1980s, the stubby-winged flightless *Patagopteryx.* Since the 1880s, the later Cretaceous of North America has been generous in providing skeletons of the flightless loonlike *Hesperornis,* and the tern-sized, big-headed flier, *Ichthyornis* ("fish bird"). These birds populated ancient shorelines and doubtless hunted over the waters, nabbing fish between their long "beaks" armed with tiny, sharp teeth. The Mesozoic diversity and relations of birds are superbly summarized by Luis Chiappe in the June 1995 "dinosaur issue" of *Natural History* magazine.

THE BEGINNINGS OF BIRDS

The fossil record of birds, from Luis' and other accounts, is rapidly improving but still sketchy. Most bird fossils are small and fragile, so the record likely treats them poorly. Thus the documentation of pre-Cretaceous birds is extremely impoverished. Of course, there is a brilliant and famous exception to this rule. In 1861 a fossilized birdlike skeleton was found in an Upper Jurassic limestone quarry near Solnhofen, in southern Germany. The skeleton was named *Archaeopteryx lithografica* and it was joined at various times over the last century and a half by a handful of other precious skeletons representing this form. Remarkably enough, *Archaeopteryx* not only had exquisitely detailed skeletal and skull parts; the limestone was also finely etched with the impression of feathers. The earliest appearance of birds was, then, as a matter of record, indicated by one of the best-preserved and most beautiful sets of fossils in the world.

Some specimens of *Archaeopteryx* show feather impressions, others don't. For some time the "featherless" skeletons from the Jurassic limestones were thought to be small theropods—coelurosaurian dinosaurs, like *Compsognathus.* But it was soon apparent that these were simply fossil skeletons deprived of feathers through less favorable preservation. Detailed work by John Ostrom at Yale in 1973 revealed a reason for the mix-up. *Archaeopteryx,* as the earliest bird, bore an uncanny resemblance to coelurosaurs. In 1973, Peter Wellenhofer, a German paleontologist whose talents extend to exquisite technical drawings of specimens, also noted strong similarities between coelurosaurs and *Archaeopteryx,* based on a featherless skeleton of the latter discovered in Eichstatt in 1951 and mislabeled *Compsognathus.*

Why were these scientists so persuaded? *Archaeopteryx,* after all, clearly departs from the standard dinosaurs in having feathers (though we have those bumpy skin impressions in hadrosaurs, we can't eliminate the possibility of feathers in other dinosaurs; we just don't have the fossil impressions to decide one way or the other). It also has those greatly elongated and modified forelimbs that closely compare with the condition found in the wings of modern birds. But differences between species, as we have seen, do not necessarily eliminate these possibilities of close bloodlines. It is necessary to look beyond the differences for similarities in special features that indicate true affinity. In the case of *Archaeopteryx* and coelurosaurs, such similarities were striking; Ostrom found over twenty anatomical characters linking the two groups. The comparison has been further refined and expanded through work on an array of early birds and their theropod relatives, including Mesozoic birds like *Sinornis, Patagopteryx,* and *Iberomesornis.* Like other maniraptors, birds primitively

Mesozoic and living birds: (from left) *Archaeopteryx, Mononykus,* hesperornithiforms, modern birds *(Ed Heck)*

---

have a crescent-shaped wristbone (semilunate carpal). In non-avian maniraptors this bone is distinctly separate, but it is in close contact with the other bones of the wrist. In birds, this bone becomes fused with the wrist, but its crescent shape is still apparent. Other features of special similarity are found in the construction of the pelvis, the development of the large airspaces in the skull, the comparatively large brain cavity, and details of the feet and vertebrae. Birds even show a connection with theropods at a broader level; just like *Tyrannosaurus,* they are characterized by an S-shaped curve of the neck, a hind foot with three ground-hugging toes, and a small fourth toe held high off the ground on the back of the foot. Birds not necessarily of a feather do flock together.

The convictions of Ostrom and Wellenhofer, as well as of Jacques Gauthier and other scientists, on the tight bird-dinosaur connection are not really new. In the 1860s and early 1870s the famous biologist Thomas Henry Huxley passionately argued for bird-dinosaur affinity. The anatomist Gerhard Heilmann took on the whole problem in a book published as early as 1926. He noted, after an astute survey of vertebrate groups, that coelurosaurs seemed to be the best candidates for the relatives of birds. Unfortunately, the latter lacked a key bird feature—a wishbone formed by the fusion of the two collarbones (clavicles). Heilmann thus concluded that the bird-coelurosaur resemblance was simply a matter of convergence, namely that the true ancestors of birds should be found among more primitive but as yet unidentified archosaurs from the Triassic. He also suggested that birds evolved from archosaurs living in trees. Allegedly, these forms first modified forelimbs into parachutes for sailing, further refined the wing for gliding, and eventually attained the flapping, powered flight seen in birds today.

Despite Heilmann's misgivings, the idea that birds and theropod dinosaurs are indeed closely related seems to stand up against most attacks. For example, some maniraptoran theropods, like *Oviraptor*, have a bird hallmark, a wishbone. The bird-dinosaur junction has been broadly accepted, but it has a few detractors. Some embryologists argue that the finger bones elongated in true birds (digits 2, 3, and 4) are not the same as those elongated in theropods (thought to be digits 1, 2, and 3). The argument raises a paradox because *Archaeopteryx*, the supposed first bird, has the 1, 2, 3 formula, not the 2, 3, 4 formula of later birds. Some workers, like Alan Fedducia of the University of North Carolina at Chapel Hill and Larry Martin at the University of Kansas, strenuously reject the bird-dinosaur hypothesis. They claim, much in the vein of Heilmann, that a transition from coelurosaurs to birds is highly unlikely because none of the former were tree-dwellers capable of evolving to the next gliding stage necessary for the bird heritage. They also stress that many of the relevant coelurosaurs with birdlike features are of Late Cretaceous age, occurring much too late to be near the roots of bird ancestry.

Two counterpoints can be raised here. First, the Heilmann scenario for transition from a non-flying to a flying animal that occurred some 200 million years ago is not a very tangible thing. In any case it should not be the basis for rejecting a pattern that is clear from the "hard" anatomical evidence. The uncertainties that come with constructing such scenarios about the appearance of adaptations in birds and other organisms will be expanded upon momentarily. Secondly, the elimination of Late Cretaceous theropods as relevant to the question of bird origins because of their later appearance in the fossil record than *Archaeopteryx* reflects a basic and frequent misuse of fossil evidence. The first fossil found in the record for a particular group is not necessarily the indication of the time of origin of that group. We could someday find fossil birds much older than *Archaeopteryx* (indeed this has been claimed, but the alleged candidate, *Protoavis* ["pre-bird"] from the Late Triassic, is highly questionable). Why then should we ignore some very birdlike theropods of the Late Cretaceous because they are younger than *Archaeopteryx*? In addition, would the discovery of an older fossil bird eliminate *Archaeopteryx* from consideration? Absolutely not. *Archaeopteryx* would still be important because it shows a wonderful mosaic of characters found in modern birds and—lo and behold—dinosaurs.

There are even some animals alive today that show such mosaics. The duck-billed platypus and the spiny echidna of Australia and New Guinea retain primitive features found in other vertebrate groups that have been lost and modified in all other mammals. For one thing, these strange creatures don't bear their young live. They lay eggs, a reproductive habit that seems to be primitive for vertebrates. By this evidence, we are sure that the duck-billed platypus should be as old as the Mesozoic hills. Until a few years ago we just didn't have a specimen to prove it. A jaw of a platypus is now known from the Cretaceous of eastern Australia. We call such organisms "living fossils," because they retain ancient characters lost or drastically modified in their relatives. *Archaeopteryx* is a "dead" fossil, but whether it is Late Jurassic or Late Cenozoic in age, it would still bear the undeniable stamp of a species that importantly retains the characters of the bird legacy. In the same way, *Oviraptor, Troodon,* and kin are not organisms to be ignored when it comes to thinking about the origins of birds. The balance of anatomical evidence linking these forms with birds through *Archaeopteryx* is undeniable.

This leads inevitably to the question of family trees. Which of the various non-avian lineages of dinosaurs are the closest kin of modern birds? Since we have nested the latter within the Maniraptora, our exploration must focus on the lineages within this group. This important search is just under way, and it is a major preoccupation of Gobi team members Mark Norell, Jim Clark, and Luis Chiappe. At this point, the closest kin to Avialae is a tossup between oviraptorids and the clade including dromaeosaurs and troodontids. But the investigation must extend beyond the usual list of suspects. It must include, for example, some *incertae sedis* members of the Maniraptora, like *Avimimus.* The phrase *incertae sedis* means "uncertain position" (the second word derives from the Latin verb *sedeo,* to sit). In other words, these taxa cannot be placed in a particular branch of the Theropoda with any confidence.

Avimimus incertae sedis, on the other hand, offers a tantalizing glimpse of what is clearly a specialized theropod, with many odd, birdlike features. This form, based on fragmentary material described by the Russian paleontologist Kurzanov, is a small delicate creature. What is known of the braincase is very birdlike, but the anterior snout (premaxilla) reminds one of the construction in hadrosaurs. The hind limb has the ultratheropod design with tridactyl foot, lacking even a small first (in other forms the most internal) toe. The pelvis is unusual in having a flange of the upper (iliac) crests medially directed in a most birdlike fashion. The most puzzling aspect of the skeleton is the forelimb. Although the upper arm (the humerus) resembles that in theropods, a bone of the lower arm, the ulna, has a long roughened ridge that Kurzanov interpreted as an attachment for feathers. Moreover, some fragments of the bones in the hand (carpometacarpals) have very much the appearance of the hand of birds. Unfortunately, the incomplete material of *Avimimus* prevents much more thought about the affinities of this creature. There is even some suspicion, expressed by our colleague Perle, that the reported skull may be that of a very young oviraptorid and not associated with the *Avimimus* skeleton at all.

The investigations of the mystery of bird origins must also embrace another theropod group, the ornithomimids. Notable among these is *Harpymimus,* first described by the Mongolian scientists Barsbold and Perle in 1984. *Harpymimus* was discovered in Hurrendoch, that Lower Cretaceous locality in the eastern Gobi that we found so bone dry in 1991. It has a highly flexible (prehensile) hand. This animal, known from a partial skull and associated pieces of the skeleton, is unique in having six or more tiny teeth in the very front of the lower jaw. It is odd for being, along with the newly described *Pelecanimimus* from Spain, the only toothed fossil among the otherwise toothless ornithomimids. Other members of this clan show conformity in the skull—long slender toothless jaws, large eye sockets with a ring of individual bones (sclerotic ring) that supported the eye—all features very reminiscent of birds.

Untangling this Gordian knot of theropod strands—the key to solving the mystery of bird origins—is a tough problem. It must be tackled through studies of every bit of anatomical minutiae, from the architecture of the limb to the intricacies of the braincase. Yet the investigation is much better served than it was only a few years ago. We now have in hand some truly spectacular skeletons of primitive birds and non-avian theropods.

ON BEING AND BECOMING A BIRD

Cladograms, phylogenetic trees, and classifications organize data for us. They provide a map that underpins our understanding of evolutionary routes among organisms. This operation is critical to considering the "what happened" questions. How did birds actually evolve? What were the steps in attaining the flapping flight in modern birds? How did feathers originally function—as a coat to insulate warm-blooded, active creatures, or as flight structures themselves? These kinds of questions are often difficult to answer, even with the knowledge of a good cladogram or classification. The fossil bird *Archaeopteryx* has been showcased as an example of transition in function as well as structure. The animal had feathers and wing structures, but, according to Ostrom, Bakker, and others, it couldn't fly very well. Feathers do have other purposes, notably the enhancement of insulation and thermoregulation critical to endothermic ("warm-blooded") animals. Hence, feathers could have originally evolved for insulation, only later to be "coopted" as a part of the wing for flight. The flight feathers of birds are in fact marvels of engineering for this purpose. The bristles along the hollow shaft form slots ideal for controlling the movement of air against the wing for lift and maneuvering.

This scenario, however, begs the question: why is the first appearance of feathers coincident, as far as we know, with some kind of wing structure? In response, Ostrom suggested that the wings in *Archaeopteryx*, and its yet unknown precursors, were handy for catching insects and were implements for predation. This argument calls upon analogy with certain modern birds (black egrets, for example), whose wings are used both in flying and in "mantling," an interesting behavior wherein the bird casts a shadow over the water with its wings in order to see its prey. Wings, then, were versatile and put to several good uses in the earliest avialians. The modification and refinement of wing design later led to its use for flight. Under this scheme, feathers would evolve through two steps: they were first structures for insulation in a winged creature where the wings were in essence "fly-catchers"; they were secondarily, and later, structures in the flight machinery.

But this scheme for evolutionary retooling has some problems. Although *Archaeopteryx* provides a striking mosaic of theropod and early avian characters, one finds that it does not satisfy all the expectations for a "proto-bird." If feathers originally aided insulation rather than flight, one would predict that some fossil, particularly a fossil filling the gap between *Archaeopteryx* and non-avian theropods, might have feathers but without modified winglike structures. That fossils rarely preserve impressions of feathers doesn't help our chances in finding this missing link. But such evidence is desperately needed. Such a fossil would be the only indication that a primitive non-winged theropod had feathers for some purpose other than flight.

The second problem with elaborate stories about the evolution of feathers concerns the supposed capabilities and limitations of *Archaeopteryx.* It is by no means evident that *Archaeopteryx* entirely lacked the ability to fly. Indeed, research published in 1991 by John Ruben suggests otherwise, even though the animal was probably incapable of the extended flapping flight. Under this view, *Archaeopteryx* could fly, but not like a bona fide modern bird. And the complications don't end here. The equation between feathers and thermoregulation relating to endothermy is not a given. The paleontologist Anusuya Chinsamy and her coauthors in 1994 published some remarkable results from detailed histology of bone in Cretaceous (and presumably feathered) birds. These bones had lines of arrested growth which possibly indicate ectothermy, a "cold-bloodedness," rather than the endothermy typical of modern birds. Hence, the origin of feathers for some kind of primitive flapping flight is just as probable as its origin for the purpose of insulation. Indeed, one can claim that *Archaeopteryx* suggests that feathers were part of the evolution of the flight structure. After all, it has both feathers and winglike structures.

More problems remain. Ostrom's argument that the wing in *Archaeopteryx* was mainly used in predation completely fails when we consider living birds. There are apparently no living birds with well-developed wings that use these structures for predation and at the same time abandon flying behavior. Do these observations, then, allow us to reject the notion that the earliest birds used wings and feathers for purposes other than flight? Unfortunately no, at least not in a rigorous fashion. Our identification of multiple possible purposes for feathers, as described in a 1994 paper by S. Randolph, and the lack of critical theropod-bird fossil intermediates suggest that the stepwise evolution in function cannot be entirely eliminated. All we can say is there are more problems with this famous story than many are willing to admit.

In discussing *Archaeopteryx,* I do not mean to suggest that fossils are useless and misleading. Although the fossil might not reveal the sequence of steps for the changing use of wings and feathers, *Archaeopteryx* does have a lot to tell us. The emergence of feathers and wings is identifiable in a form that shows clear affinities with other coelurosaurian theropod dinosaurs, which allows us to look broadly for the clues to the origin of bird structure and function. Accordingly, we may look to theropod diversity for the many plans that anticipated redesign of the bird skeleton. Fossils, in this sense, do offer such insights and, in some cases, these insights cannot be extracted from the living world. The notion that feathers were not originally used in flight, for example, cannot be established by examining living creatures. (There are flightless, feathered birds today but these are readily identified as cases of secondary loss of flight.) If a further fossil discovery offered evidence of feathers with an absence of wing structures and a combination of other structures that suggested it was more primitive than any other fossil or living bird, including *Archaeopteryx,* such a skeleton would resoundingly underscore the power of the fossil record.

A Bird by Any Other Name?

It may seem rather odd to call birds dinosaurs. Or, more explicitly, "birds are a kind of dinosaur." But that is just what we mean by weaving the intricate tale of bird-theropod relationships. Classifications involving names tell us things about the history of life. As I discussed in Chapter 2, they do this by giving us a hierarchy—a system of groups within groups—that can be "read out" in the geometry of a big tree of life. As Flaubert wrote to Turgenev in one of his famous letters, "Draw a tree so no other tree can be mistaken for it." That is precisely the aim of biological classification. "Birds are dinosaurs" is a correct statement. But "dinosaurs are birds" is technically incorrect. Why? Because the latter statement would force us to draw a tree wherein all the various branches of dinosaurs came off some branch of birds. As we have seen, the converse seems to be true; all birds branch from one of the dinosaur lineages, the Maniraptoran theropods. A classification would read:

DINOSAURIA
  SAURISCHIA
    THEROPODA
      COELUROSAURIA
        MANIRAPTORA
          AVIALAE
            AVES

Where does something like *Archaeopteryx* go in this classification? It might be tempting simply to say that *Archaeopteryx* is a bird and should naturally go under the formal category Aves. But the problem is trickier because of some recent developments in the theory of classification. To dispatch *Archaeopteryx* properly, I must first digress a bit and mention those recent arguments. Traditionally, ranks of the Linnaean hierarchy call for a general name—a species belongs to a genus, a genus to a family, a family to an order, an order to a class, a class to a phylum, and a phylum to a kingdom. In some cases, the names are easily applied—we recognize a five-kingdom system for life, one of which includes the Animalia. But naming ranks also leads to problems. As we become more knowledgeable about the relationships of groups, we can be more specific about various levels in the classification, and we can proliferate greatly the number of ranks. Likewise, proliferation of general rank names then becomes cumbersome; we start designating things as superfamilies or super-superfamilies, as more levels are recognized.

There is another problem with rank names. A rank like family suggests more knowledge than we actually have. It suggests somehow that the family Protoceratopsidae and the family Hominidae (the group including us—*Homo sapiens*) are equivalent in some biological way. But this equivalence in ranks means nothing really. Protoceratopsidae and Hominidae simply happened to "fall out" as families as we moved up along different branches on the tree of life. Certain families, like Protoceratopsidae or Oviraptoridae, may have only a few species within them. Others, like many of the prolific families of insects, may have huge numbers of species. Why, then, should the two groups be given equivalent rank names? These annoyances can be avoided if we simply rid ourselves of these rank names. After all, the indented list above doesn't need all these rank names; it tells clearly where the various groups belong. It also allows the insertion of other groups in the hierarchy at some later time. For this reason, I have throughout this book avoided referring to the groups as families, orders, classes, or superclasses.

This provision is important because we would like names of groups to have an explicit meaning. The Maniraptora should mean "a group including the maniraptoran ancestor and all its descendants." It should not simply mean "a group that, because it is *somehow* different, isn't another kind of dinosaur." Unfortunately, the traditional scientific literature is plagued with names whose original meanings are very foggy. Invertebrata, for instance, the great suite of creatures that includes flatworms, sponges, worms, insects, snails, and starfish, are simply those animals that for some reason are not vertebrates. There is nothing that makes invertebrates a real, i.e., monophyletic, group. Some so-called invertebrates are on the same evolutionary spine leading to vertebrates, while other so-called invertebrates are not. But all "invertebrates" cannot claim to share an ancestor unique to themselves, in other words one that excludes vertebrates. In this sense, the common textbook and encyclopedia category Invertebrata is no better than Linnaeus' primitively obsolete term Beastiae.

Despite the long history of taxonomy, this attempt to use group names consistently to indicate ancestry and descent is rather new. The approach was clarified in the works of the German zoologist Willi Hennig in the late 1950s and 1960s, some two hundred and twenty years after the publication of Linnaeus' *Systema Naturae.* Hennig was also responsible for the development of the methods later called cladistics. The rules of classification have been further refined by two brilliant colleagues, Jacques Gauthier, a dinosaur and reptile specialist in his own right and one responsible for many of the clarifications of dinosaur groupings noted above, and Kevin de Queiroz, a curator at the Smithsonian Natural History Museum. Jacques and Kevin have outlined a set of rules they call phylogenetic taxonomy. First, the names of ranks are abandoned; no more families, superfamilies, or super-superfamilies. Secondly, names of groups are defined by their genealogical connections, not by their characters or contents per se. Thus Dinosauria is defined as the clade stemming from the last common ancestor shared by the two dinosaur subgroups, the Ornithischia and the Saurischia. It so happens that dinosaurs also have certain features that enable us to recognize them—such as the hole in the hip socket mentioned in Chapter 2. This trait is then a handy way of *diagnosing* the group; it is the shared-derived feature that demonstrates the monophyly of Dinosauria. But Dinosauria is *defined* by its basic connection—ornithischians and saurischians through their last common ancestor. This way a group can change in content (by the inclusion of more taxa) or its diagnostic traits may be modified, but the group's name can be retained. Such a method offers more stability in naming biological groups.

Jacques and Kevin have also sought to retain some of the familiar names. They use them for clades (called *crown clades*) issuing from two or more extant taxa. Mammalia is the clade issuing from the last common ancestor of the three major branches of living mammals: monotremes (duck-billed platypus, echidna), marsupials (pouched kangaroos and kin), and placentals (our own group as well as that for many other mammals). Their ideas are constructive, especially if some of the original group names—the Mammalia, Aves, Crocodylia, and so forth—can be retained at levels that make sense in the hierarchy of life. But phylogenetic taxonomy is relatively new, being first described in papers dating back to the late 1980s, and it is controversial. It has not been widely or enthusiastically embraced by all taxonomists.

It is easy to see how this new system impacts on our designations of birds and dinosaurs. As noted, the formal equivalent of birds is Aves. As Jacques and Kevin would have it, Aves is defined as the group that includes the last common ancestor of all the living groups of birds. Under this scheme *Archaeopteryx* is not a member of Aves. There are other fossil birds, including *Gobipteryx*, that are more intimately nested within the clade including all living birds. In other words, there are some fossil taxa that are closer to the ancestry of all living birds than *Archaeopteryx*. A phylogenetic classification for these players might then read:

THEROPODA
    COELUROSAURIA
        MANIRAPTORA
            AVIALAE
                Archaeopteryx
                    AVES
                        Gobipteryx
                        "living birds"

This change is a little uncomfortable; it means that calling *Archaeopteryx* a bird, as I have throughout this chapter, is only informal usage. Nonetheless, the above scheme serves several good purposes. It retains as a stable point the familiar (indeed Linnaean) word Aves for the crown clade that contains all living birds. It explicitly puts the relationships of living birds in context with fossil birds and other theropods. Finally, it gives a proper place in the hierarchy to *Archaeopteryx*. Our Jurassic friend is much too important to just be put in the crown clade Aves. It is a critical clade outside Aves, one that points the way to modern birds from dinosaurs.

The Odd Bird

There is yet another of our Gobi discoveries that bears on the origins and early evolution of birds. This is *Mononykus,* the new animal first discovered by the Soviet-Mongolian expeditions, then by Malcolm at Tugrugeen, and subsequently by other members of our team from Ukhaa Tolgod and various localities. As I noted, *Mononykus* appears to be a bird. It was, nonetheless, a very odd bird. Although it has no wings, it has several features that suggest a closer relationship with modern birds than the famous primitive bird *Archaeopteryx.* In addition to the enlarged sternum, these features include an antitrochanter (a small knob on the pelvis at the hip joint), a continuous crest on the femur for attachment of the limb muscles, and a greatly shortened fibula, the more delicate of the lower hind limb bones. True, our *Mononykus* fossils do not show evidence of feathers, but it is only by some miracle of preservation that the fine Jurassic limestone entombing *Archaeopteryx* preserves impressions of tiny feathers. *Mononykus,* like most fossils, is not preserved in such unusual rock, so any trace of feathers, even if originally present, would be highly improbable. A detailed analysis of *Mononykus* published in an April 1993 issue of the scientific journal *Nature* favors the view that this creature was a flightless relative of modern birds.

The argument has attracted some controversy; certain specialists claim *Mononykus* is simply a small dinosaur whose birdlike features are a product of *convergent evolution.* In other words, *Mononykus* evolved an enlarged sternum, antitrochanteric process, and femoral crests along an independent evolutionary pathway. Convergence is doubtless a widespread phenomenon in evolution. We can see it in the remarkable but clearly independently derived similarities between fishes and the mammalian whales. There were marsupial wolves (now extinct) from Australia that looked in an uncanny way like the "true" wolves of the northern continents. And a preference for ants has refashioned some unrelated creatures, like aardvarks and anteaters, in striking ways. An elongated palate (roof of the mouth), extremely complicated nostril passageways, a powerful and highly mobile tongue, and

Mononykus (Ed Heck)

sharp front claws were acquired independently in the evolutionary backgrounds of aardvarks and "true" anteaters. However, a similarity in form does not necessarily indicate convergence, despite the frequent leap to this conclusion by some scientists. The evidence at hand for *Mononykus* does not force us to the conclusion of convergence, as long as one recognizes that flight might have been lost or regained several times in bird evolution. The loss of flight is indeed a reality of bird history, as the existence of the flightless ostrich and other birds today clearly shows.

There are other mysteries with *Mononykus.* What were the curious can openers on those stubby front limbs for? They certainly did not have a long reach. Their use in defense seems doubtful. It is likely that *Velociraptor* would have little trouble dispatching one of these stub-armed turkeys. But that short arm is not puny; it has very robust bones and presumably large muscles, and that big claw. Perhaps, if the beast took on the right posture, the claws would be useful for digging termite mounds and ant nests—an anteater of the Cretaceous. Perhaps the forelimbs were used to grasp a mating partner. Perhaps they were employed for digging deep nests or burrows. Any, all, or none may be true.

The story of *Mononykus* cannot be completely told without mention of one of the most coincidental and fortunate discoveries of the entire expedition. This find was made not on the blinding white outcrops of Tugrugeen Shireh but in the bowels of the American Museum's storage area. A few months after our return from the Gobi with *Mononykus,* Bryn Mader, a scientific assistant at the museum, happened by to look at the specimen in Mark's office. "Hey!" he exclaimed. "I think there's something in the C.A.E. collections that you might be interested in." Bryn is an expert on mammals, particularly the big rumbling rhinolike Brontotheres (the thunder beasts) that lived about thirty million years ago, halfway between the Cretaceous extinction event and the present day. At the time, though, he was in charge of taking care of various collections in the Department of Vertebrate Paleontology. When rummaging through the Central Asiatic Expeditions collections, Bryn spotted an inconspicuous and undramatic piece of limb fragment in a chunk of red sandstone from the Flaming Cliffs. He brought the specimen, described on its label as "birdlike dinosaur" and numbered "Am. Mus. Nat. Hist. #6542," back to Mark.

AMNH 6542 was clearly a third specimen which to that point could be identified as *Mononykus.* In their urge to get "bigger and better eggs" the 1923 expedition was not impressed with this specimen. One can hardly blame them, considering the precious *Oviraptor, Saurornithoides,* and mammal skulls discovered at the base of the cliffs. It took seventy years to demonstrate that Andrews' team had recovered a strange flightless member of the clade that led to modern birds. And it was a riveting reminder to all that great museum collections are great terrains for discovery.

EMBRYOS AND ANCESTORS

The origin of birds is one of the outstanding problems in vertebrate paleontology. It draws on the evidence of critical fossils as well as a vast array of data on the form in living creatures. Like other such problems, it is also greatly clarified with certain information of younger stages in both fossils and living species. Our dinosaur embryo was therefore regarded as an extraordinary prize. Likewise, such embryos and juveniles, whether found for fossil birds, non-avian theropods, or any dinosaurs for that matter, are a cause for great rejoicing in the paleontological community. Why all the fuss? Because there is much these tiny bones can tell us. The exquisite series of *Protoceratops* growth stages in the American Museum collections show clearly how the skull of these creatures changes with age. As these individuals get older and larger the head is distinctly remodeled. The plate-like frill at the back of the skull becomes more pronounced, as do the protuberances of the cheekbones (jugals) below the eye socket, and the hooklike beak or rostral element. You may recall that *Protoceratops* and its near relatives can be distinguished by different expressions of these very features: frills, cheekbones, and beaks. *Protoceratops andrewsi,* for example, has a frill that is much larger than that in other members of the group. Juvenile skulls of *Protoceratops* and its close relative *Bagaceratops* are more alike to each other than are the adults of either species. Moreover, it seems reasonable that the primitive condition for the several genera in Protoceratopsidae approached the more juvenile state. In other words, it is likely that the ancestor of this group had a smaller beak and frill.

This is what we would expect according to the rules of phylogeny—the young ones do converge on the conditions ancestral to the group as a whole (as I'll note later, there are often anomalies during development). We would predict that, if juveniles were known for all the species within the Protoceratopsidae, they would all have relatively smaller beaks and frills. Think of the possibilities of extending these conclusions if juveniles and hatchlings were known for a wide variety of dinosaurs. There might be a prototype for many of these groups—theropods, sauropods, and so forth, based not only on data from adults but on features known from juveniles, hatchlings, and embryos. There has actually been some notable progress on this front in recent years, but the search for dinosaur embryos and young ones should carry on; much more information is needed. And the results of this mission could be very important: youth may be the key to understanding the outstanding theories of vertebrate phylogeny.

Indeed, dinosaur embryos and juveniles themselves are a fossil record of a life cycle. They represent particular stages of growth, milestones in the process of development from unfertilized egg to embryo, to young, to full-scale adult. This process of growth and change is what biologists call *on-*

togeny. An ontogeny is effectively a history within a history. Each individual organism has its own ontogeny, which is a component of a larger plot—the *evolutionary* history of the species comprising this organism and its relatives. When we consider the pattern of relationships we call phylogeny we don't really just compare static items. The best phylogenies are phylogenies of ontogenies—that is, they account, to the greatest extent possible, for information on development of the species being compared. Such ontogenic data are hard to get, certainly in fossils but even in many living organisms, where either collecting embryos or rearing organisms from the point of fertilization is challenging. When available, however, such data often bring great insight to evolutionary problems.

Consider these points at a broader level of life's diversity. We know that some animals, like insects, change at various stages to an extraordinary degree: eggs become larvae, larvae become pupae, pupae become adults—the kind of ontogeny we call *metamorphosis.* In other organisms the process is not so punctuated. Vertebrates change more smoothly, although the first stages as an embryo may bear little superficial resemblance to the adult product. A tiny human embryo, for example, looks at first glance something more like a shrimp than a bipedal human. Only closer inspection reveals its essential chordate features, an enlarged head region with a bulging eye, gill slits, small appendages or limb buds, the ghost of segmentation that foreshadows the development of vertebrae, a vestigial notochord (a rodlike structure made of connective tissue extended along the back that was also an important development in the chordate ancestor), and a dorsal nerve cord that becomes the spinal nerve. Knowing the signature of ontogeny for a particular group allows us to consider their evolutionary potentials, namely what these organisms are likely to be or not to be during the course of their evolutionary history. For example, we know it is highly unlikely that tetrapods, the four-limbed groups to which we belong, will become pentapods—the program of ontogeny that is common to all vertebrates seems to prohibit except in the most extraordinary cases the emergence of a fifth limb. Nor are we likely to mimic the insects in development of a third pair of legs.

This stereotyped design and pattern of growth is very important in another sense—it brings diversity of species and their wildly varied architecture together as part of a single history. For over a century it has been appreciated that embryos share a greater similarity among themselves than do the adults they are destined to become. At some stages of growth, the embryos of fish, amphibians, turtles, lizards, birds, and mammals all look strikingly alike, an odd juxtaposition in light of their adult design. This similarity in form naturally helps us understand how characters diverge from one another. Thus the wings of bats and the front limbs of a human at some stage look very nearly alike. During ontogeny of the bat, the upper arm (humerus), lower arm (radius and ulna), and finger bones elongate extravagantly to form the struts of the wings. This does not mean, however, that wings evolved directly from the kind of front limbs seen in an adult human. Rather it means that both human forelimbs and bat wings diverged from a common plan seen at some stage of their respective ontogenies. In a loose sense, embryos serve as keys to our ancestry; at least they show common sets of conditions that might be expected in our ancestors. Ontogeny is then a critical key to the pathway of evolution.

Ontogeny, Mutations, and Hopeful Monsters

It would be unfair to ignore the fact that ontogeny does not always perfectly mirror this pathway. There are special cases of anomalous events occurring in an ontogeny—development slows down or stops entirely—leaving the adults rather juvenile-looking. Experiments with living salamanders show that certain species look like the juveniles of other species because they don't fully metamorphose by the time they reach sexual maturity. In other words, the adult resembles the juveniles of its relatives and presumably of its ancestor. The remodeling of ontogeny to produce this situation is called *paedomorphosis,* with a special process, *neoteny,* occurring when the *rate* of change in ontogeny slows down. A growth phase in ontogeny can also be delayed at onset or truncated too early to produce the same result. Now remember that juvenile characters are often attributed to the most primitive adult conditions, the blueprint from which other characters evolve. But here evolution seems to be reversing itself. Descendants seem to take back characters that belonged to their remote ancestors but were refashioned in their immediate ancestors.

This may seem rather cumbersome and highly improbable, but the mechanism behind paedomorphosis is really very simple. What may seem like a strange reversal in the adult state is in actuality simply a resetting of the clock of ontogeny. Suppose (it is actually probable) that the protoceratopsid ancestor, as an adult, had only a weak frill and cheek flaring. Its ontogeny did not carry through or change fast enough to allow the development of large frills and prominent beaks. Subsequently, ontogenies in the descendant protos were modified, sped up, or lengthened and frills and large beaks were part of the adult equipment. But later, one of the species that arose from the frilly lineages experienced a change in the ontogenetic program, resulting in adults with small frills and beaks. Its developmental clock simply ticked more slowly. We don't in fact know whether this happened in the evolution of protos, but such a discovery might not be surprising. Paedomorphosis likely accounts for some dramatic evolutionary events in the history of dinosaurs and other biological groups.

Such phenomena are, however, excruciatingly difficult to document. Paedomorphosis would seem, for instance, to be the response to a major genetic mutation, one that altered not only genes that code for a structure but also genes that control development. In the last few years exciting new experiments have identified genes that code for the development of particular structures—gill arches, head segments, and nerve branches. These findings are at last opening that black box of evolution—the mysterious connection between the gene makeup, called the *genotype,* of the organisms, and the expressions of body parts and functions, called the *phenotype,* which those genes control. Still we are a long way from explaining evolution in ways that link genes to structure—patterns on a broad scale that involve phenomena like paedomorphosis. Whatever the refinements of such patterns—and they will come—there is no doubt about one thing. Ontogeny is a logical place to look for them. Much of what we will find in that black box will be the clues to how mutations reshape genetic codes, and how these codes provide the "software" for rates of growth, remodeling, and part differentiation.

Mutations that affect developmental genes are very impressive. A change in such genes may switch on or off sweeping ontogenetic changes. They have a cascade effect, spilling over one level of organization to another. The power of such genes resides in the pliability of growth. At early stages of ontogeny, cells are rather more flexible in their choices. Cells that produce limb buds are quite like other cells that produce vertebrae or abdominal muscles. At some point early in development, however, those cells take their own pathways—ones that are fated for a variety of body structures. This branching of directions for cells in ontogeny is appropriately called differentiation. Differentiation is controlled by genes whose mutations can cause dramatic changes in body structure. This has been experimentally demonstrated in myriad cases. Laboratory-induced mutations of certain genes may lead to poor limb development, an uncoiled cochlea (the tiny snaillike organ responsible for hearing), extra teeth, and grotesque reshaping of the jaws. Likewise, in developmental experiments, one can place cells that program for limb development in another region of the embryo, say the head region, as long as the stage of differentiation is early enough so the alien cells are not rejected. The results, as one can imagine, are also grotesque: creatures with new limbs sticking out of their heads and other monstrosities.

The drama of these experimental manipulations much impressed biologists in the late nineteenth and early twentieth centuries. They proposed such anomalies as a key to major changes in evolution. Slight differences among species were one thing, but the major shifts that put tentacles on octopuses, feathers and wings on birds, hair on mammals, or took away limbs from snakes had to reflect a major and very drastic mucking around with development. And this was likely caused by very powerful mutations, ones that produced "hopeful monsters." This was actually the term applied to such products, with the implication that the monstrosities that might arise from mutation could, on rare occasions, point the way toward new ("hopeful") evolutionary pathways. The strongest advocates of the theory tied hopeful monsters to the origins of the major groups, certainly including such things as dramatically and grotesquely specialized as dinosaurs.

The problem with this theory is twofold. First, the vast majority of genetic or developmental manipulations have tragic consequences. Mutations that affect such gross changes in phenotypes are in overwhelming cases lethal. The organisms affected by those mutations simply don't survive, unless, in the case of some human disorders, the wonders of medical technology and science intervene. Evolution involves natural selection, the winnowing away at new genotypes and phenotypes, leaving only those that offer the best chances for the survival and reproductive potential of the individual. Because of selection there is a strong resistance to the massive disruption of the genotype. After all, the surviving genetic blueprints may be the products of millions or even billions of years' refinement—if it ain't broke don't fix it. Evolution, especially out in nature, won't allow hopeful monsters to endure.

A second problem with the monster theory is that it presumes that such rapid and drastic alterations are necessary for the kind of major changes that led to birds or mammals or octopuses. This ignores the fact that thousands of generations through time provide plenty of opportunity for evolutionary change. These changes can culminate in a major transition from one form to another. One of the great contributions of the fossil record is in illustrating some of those steps. This doesn't mean fossils preserve the very ancestor of a living form. But they do preserve species whose combinations of characters represent intermediate steps. Living horses, with their high-crowned grinding teeth and single hooves, bear little resemblance to the first horses from 55 million years ago, those poodle-sized creatures with four toes and low-crowned, bumpy teeth. Yet the fossil history of horses is replete with examples of transitional states—the missing links have been found. The horse record is paleontology at its best, but many other examples abound. There are even recent studies of small aquatic one-celled plankton (called diatoms) that document a near perfect record of transition from one species to another over several thousand years. Our much better understanding of the origins of birds and the modern mammals reflects the triumphs of paleontology.

Thankfully our work in the Gobi, especially the discovery of key fossils at Ukhaa Tolgod, are part of this success story. Some of these Gobi forms, like *Mononykus* and *Oviraptor,* are bizarre creatures indeed—they look for all the world like "hopeful monsters." But looks can be deceiving. These skeletons do harbor codes that, when combined with other codes in other species, retrace the evolutionary tree, through dinosaurs, birds, mammals, vertebrates, all life. Breaking that code is one of life's great professions.

CHAPTER 9

1994—BACK TO THE BONANZA

Bones picked an age ago,
And the bones rise up and go.
EDWIN MUIR.
"The Road."

One day in the spring of 1994 my assistant Barbara Werscheck (some of my colleagues call her my boss) alerted me to a message from Mark Norell, whose office at the American Museum of Natural History is in the Childs Frick Wing, the edifice for the Department of Vertebrate Paleontology.

"Mark called. He wants you to come over to his office."

My research office is actually next to Mark's, but at that moment I was buried in museum business at the desk of my administrative office in one of the museum's stately outer turrets. It was a bad time for interruptions. Mark, however, wouldn't call for trivialities so I made my way over there as soon as I could. During the year since the discovery of Ukhaa Tolgod we had worked feverishly on some of the best material from the site. High on the priority list was, of course, the egg nest with the embryo in the half shell and the tiny dromaeosaur skulls. For months, Amy Davidson had with extraordinary care and skill picked away at the matrix around the delicate embryo skeleton with a sharp steel needle. By spring, when Rinchen Barsbold from the Mongolian Academy of Sciences visited us in New York, Mark was laboring over the specimens, trying to make sense of them. When I arrived at Mark's office, Barsbold and Jim Clark were with him. All three excitedly began sharing with me their unusual observation. The tiny "hatchlings" outside the eggs associated with the nests were indeed tiny dromaeosaurs, perhaps very young *Velociraptor.* But the embryo in the half shell was something completely different. Yes, it was the exquisite skeleton of a theropod all right, but not a *Velociraptor*-like dromaeosaur, as had been suspected. Instead, it was an oviraptorid, complete with the skull.

So why were these different species found in the same nest? The preservation was too exquisite, too undisturbed, to be accepted as a random association. There were at least three possibilities. Perhaps the little "dromaeos" were raiding the nest, snatching up succulent oviraptorid hatchlings. Alternatively, the dromaeos could be part of the nest too, the hatchlings of eggs belonging to another species. This would draw on the example of some living birds, like the cuckoo, which places its eggs in the nests of other birds in order to get some free brooding time. Finally, the parental oviraptorid could simply have provided her hatchlings with some dromaeosaur food. It was impossible to decide which of these scenarios was true. The first explanation, however, did not seem altogether likely, as the oviraptorid embryos were not much smaller or less developed than the tiny dromaeosaurs in the nest. The rarity of cuckoo behavior in living birds suggests that dromaeosaurs as nest parasites was improbable if not impossible. In addition, the outside, rather than inside, of bits and pieces of the eggshell were stuck to the skulls of the dromaeosaurs. The last explanation—dromaeosaurs as baby food—was slightly favored. It is fair to say we may never know what exactly happened and why the two species were associated in one nest. Nonetheless, the simple observation that oviraptorids and dromaeosaurs were in the same basket was a riveting revelation. Barsbold called it "the greatest dinosaur discovery since the Andrews expeditions first found dinosaur eggs!"

The discovery of the true identity of the dinosaur embryo also clarified another critical matter. It was increasingly apparent that the most common theropods at Ukhaa Tolgod were not *Velociraptor* and certainly not troodontids and *Mononykus.* Oviraptorids were by far the dominant theropod here, with nearly a score of skeletons representing the group found in the 1993 season. This was striking because these forms are so rare at other localities. Most of the numerous eggs at Ukhaa were oblong, several inches long with a wrinkled external shell—exactly the type that encased the oviraptorid embryo. It was safe to reason that most of these eggs at Ukhaa therefore were laid by oviraptorids. There was even some evidence that bore on this connection. The 1923 American Museum expeditions found an *Oviraptor* at Flaming Cliffs lying on top of a nest of eggs, the only such association that had been discovered. Already convinced that these kinds of eggs belonged to the most common animal around, *Protoceratops,* Osborn and others naturally concluded that *Oviraptor,* as their name for the theropod suggested, was raiding rather than laying eggs.

Until Ukhaa Tolgod, we too were duped by the abundance of *Protoceratops* into assuming that it was the likely candidate for the nesting dinosaur. This was despite some nagging deficiencies of the evidence. None of the hundreds of dinosaur eggs collected have clearly identifiable *Protoceratops* embryos within them. Even some tiny skulls of *Protoceratops* recently discovered cannot be associated with an egg of a particular type. Now we had evidence that all but eradicated this connection between protos and eggs. The eggs under the Flaming Cliffs *Oviraptor* were exactly the same as the Ukhaa Tolgod egg containing the embryo. *Oviraptor* seemed to be more the good mother (or father) than an egg snitcher. Of course, it is possible that very similar or nearly identical eggs do belong to *Protoceratops.* But here there is no positive evidence for a linkage, no evidence that in any way measures up to the proof that at least one egg of this type belongs to oviraptorids.

More evidence bearing on this mystery was soon to emerge. The months following the 1993 season saw a great deal of work in the lab on the fossils we recovered in the Gobi. With National Science Foundation grant support, we hired Amy's mentor, Bill Ameral, to prepare some of the intricate mammals and small dinosaurs. Amy divided her time between the delicate bones of dinosaur embryos and the bulkier items, particularly that headless oviraptorid skeleton that Mark found above my troodontid pocket. As she flicked away the dull red sand between the bent, clawed appendages of the oviraptorid, a series of oblong objects revealed themselves. They were covered with a hard prismatic substance of pale purple-blue. There were not simply one or two objects but a large number of them, arranged in some definite pattern. The objects and the skeleton around them would not be studied until a year later. But at this early stage there was no mistaking their identity. Our hasty identification in the field was confirmed. Quite unintentionally, Mark had found the perfect complement to the puzzle piece represented by the oviraptorid embryo. His "big mamma (or papa)" oviraptorid was crouched on a magnificent tier of eggs!

We had also brought back an amazing bounty of tiny mammal skulls. Many of these were in excellent shape—with bones, sutures, and associated limbs and finger bones exquisitely preserved. We were certain that many of the more than ninety mammal skulls represented new animals and new information on crucial parts of the anatomy of these creatures. There was only one slight disappointment to the sample. Overwhelmingly, the Ukhaa mammals were represented by multituberculates, those extinct herbivorous forms that resembled rodents in habits. We were more eager of course to retrieve some of those mammals that related directly to the living groups like the placentals and marsupials—the ancestors of our own ancestors, as one of my professors at Berkeley used to call them. But our sample included only a few table scraps, some tantalizing partial jaws and fragments of skulls. Where were those placental gems like the *Zalambdalestes* skull that Jim had found in Tugrugeen in 1991? It seemed that such important targets would require relentless hunting, even at a place as bountiful as Ukhaa Tolgod.

Despite this shortcoming, our abundant sampling of mammals—not to mention our impressive pile of lizard skulls, many with skeletons—delighted us and awed our colleagues. One visiting scientist, upon seeing the small, skull-filled nuggets in the specimen trays in my research office, remarked, "I've been working on these things all my life—but only bits of teeth and jaws: I never thought I'd see anything like this."

THE RETURN TO XANADU

After a glorious year of sifting through our treasures from Ukhaa Tolgod, we were hot to return to the site in the 1994 season. By the late spring of that year the locality had become rather famous. John Wilford had written a detailed account of the discovery for the *New York Times* which was widely released by papers and media internationally. Now we were ready for a more heralded return. Our team was bigger than the 1993 crew, but not unwieldy. Lowell rejoined us, this time with our friend Carl Swisher, from the Berkeley Geochronology Laboratory. We needed a good geologic section of the world's best new Mesozoic site, and Lowell and Carl were to provide it. In addition, Carl hoped to do a detailed sampling of the rocks in Ukhaa and elsewhere for a magnetic signal. As I discussed in Chapter 4, some minerals preserve the orientation of the earth's past magnetic field by virtue of their orientation in the rock. The pattern of flipflops in this field through millions of years is its own earth calendar. We were anxious to get some data on paleomagnetism, because the rocks at Ukhaa were devoid of any radioactive elements that might provide other clues to their age. The only thing we had to go on—besides the fossils themselves—was the possibility that paleomagnetic signals in the minerals might give us a chance to date the age of these dinosaur beds.

In addition to the geological team, our group comprised the usual suspects—Mark Norell, Jim Clark, Amy Davidson, Lüis Chiappe, and me. Amy's husband, Jim Carpenter, an entomologist from the museum, would accompany us in an effort to build an important insect collection from the Gobi. Our mechanic John Lanns would also join us, this time for nearly the duration. As in the previous season, Malcolm and Priscilla opted to rendezvous with us at a later date.

By now our launch of the expedition out of Ulaan Baatar was becoming routine. The operation for the most part ran smoothly. The shipment was not only on time; it was early. We purchased another GAZ, one in fairly decent condition. We did have one big problem though. On the flight from Beijing to Ulaan Baatar I noticed something unusual out the window. We seemed to be flying over endless plains of a pale green color. Where was the rich, red Gobi? Were we off course? When we arrived at the Ulaan Baatar airport, Dashzeveg explained the mystery.

"Beeg problem this year. Many rains!" he laughed.

"How do we get down to Flaming Cliffs and beyond?" I asked.

"*Mitqua,* maybe we will go around farther west and then south."

The Mongolian contingent for the season included Dash, Perle, and Batsuk. There was also Gunbold, a very young (seventeen-year-old) tenderfoot. Gunbold was serene and shy, but fiercely intelligent. His father was Dr. Dumaajavyn Baatar, president of the Mongolian Academy of Sciences. There were two other members of the team—Sota, our gas tank driver, and Temer, the animated, elfin driver of our additional GAZ. It was not long before we developed a very close and trusting bond with our new comrades.

We had another critical colleague who was not to join us in the desert. The youthful secretary to the president of the academy, Dr. Tsagaany Boldsuch, helped us immeasurably. Boldsuch had also been involved with assisting us on the 1993 expedition and was even more enthusiastically engaged in the project this year, despite his work coordinating several big international projects with the academy. Mongolia had become popular and more accessible for such work, but its capacity was seriously stretched to accommodate this converging attention. Fortunately, Boldsuch had an uncanny skill with English and a second sense about the politics and intrigue of the Mongolian scientific community. With his assistance, I navigated the tricky course of conferences, parleys, and protocols.

We struck out for the Gobi on July 1. As Dash had predicted, though, our way was not direct. The river valleys south of town on our familiar road to the Flaming Cliffs were completely flooded. We made a flanking movement directly west on the paved but crater-filled highway to Aarvaheer. This allowed us to cross the Tule River on a firm bridge and then head south again toward the Gobi on the usual unpaved route.

This season I had my own personal struggles. For nearly a year I had suffered a stubborn sinus, throat, and lung infection, one I might have contracted in Mongolia or China during our last summer. A bout of pneumonia seriously impaired a trip to Egypt in the fall of 1993. I had improved but was not yet healthy. My congestion and coughing fits were severe, especially at night. Knowing that my illness would worry my comrades, I tried to suppress any sign of how sick I was.

Our new route took more time and gasoline, but it followed a dry and unimpeded pathway until we reached the usually dry lake just south of the outpost of Mandal Obo. This flat spread across the road leading through the old volcanics. Beyond the volcanics were, of course, the Flaming Cliffs, only a tantalizing fifty miles away. We reached the lake area by dusk and rather brashly pushed on. After a few hours of misery skating through the mud and towing the tanker out of the troughs, we gave up. In the light of day, we found the proper route south of the volcanics and on a straight line to the Cliffs.

We were so overwhelmed with the desire to get back to Ukhaa that one night's bivouac at Flaming Cliffs seemed sufficient. But after some discussion it was clear that Lowell and Carl, along with Dash, had some important geological problems to settle that had bearing on our work at Ukhaa. There was a need to carefully describe the layers of rock in each section at Flaming Cliffs, in order to see whether the sequence was repeated at Ukhaa and elsewhere. At the bottom of the cliffs were layers of sandstone with numerous nodules. These were the slopes first crawled by Granger and his patrol in the 1920s. The nodules were a hot pink in color, varying in size from pebbles to nearly boulders, with an average size about that of a potato. One out of every million or so of these nodules encased a small, astonishingly perfect fossil skull. Above the nodule layer were the great bands of cliff-forming sandstones, with their waves of cross-beds, the remnants of ancient sand dunes. This main escarpment was topped with a series of chunkier sands, thinner nodule beds, and shaley layers that indicated ancient stream channels. Bone was found throughout the section, but Dashzeveg claimed that even the mammals found at the base of the Cliffs had weathered out from nodules much higher up. We were skeptical of this view, and Carl, Lowell, and Dash entered into prolonged controversy on the issue. The rest of us stood on the sidelines, resigned to the prospect of searching for some table scraps overlooked by years of paleontological exploration. I reflected on how mundane this visit to the most famous dinosaur site in the world had now become.

But our 1994 stop at the Cliffs turned out to be unexpectedly productive. On the very evening of our first day Jim Clark nabbed a beauty of a *Zalambdalestes* skull. On the next day, Independence Day, I returned the favor. A few of us struck out for some very undramatic low gentle slopes far to the west, near an area originally called the volcano locality, in honor of an evocative cone-shaped hill nearby. Walking in a line with the others up one of these slopes, I nearly tripped on a skeleton of a dromaeosaur, possibly a *Velociraptor.* A sinuous string of neck vertebrae bent backward in U-shaped bend over the top of the trunk vertebrae. To our delight, this twisted neck ended in a skull. It was one of but a handful of theropods collected over many decades at Flaming Cliffs. By late that evening Amy had scooped out the conveniently soft sand around the skeleton and applied the requisite plaster jacket, and we lovingly nestled the block in the bed of a GAZ. But even the successes of these two days did not induce us to stay. On to Ukhaa Tolgod, we decided, and even better fossils!

By July 6 we were camping at the opening of the gulch just south of Ukhaa Tolgod. A brief afternoon hike affirmed that the locality was still strikingly rich with fresh skeletons. The rains of winter and spring that made for such nagging problems on the road were a boon to our prospecting effort. Many other dinosaur skeletons and nearly one hundred more mammal and lizard skulls were weathering out on the rock surface. By this time Ukhaa Tolgod, a small bowl in the Gobi Desert, a pocket of sand less than three square miles in area, was probably mapped in greater detail than any other Gobi locality. Lowell and Carl made a detailed geological description of the rock sections exposed, placing our individual fossil sites in the proper stratigraphic position. The sequence here, from bottom to top—a thin layer of shales, a nodule bed, cross-bedded, cliff-forming sandstones, and layers of sandstone alternating with beds of shale and pebble-choked rock called conglomerate—was remarkably like that at the Flaming Cliffs. I also had in hand a color-enhanced "map" derived from LANDSAT satellite images. The red pixels in the image exactly matched the outcrops of red rock, giving a near perfect outline of the site. This map

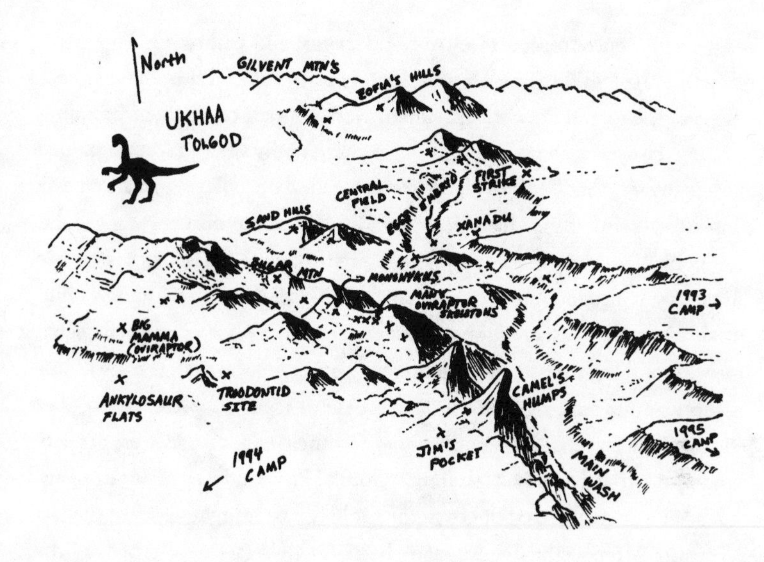

Field sketch, Ukhaa Tolgod, showing fossil sites *(Novacek)*

had been expertly prepared by a young undergraduate student, Evan Smith, working at Yale University's new Center for Earth Observation. On our field maps and our sketches in notebooks, Ukhaa was portrayed as a mosaic of dots and accompanying names—First Strike, Xanadu, Zofia's Hills, Granger's Flats, Camel's Humps, Ankylosaur Flats, Jim's Pocket, Mike's *Mononykus,* Mark's Egg Site, Central Field, Gilvent Wash, and several others. These were the sublocalities, areas varying in size from the surface of a baseball infield to a plot of only a few square feet. Each sublocality indicated its own triumphant event.

That summer we were further impressed by the abundance of *Oviraptor* and its smaller relative *Ingenia.* I found a nearly complete skeleton of a fifteen-foot beast in the sand, not far from the spot where in 1993 Mark had found the headless skeleton. This fossil proved to have a beautiful skull, perhaps the best preserved of any oviraptorid, of the high-crested adult *Oviraptor.* Perle also found an excellent skeleton. Several other oviraptorids—more than we could efficiently extract in one season—were discovered by Luis, Jim, and other members of the team.

This concentration of oviraptorids, eggs, and embryos is impressive and odd. Not all these fossils were found at the same level in the rock section, so the total number of specimens were not part of one huge brooding colony. However, the concentration of these items at several stratigraphic levels suggest that Ukhaa Tolgod was some kind of collecting area for nesting sites, one that may have been a traditional oviraptorid nesting ground at regular intervals for a fairly long stretch of time. Was this a strategic hatchery, a destination for seasonal migration? Were these animals and their nests the product of an unusual circumstance—for example, repeated periods of drought—that forced them to gather in local refuge? Or was this simply an unusual, nay, extraordinary, case of localized preservation, that distorts the fact that such populations and their nesting sites were actually widely distributed in the Cretaceous Gobi? The abundance of sand and evidence of extensive sand dunes in the Gobi red beds argues against the last scenario. Moreover, there are many handsome Cretaceous red beds of the Gobi that have little, if any, fossil bone. Ukhaa Tolgod and other rich sites may more likely represent concentrated areas of Cretaceous flora and fauna—oases in the dunes. Beyond that it is difficult to say what was going on. We can, however, recognize there is evidence that oviraptorids were responsible parents; they stuck near or rested on their nests, even—as suggested by the Flaming Cliffs *Oviraptor* and our "big mamma"—in the face of death.

VERMIN

We again found a great many lizard skulls at Ukhaa. In raw numbers, the abundance of these vermin was astounding. Nonetheless, we expected that such fossils would be relatively more common than dinosaurs and mammals. Indeed, lizards rank first in abundance among the vertebrates thus far collected from the Gobi Cretaceous. These are part of a Mesozoic renaissance that endured. Some of the lizard species are extremely well preserved and show anatomical features that offer clues to the relationships among the major lizard families. This was apparent with the 1990 discovery of *Estesia,* our Komodo-like lizard, near Eldorado.

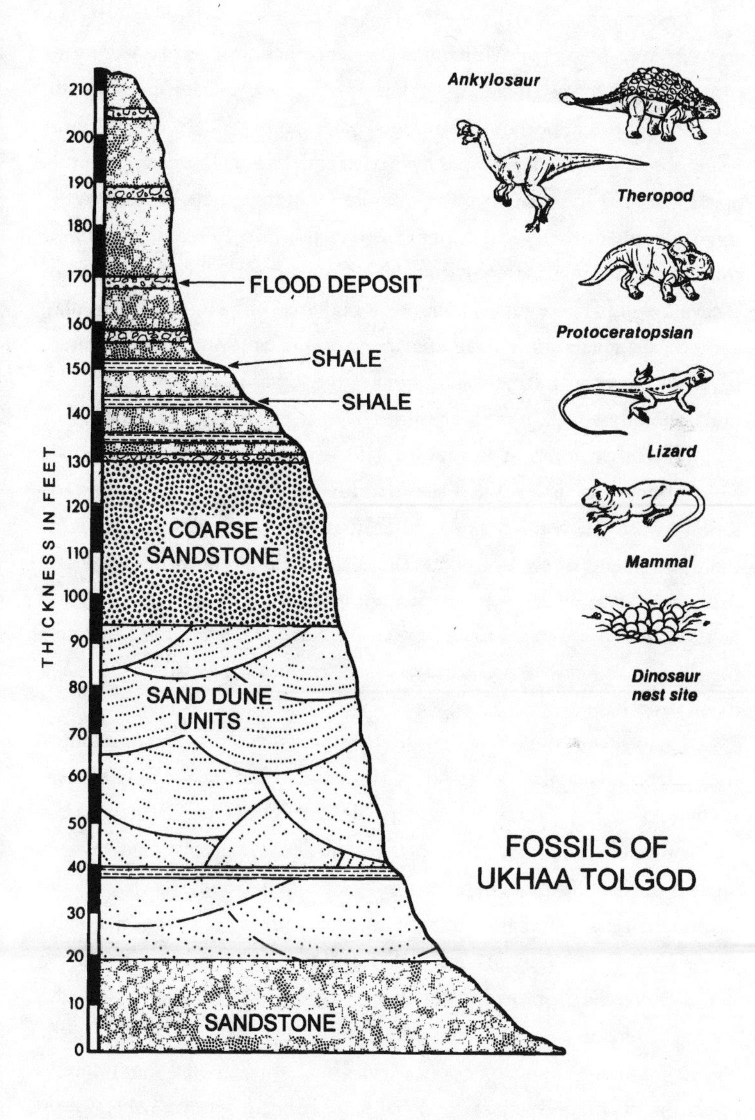

Rock strata at Ukhaa Tolgod (numbers = thickness in feet), showing the levels where fossils were found *(Ed Heck)*

Less abundant and diverse than the lizards from the Gobi Cretaceous are the crocodiles. The Mongolian sequence has produced a variety of crocodile-like forms, including a key species *Shamosuchus djadochtensis.* This animal has important bearing on the origin of modern crocodylians. Many specimens of *Shamosuchus* have been collected by Russian paleontologists, but only one specimen, the type skull collected in the 1920s, is available for study in the North American collections. In 1992 our expedition collected additional material from Zos Canyon, a locality not far from Ukhaa Tolgod. These fossils included a braincase of a crocodylian similar, and perhaps related to, *Shamosuchus djadochtensis.* Study of the new material, as well as comparative examination of the fossils in the Russian collections, should help clarify the mysteries of early crocodylian evolution.

The other group of animals usually associated with aquatic habitats are of course the turtles. These are found in various Gobi localities, in addition to the death pond at Bugin Tsav. In the North American Cretaceous, fragments or even complete shells and skeletons of turtles are abundant, indicating ancient environments of streams, ponds, mud flats, and deltas. The sporadic occurrence of such creatures in the Gobi beds bolsters the view that many of the sandy rock layers bearing fossils represent a much drier habitat.

I have given these reptilian groups unfairly short treatment, admittedly preoccupied with the wonders of dinosaurs and mammals. This bias is aggravating to paleontologists. My colleague, Dr. Gene Gaffney, one of the world's foremost experts on the fossil turtles, was deeply annoyed with a paper published several decades ago by C. Lewis Gazin, the well-known mammalian paleontologist. Gazin had been working an important Paleocene locality in Wyoming called Bison Basin. In writing up the Bison Basin fauna, Gazin provided detailed descriptions of teeth and jaws of a variety of fossil mammals, including important early primates. In the opening remarks of his paper on the site, Gazin announced that vertebrates other than mammals were virtually absent. This is a very odd observation because anyone who has ever worked this locality—as I have many times—knows that the gray-white sandstone surface of Bison Basin is riddled with pieces of skeletons, skulls, and myriad shards of turtles. Some of these are respectably preserved. Apparently, though, they didn't count to a mammalian paleontologist who early on worked this site.

That is blatantly unfair. While animals like turtles are intrinsically interesting from an evolutionary standpoint, they also have other critical uses. Bones of many turtle species are so abundant and well distributed in layers of Mesozoic and Cenozoic strata that they are very useful for age-dating beds and correlating fossil faunas across widely separate regions, even different continents. This is the forte of Dr. Howard Hutchison, a paleontologist from the University of California at Berkeley, who objectively extended his mammal expertise to other vertebrates, including turtles. Howard can spy a piece of turtle shell on a hillside and identify it often to a particular species. These skills, coupled with a prospector's eye that rivaled those of various members of our Gobi team, put Howard in great demand. He hops all over North America or joins teams on their way to Ellesmere Island in the Arctic and elsewhere to find and analyze turtles as well as other vertebrates.

There are now hundreds of choice specimens of turtles and lizards from the Cretaceous Gobi of both Mongolia and China. These have not been ignored. A number of important papers have been offered by Polish and Chinese scientists, as well as earlier scientists working at the American Museum. Our paper on *Estesia* marks the first and as yet only contribution of our team to this part of the collection. There is still much work to be done, especially on the embarrassment of riches confronting us in the collections of fossil lizard skeletons from Ukhaa Tolgod. Field science has been productive here, but a proper story on these important members of the Cretaceous ecosystem must wait for laboratory science to catch up.

SUGAR MOUNTAIN

On July 9, 1994, Amy and I found ourselves crouching over a little mound of red rock at Ukhaa Tolgod, high in the saddle between the saggy Camel's Humps and the steep sandy rampart above Ankylosaur Flats. The day was as beautiful as any I've ever seen in the Gobi. The sky was deep blue and converging on black, with cumulus clouds so white and sharply shadowed that they looked as if they were about to land. The wind was gentle, and rather cool and bracing.

"That small skull came out of here," Amy said, carefully teasing the edge of the sand with a small camel's hair brush.

"Well, the one Mark found below us was directly downslope," I replied.

Earlier, Mark had picked up a delicate little sharp-nosed skull of a tiny shrewlike mammal—an "ancestor of man's ancestor" from the age of the (non-avian) dinosaurs.

"This could be more of it," Amy continued. There were small limb bones sticking out of the broken surface of the sand.

We brushed gingerly at the surface.

"Wait a minute, that's another one!" Now I could see the outlines of a skull and jaws, narrow arcs of white bone dulled by the impregnation of red sand.

All we usually find of Mesozoic mammals are isolated teeth. On rare occasions, if one is lucky, a skull might be found. More rarely, one might find a skull and a skeleton. But almost never, even in the bounteous Cretaceous rocks of the Gobi, are there more than one skull and skeleton.

Amy and I just looked at each other for a moment, a bit stunned.

"Let's dig a little back here." Amy pointed to the bench of sand going into the hill above and behind our second mammal skeleton.

This was close-in work for one person only. I wandered about on the slope beneath our red mound, hoping to find more mammal skeletons in the float. After half an hour I returned to Amy. As always, she had been painstakingly careful but efficient. A flat table of sand, level with our exposed skeleton, had been carved out. Amy was now lying facedown with her legs stretched out behind her, staring at her ledge, her head only inches from the tiny quarry face.

"Mike?" she asked as I approached. Her voice was weak, almost muted by the soft breeze, but I heard the quiver of excitement in it.

Before she could say anything more I saw it. There was another skull, its roof perfectly symmetrical and undistorted, less than a couple of inches long, etched by the sand. There was also the ghost of a skeleton, white bits of bone, that trailed behind the skull—a skeleton.

"And . . ." She pointed to a splotch of white bones no bigger than matchsticks about ten inches from the other. Another one.

We hugged each other in elation.

For the rest of the afternoon we talked excitedly and carefully brushed the sandy matrix covering the fossils. Amy kept reworking the strategy for removing this incredible little aggregate of mammal skeletons. There would be separate plaster blocks, each no bigger than a cigar box, that would be carefully marked and fitted together back in the lab. Each skeleton was so tiny, about eight inches long, that we could give it its own block. But these mammals obviously belonged together. They were the same species of those all-important and rare placental mammals, our own antecedent stock. Were they a family? In a nest? Were they residents in a burrow? They were possibly the first discovered family on the pathway to our family. They were, more certainly, the first such congregation of such critical creatures ever found.

This "family" appears to represent a completely new kind of early placental mammal. It is doubtless closely related to a form first described by Zofia Kielan-Jaworowska, the tiny shrewlike *Asioryctes.* But the new ones have odd front incisors with small hooklike cusps, and a relatively much larger upper canine tooth with a single root (the upper canine in *Asioryctes* has two roots). A real bonus was revealed only after Amy's preparation back at the lab. Near the back of the skull of one specimen, in the ear region, we can see tiny remnants of not only the stapes but the minuscule parts of the incus and perhaps the malleus. These critical bones for mammalian evolution—the hammer, anvil, and stirrup, each no larger than a grain of sand—represent the first known set for a Mesozoic placental. Heretofore only bits of the stapes had been preserved in our archaic antecedents. The disclosure is exciting and potentially illuminating, at least once the study of this swarm of skeletons is completed.

"Well, our day is almost perfect," Amy said as we continued working our mammal pocket.

"What do you mean, almost?"

"We don't have a *Mononykus* skull yet."

It was true—despite the dramatic skeletons of the bizarre *Mononykus* thus far recovered, we had only some very small bits of the skull. The head of the beast, as reconstructed on the cover of *Time* magazine back in April 1993, was more a matter of art than science. We needed a skull because many of the issues concerning bird-dinosaur relationships apply to detailed architecture of the cranium. If, for example, the skull had a characteristic system of loose attachments between certain bones, it would offer powerful evidence for the bird affinities of *Mononykus.* Moreover, details of the skull had an added advantage. They might provide evidence of bird kinship independent of structures associated with the limbs and the back of the skeleton.

At that moment we saw the red Mitsu approach from the west corner next to Camel's Humps. The vehicle came directly toward our new locality, named after Neil Young's "Sugar Mountain."

Mark stopped the car fifty yards below us. He and Luis got out and walked slowly toward us; they were smiling.

"We got one," Mark said.

Mark and Luis had discovered a *Mononykus* skull that glorious July 9 that was complete, exquisitely preserved, and attached to a skeleton. It further bolsters the case for the close affinities of *Mononykus* with other birds (though some critics remain unconvinced). At this point, the work on this priceless specimen is ongoing; this is simply an early and very incomplete report. Meanwhile these discoveries taught us something about our generous bone bed. Ukhaa was so rich that only a few rains and windstorms each year were necessary for us to pick up dozens of the best-preserved fossils in the Gobi Cretaceous. Come hell or *Allergorhai horhai,* we made a solemn vow to keep working this astounding site.

The Rat Patrol

Over the next several days a team worked at removing the very large oviraptorid that I had found; the skeleton was complete with skull. Perle and Batsuk also chose the best of their oviraptorid finds for excavation. We made Ukhaa our home and went off prospecting on a daily and fulfilling routine—finding exquisite fossils, excavating them, loading them up, socializing around the campfire. It is hard to describe the pleasure of working so productively with a group of friends enveloped in this spectacular rhythm. No effort seemed wasted, and our obvious success minimized any tensions that come with an impoverishment of fossils. We were grabbing up everything Ukhaa had to offer, and were having a lot of fun in the process.

Not every aspect of 1994 was a success story, however. I had a hunch that we might also do well locating fossils much farther south near the Mongolian-Chinese border. On the Chinese side of the border there were some red beds at a place called Bayan Mandahu that looked very much like the Djadokhta Formation at Flaming Cliffs. These were found by a Sino-Canadian team to be rich with dinosaurs—particularly protos and ankylosaurs—as well as some lizard and mammal skulls. Although aerial and LANDSAT images of this region were poor, there was sketchy information of some red rocks on the Mongolian side of the frontier. Dashzeveg himself endorsed this story, though he had never been there. He went through considerable trouble to obtain military clearance to work near this border. On July 13 five of us—an itinerant rat patrol—left the others at Ukhaa to strike out for the frontier.

We came back to our favorite site a few days later, exhausted and empty-handed. Despite miles of marvelous, empty country it had proved a wasted effort. We had found nary a Cretaceous red bed. The commander of the border station we camped near, an impressive young man of obvious intelligence, had pointed at a spot on our map, however, almost sixty miles directly east of the station.

"This, he says, is a valley with some red cliffs near a border station, Ulaan Ushu," Dashzeveg translated.

Our eyes widened. Ulaan Ushu means "red point."

"How far is it by road?" I asked.

"A hundred and eighty kilometers, but very bad road," Dash replied.

We were not up to 112 miles of bushwhacking, nor did we have enough gas to make it to Ulaan Ushu and back to our tanker at Ukhaa.

And so the frontier region of the central Gobi remains ripe for exploration, although one may need a squadron of helicopters to do so.

By the time our rat patrol returned, the team was wrapping up its work at Ukhaa. They had finished with the major excavation projects, including the crowned *Mononykus,* several extraordinary oviraptorids, the Sugar Mountain mammals, and incredible numbers of mammal and lizard skulls, even more than collected during 1993. We pulled out and settled down for a bit of recuperation at Naran Bulak, "Sun Spring."

The rains that spring had brought death and destruction to the Gobi. A great mudflow loosened by heavy spring rains had engulfed several houses and their families at the base of a mountain that flanked Gurvan Tes, the village in the rugged highlands some sixty miles southeast of Naran Bulak. The news was related by the town officials with muted emotion. Mongolians, by tradition, understate the pain of death or sickness. But I was sad to hear of a tragedy that had doubtless affected such a wide circle of this small, hearty community. Perle was more philosophical. "The Gobi is really not a place for people," he said.

As in prior years, from Naran Bulak we headed out to Kheerman Tsav, Khulsan, and Altan Ula. Not all these jaunts were as unsatisfying as our frontier folly. On July 25, Mark, Dashzeveg, and I struggled out to a small set of outcrops several miles and three very tough gullies southwest of the ramparts of Kheerman Tsav, new territory for us. Several feet from where we stopped the jeep I saw a number of eggs sticking out of the cliffs, looking like dull white sand pebbles. All in all, we found some twenty eggs scattered about. To our elation, just about every one of these eggs had the delicate bones of a chick, possibly the offspring of nesting groups of the Cretaceous bird *Gobipteryx.* At last, Luis had some of those precious little birds to study.

Throughout the rest of that day we crossed virgin badlands and found various tantalizing bits of bone—pieces of a *Mononykus, Velociraptor,* and probably that Komodo-like lizard *Estesia.* None of these fossils rivaled the wonders of Ukhaa Tolgod, but they came from new localities that would warrant more work in seasons to come. It seemed rather eerie that discovery was now coming so easy to us. Maybe we were getting to know the texture of the Gobi much better; maybe we were just lucky. Whatever the reason, we enjoyed ourselves immensely. Toward the end of that beautiful, nearly windless day we picked out a few small outcrops of red rock on a distant mountain, separated from us by a maze of canyonlands and sand dunes. More targets for future years? We had now learned not to treat such sightings so dismissively.

SCIENTIFIC SUPERLATIVES

In the late fall of 1994, after our return from the Gobi, the description of the oviraptorid embryo hit the cover of the international journal *Science,* the front page of the *New York Times,* and papers and television everywhere. A photo of Mark was printed, showing him clutching his precious little embryo on a half shell. A friend wrote and said he had seen Mark and me on TV in a bar in Katmandu. Another friend caught our spot in his hotel room in Bolivia.

I labored on a paper summarizing the discoveries at the site, trying to keep up with the daily intrigue and surprises that confronted us. A detailed report of our discoveries appeared in the prominent scientific journal *Nature* in the spring of 1995. We had originally called our paper "Unusual preservation in a new vertebrate assemblage from the Late Cretaceous of Mongolia." The only major criticism voiced by reviewers was that the word "unusual" in the title was too subdued, which I then replaced with "extraordinary."

The paper ended with the proclamation:

"Ukhaa Tolgod exceeds all other known Cretaceous localities in the abundance, concentration, diversity and cumulative quality of the fossil terrestrial vertebrate remains."

It is fashionable in science to avoid being labeled, even if unfairly, for hyperbole. But there was no denying, as the reviewers insisted, the extraordinary nature of our discovery. With respect to the variety of the beasts—not just dinosaurs but the whole suite of land vertebrates—as well as the quality of their preservation, we had found the richest locality from the late age of the dinosaurs, perhaps from the entire sweep of the Mesozoic. By the end of the 1994 season, just a few weeks over two summers at Ukhaa Tolgod had yielded 39 theropod skeletons. Seventy years of collecting at the Flaming Cliffs had produced only half a dozen theropods. Many years at Khulsan yielded perhaps less than ten skeletons. It was difficult to get exact figures because much of the data on the Russian collections were not available, but it was certain that these did not add big numbers to the samples. We had found roughly 260 lizard skulls, many of these attached to their skeletons. Moreover, the 187 mammal skulls from Ukhaa Tolgod matched or even exceeded the total number of mammal skulls collected over seven decades from *all* of the Gobi Cretaceous localities. Indeed, during the first two days of discovery of Ukhaa Tolgod in 1993, over a terrain about as large as the great lawn in Central Park, we had found more mammals than the Polish-Mongolian expedition had extracted over several seasons from the miles of sculpted badlands at Khulsan, up to that time probably the richest locality for high-quality Mesozoic mammals in the entire world.

But superlative samples aren't the only quest for science. Ukhaa Tolgod's magnificent assemblage wasn't simply a matter of big numbers. An ancient terrain swarming with colonies of tiny mammals and crowds of lizards, a congregation ground for scores of oviraptorids, some of which were obviously caring for their numerous nests, the evidence of other dinosaurs—protos and dromaeosaurs and newly recognized troodontids—all these wondrous things together suggested, to say the least, an unusual set of circumstances. Something indeed remarkable happened there some eighty million years ago. And just what happened was one of the most intriguing puzzles of all.

CHAPTER 10

DISASTERS, VICTIMS, AND SURVIVORS

DATE: LATE CRETACEOUS, ABOUT EIGHTY MILLION YEARS AGO

LOCALITY: AN OASIS IN THE DUNES, IN WHAT IS NOW UKHAA TOLGOD, EASTERN NEMEGT VALLEY, GOBI DESERT, MONGOLIA

The sand is their world. It sweeps in great rippling waves across the tawny oceans between the mountain ranges, trackless and mercilessly blank. Some dunes are hundreds of feet high, others look like enormous swells traveling over the desert plain. There is no relief from the sun; the landscape is waterless, and virtually lifeless. Yet there are telltale signs of living creatures. Delicate three-toed footprints form long sinuous trails over the sand dunes. At night a distant roar or a scream of terror sometimes travels over the dunes with the wind. Occasionally the small black specks of birds can be seen silhouetted against the sky. On the horizon, mushroom clouds build with moisture that must drop to the earth somewhere on this parched sand planet.

Ahead lies an oasis in the dunes, where a river has edged its way down a gentle flood plain, its borders studded with small trees. In other places trees are clumped in groves that are virtually surrounded by sand dunes. A large patchy

field of shrubs and thickets spreads from one end of the crypt to the other. Beyond the bushland stands a small shallow lake. The sky above the lake is peppered with birds. An enormous grouping of shield-headed Protoceratops *moves in a long column on the northern edge of the marsh. In the brush, small lizards and mammals dart here and there in a flash of movement. Rising dust from herds of ankylosaurs hidden by the thickets clouds the air of the oasis in places. From a high sand promontory at the edge of the oasis, two* Velociraptor *inspect the scene before them greedily. It will be a good hunting season.*

The raptors quickly move to the south in order to sweep around the graceful golden shore of the small lake. The path this way is unencumbered by tall reeds and the danger of camouflaged crocodiles. Ahead, high dunes set back an appreciable distance from the shoreline provide a good vantage point for launching an attack. Bent on reaching the high dunes, they miss discovering a spectacular city of egg nests on low cliffs at the water's edge on the northern side of the lake. Over one hundred of these brooding sites are all arranged in a neat circular mosaic.

At one nest located in a small depression at the base of a sand dune, an egg, newly cracked, reveals the grotesque head of a small beaked Oviraptor. *Suddenly, the large tongue of the lizard* Estesia *appears, its forked tip maneuvering skittishly over the skin of the newborn* Oviraptor. *As the tongue retracts, a set of razor teeth takes its place. The young hatchling is snatched up, crushed, and swallowed as the shadow of a cloud darkens the nest.*

The lizard begins to crack the other eggs with a simple thrust downward of its upper jaw. A full English breakfast proceeds, interrupted only by an occasional look about for intruders.

A thump and a honking noise bring an adult Oviraptor, *fully twelve feet long, with powerful legs appointed with a formidable set of claws and a jaw that could easily slice a lizard limb, to the rescue. The creature lowers its head and propels itself into the* Estesia, *kicking furiously with its hind limbs. The lizard backs away, snarling, moving in a slow circle with its head always facing the* Oviraptor. *With some struggle, the eight-foot-long lizard could hold its own. The claws of the* Oviraptor *are no match for the row of scimitars and the deadly venom that flows from those teeth. Yet the* Estesia *is at least partially satisfied with its hurried meal, and in no mood to fight for more. It slides away into the okra-colored grass.*

The parent Oviraptor *returns to the huge nest, squatting over the three-tiered tower of eggs with enough bulk to provide protection. Its hind limbs are deeply flexed at the knee; the long forelimbs extend around the clutch of eggs as if the parent were wrapping the nest in a protective embrace, remaining in this position for hours. Perhaps its mate will return with some food for both the brooder and the young hatchlings.*

Eventually the brooding parent grows restless; there is something in the air. The brilliant blue of the sky has been smudged by the faint haze of desert dust and tinged with the smell of rain. The west wind picks up. Other members of the colony react to the scent of the storm in the wind. A couple of large adults break out into a brawl over the ripped carcass of a stubby-limbed bird Mononykus, *kicking, slashing, and screaming. The realization that a storm is approaching is enough to agitate the colony.*

The starving raptors have given up the hunt on the south side of the lake. A herd of frill-headed Protoceratops *had no elderly or sick individuals that might have been fair game for the tired stalkers. Moving north toward the nesting site, they home in on the brooding parent near the edge of the colony, who only hours before had repelled the* Estesia.

But the attack is intercepted as its mate returns. With a body length of seventeen feet, and claws several inches in length, it is not intimidated by a couple of six-foot raptors. The raptors in their half-starved state are no match for the Oviraptor. *They launch a suicidal attack, scoring a few vicious cuts in the long neck of their target, but they too are slashed and torn. The raging kick of the defending* Oviraptor *leaves one* Velociraptor *with a yawning gash in the abdominal region, releasing an explosion of blood and entrails. Moments later, it is dead. The surviving* Velociraptor *is more dogged in the attack and scores a number of vicious cuts. But it will be the last fight for both attacker and defender. The beaked defender, the voracious raptor, the nestlings, the other members of the colony, the shield-headed beasts, and much of the life of this oasis are about to be decimated.*

The storm, when it comes, arrives from the west with an abrupt bolt of

A nesting *Oviraptor* confronts *Velociraptor* as a dust storm approaches

electricity and a crack of thunder that compresses the thick air. Great whirling columns of dust are kicked up in front of the black curtain of rain. The storm is awesome in its power, but it is only the early herald of a greater disaster. The winds sucked up as the rainstorm moves over the oasis take on a ferocity never before experienced by the dinosaurs. For a moment it looks as if the storm has passed and the danger is over. A brilliant stabbing light emanates from the western sky, an afternoon sun attempting to cut through the dust-choked atmosphere. But this stillness ends as a great wall of dark sand fills the western sky and approaches with unprecedented fury. The sand is several hundred feet high, too solid for air and cloud, too vaporous for the earth. Sky and earth have now become one. As the sand wall converges on the lake and the colony, it throws up geysers of dust, as if the earth itself had just been shattered by the impact of a huge projectile. The wind and the stinging sand are relentless; the light of the day is blocked. There is no air to breathe. No place to escape. The dark, moving wall kills everything in its path. The Oviraptor *do not move from their nests.*

The storm persists, fed by the very winds it created. Dunes near the lake shore are remolded with extraordinary speed. Avalanches from the dunes combine with the suffocating wall of sand to convert the oasis to an aeolian hell. It is hours later before the winds abate slightly and the sky begins to materialize through cracks in the wall of dust. Now the outlines of a towering white cloud—not one of dust but of water vapor—grows in the western sky. A faint cry of a bird issues from some remote stratosphere. But there is no life below. The colony of life at the oasis is a vast sand cemetery. In some places, the outlines of Oviraptor *can be seen, their corpses covered with a thick coating of sand. In other places, particularly near the sand dunes, the animals are buried in deep piles of sand caused by the avalanches from the dunes. The sand has no luminescence; it is the ugly, brown, and non-reflective earth of a burial ground.*

Among the corpses can be seen an Oviraptor *in a strange contorted grapple with a smaller sharp-toothed* Velociraptor. *The maimed body of another* Velociraptor *lies nearby. Not far from this trio is the body of yet another* Oviraptor *squatting, head down, its front limbs wrapped in a powerful embrace around a clutch of sand-covered eggs.*

ARID LANDS

The carnage described above is one scenario for what happened at an oasis in the dunes some 80 million years ago in the vast sand deserts of Central Asia. We can reconstruct the rigors and dangers of an ancient desert based on our present-day knowledge of these arid lands. Deserts today make up only about five percent of the world's land surface. To be in one, though, is to be lost in a universe of infinite space and emptiness. Deserts—especially deserts like the Sahara or the Gobi—inspire stereotyped images, a great range of sand dunes, camels, the silhouette of a date palm in the sunset, a rider on a white Arabian stallion swooping in on a lithe maiden with an exposed jewel-encrusted navel and a veiled face. But in reality deserts are extraordinarily varied in conditions and form. So how do we define them?

There is a formal meteorological definition of a desert: any place on earth that receives less than ten inches of rain a year. But this definition isn't very useful; it lumps together a lot of varied terrain. Meteorologically speaking, Antarctica is a desert; so, during some parched years, is my hometown of Los Angeles. Moreover, some deserts hardly have much sand at all—only about two percent in North American deserts, compared to ten percent in the Sahara and an impressive thirty percent in the Arabian Desert. Some deserts, like the Gobi today, are strikingly heterogeneous. Temperatures here range from 113° F. in the summer to –40° F. in the winter. Varied landforms and seasonal conditions leave the Gobi with a broad spectrum of habitats. There are great sand seas in the Gobi, but there are also large shallow lakes, snow-capped mountains, and scraggly bush country. The Gobi is also comparatively lush. This 450,000-square-mile parcel of property receives on average between three and eight inches of rainfall. By contrast, deserts like the Atacama of northern Chile may receive only traces of rain, and rarely more than three inches annually. The contrast can be intimidating. In 1988, I worked in Yemen on the Arabian Peninsula with a colleague, Ian Tattersall, from the American Museum. Our search for any kind of decent vertebrate fossil—dinosaurs from the

Jurassic or fossil mammals, hopefully including hominids, from the Cenozoic—took us over miles of terrain. To the east, Yemen borders the great Rub' al Khali, the "Empty Quarter" of the Arabian Desert. We could not navigate this sand ocean without the help of drivers from an American oil company. The dunes seemingly expanded to infinity to the east and soon closed in behind us. The dust-laden sky merged with the color of the dunes and obliterated the horizon. It was a hostile, disorienting place.

Sand, where it does occur, is a dynamic element of the desert. Dune fields can be localized or expansive. Some sand seas or ergs, like those in the Sahara or the Rub' al Khali, are as big as France or Texas. Others, like the isolated but magnificent dunes next to our camp at the Gobi locality of Tugrugeen Shireh, are no larger than a baseball field. Relief is comparably variable. Some dunes are truly mountains, reaching more than six hundred feet in height, but many sandy areas in deserts are actually called sand sheets, where relief is minimal. The migration, expansion, or contraction of sand of course depends on the forces of wind. Dunes are aligned in waves that correspond to the prevailing vector of winds, just like lines of swells on the oceans.

Dune formation is complex, but the processes shaping dunes on hard desert pavement share some basic qualities. Most of the sand accumulating to form the dune moves as a mass of jumping (saltating) grains. Some of the sand accumulates by creeping more slowly along the surface. Bouncing grains move readily on hard pebbly surfaces but slow down or come to rest in hollows or softer ground, where a small embryonic dune, a patch of sand, will form. The accumulation of sand in such a spot is contagious. The sand itself creates surface friction that impedes the velocity of the wind, and more grains drop on the sand patch. This process continues until the sand dune reaches a certain critical height. As the dune grows, the smooth leeward side steepens. But at a certain angle the wind cannot be deflected down sharply enough to track the profile of the slope; it shears off the crest of the dune, creating a dead zone in which the sand simply falls from the windward side. When the angle of slope of the lee side of the dune steepens to 32 degrees, this angle, called the *angle of repose,* is maintained. Sand is lost at the bottom of the slip face as it is added at the top. The dune is then stabilized; it can move as a whole because sand eroded from the windward side is simply deposited on the lee.

Dune migration can be spectacular. In Peru there are records of dunes moving a hundred feet per year. Our isolated dune at Tugrugeen Shireh, first visited in 1991, has since covered some choice latrine spots on its far, lee side.

Animals and plants flourish among the dunes of desert regions because there are places of relief—oases—that effectively sustain life. Oases, like the deserts encompassing them, vary greatly in size and character. Many are no larger than a hectare (2.5 acres); others stretch for hundreds of square miles, fed by the natural irrigation of a river or water percolating from artesian springs. Oases are often found in depressions scooped out of the sand and scree surface of a desert by winds—a process called deflation. In the Sahara, deflation has produced verdant depressions, the largest of which have an area some 50 miles by 20 miles, with depths of 330–525 feet. The floors of many of these pockets lie below sea level, and the depressions often have large sand sheets and dune fields on the downwind side. Accumulated rainfall with poor drainage, or water rising to the surface of these stripped depressions, forms shallow playas, highly alkaline lakes. But not all oases conform to this pattern. In a sense, an oasis is properly defined by its poetic meaning—any slice of comparative paradise offering sustenance and protection from the bleakness all around. Functionally speaking, the Nile River is perhaps the world's greatest oasis, and is certainly its most extended one. A good portion of its winding 4,160 miles dissects the blazing Sahara, providing a finger hold for life and culture.

Oases are not, however, permanent fixtures. They may emerge miraculously from several years of heavy rain or the steady rise of the water table. During some of our summers in Mongolia—like the wet 1994 season—the comparatively heavy rains gave the Gobi too many oases. The flat playa lakes expanded and the mud and water ruined some of our favorite routes. Worse yet for us, the rains at some favorite localities created a veneer of vegetation that frustrated our prospecting of the ground surface. At the other end of the extreme, oases can shrivel up and disappear with cumulative periods of drought, taking their vital gifts with them. The effect on the biota, including humans, can be disastrous. Deserts that have formed even in the last few thousand years have buried well-irrigated kingdoms of animal and plant life. Droughts have, of course, disastrous effects on a shorter time scale. Recent famine, precipitated by drought, as well as political and economic problems, in arid Somalia killed 120,000 people. It is estimated that starvation during a horrific drought in India between 1876 and 1878 killed more than 5 million. In northern China, including parts of the Gobi, between 1876 and 1879 the drought was even more devastating. Estimated deaths range between 9 million and 13 million. Slavery, the sale of children, and cannibalism have been documented during these times of massive decimation.

What was the Gobi like during "Djadokhta times" eighty million years ago? It is not certain that the late Cretaceous Gobi meets our formal definition of a desert. If it approached desert conditions, it was certainly not a Rub' al Khali. Reconstructions by Russian geologists like Shuvalov (published in 1982) minimized the influence of the dune environments, suggesting that the Gobi during "Djadokhta times" was well irrigated by extensive lake systems. More recent analyses by Jerzykiewicz and Eberth, however, as well as work of our team geologist Lowell Dingus, opt instead for a habitat of extensive dunes, sprinkled with small intermittent streams, restricted ponds, and shallow lakes. We see this situation today when the rains raise ponds and lakes near the sand hills of western Nebraska, or the Great Sand Dunes of Michigan. In the Cretaceous rock sequence of the Gobi, cycles of rising and falling water levels are documented by clays and shales that alternate with calcite-dominated rocks—caliches—indicating a leaching out of minerals in the soil horizon as it loses water. It is possible, Jerzykiewicz and Eberth argue, that the Barun Goyot beds indicate a somewhat more extensive lake system than the Djadokhta facies, but the evidence for this is not clear-cut. It is at this time safe to say that fossils from both rock sequences represent the stubborn resistance of life to the rigors of aridity, sand, and wind—animals that lived and died at or near oases in the dunes.

DESERT STORM

Rigors of the Gobi, both present and past, doubtless include desert storms. To humans, even those accustomed to the desert, such a storm can be a frightening thing. In the memoirs of the C.A.E., Andrews related some harrowing encounters:

> In the gray light of dawn we could see an ominous bronze cloud hanging over the rim of the basin to the south. Evidently, there was more of the storm to come, but, since it might miss us, we decided not to wake the camp. Ten minutes later the air shook with a roar louder than the first, and the gale struck like the burst of a high-explosive shell. Even with my head covered I heard the crash and rip of falling tents. It was impossible to see, but I felt for Granger with one foot. He was lying across a green suitcase, his face protected by a shirt. As our tent swept away, he had leaped to save the box that contained six tiny fossil Cretaceous mammal skulls, the most precious treasures of all our collections. For fifteen minutes we could only lie and take it. While I was feeling for Granger, the sleeping bag had been torn from under me and my pajamas stripped off. The sand and the gravel lashed my back until it bled, and poor Granger on the mammal skulls fared no better.
>
> Suddenly the gale ceased, leaving a flat calm. The camp was completely wrecked. (*The New Conquest* . . . , p. 308.)

Although we have not experienced extensive destruction from one of these miserable storms, we have had a taste of their fury. At the end of June during the 1991 season, our caravan was heading out of the town of Dalan Dzadgad toward the Flaming Cliffs. A storm seemed to be taking form somewhere in the west, in the direction of travel. It appeared remote and intangible enough not to stop us. Besides, Mark and I were anxious to reach our destination; we were carrying an evil-smelling and dangerous 50-gallon barrel of gasoline in the back of the red Mitsu. Unfortunately, like Andrews, we underestimated the velocity and power of bad weather in the Gobi. In front of us, slightly to the left, we could see a row of white gers, the Gobi tourist camp, less than a mile away. Suddenly the gers disappeared. The storm issued its first volley, pulverizing the glass surface of our windshield to a milky haze. The sand came into the cab of our jeep like a hideous vapor, through the door jambs and the wind wings, up our nostrils, in our eyes. We swathed our heads in bandannas. I could feel the slow panic of suffocation. Could we die in something like this? We had to stop the car and turn its rear end toward the onslaught of the horizontal wind. As we rotated, the wind hit us broadside; the car seemed ready to launch and roll like a tumbleweed. I envisioned the end of an expedition; an incinerated Mitsu exploding with its gasoline cargo, the boom of an explosion in the dusty darkness, terrifying the people in the gers nearby.

Finally, we resumed a slow crawl forward. There were no gers in sight. It was a raging Martian landscape outside the windows of the car. After an hour of this groping in the blackness of the storm, we suddenly saw a chain-link fence. It was the only fence in the Gobi, the one meant to keep the tourists in, or offer them rather tenuous protection from the wilderness all around. We practically collided with it in the blinding whirlwind. On the other side, not much closer than we had seen them an hour ago, were the gers. Mark and I decided to change plans. We paid for a stay in sturdy gers, while the storm raged throughout the night.

Sand and wind combine in deserts to create their own mix of sound and fury. The power of this combination has been well documented. Wind has carried dust over 2,000 miles from North Africa to Germany. Likewise, storms generated on the sizzling outback of Australia produce hammer-headed dust clouds that drift to New Zealand. Individual storms have transported as much as 100 million tons of dust and sand. Late on a June day in 1988 we reached the 12,000-foot summit of the Yemen highlands. Our perch provided a spectacular view of the craggy peaks and bumpy foothills flowing like a bed quilt toward the Red Sea. Beyond, there was a haze of yellow-brown that blocked the light of the sun.

"Storm, over Khartoum, in the Sahara," said Abdul.

I shivered at the reality of a dusty hell, making itself known some six hundred miles from where we sat in green meadows and ate fresh bananas.

PALEO-AUTOPSY

The scenario offered in the opening of this chapter, like others offered in this book, and for that matter numerous other books about prehistoric life, is primarily a product of imagination. On the other hand, there are clues left by the rocks and the bones, undeniable signs, that compel us to at least consider the possibility that wind and sand had a profound impact on the life, death, and certainly the entombment of these Cretaceous creatures.

The first clue resides in the spectacular preservation of fossils in the Cretaceous "red sands" of the Gobi. We think that the superb state of interment relates to the rather peculiar degree of aridity represented by a habitat where protos grazed and raptors stalked. Although this Cretaceous habitat was irrigated by small streams and ephemeral lakes, it was also doubtless appointed with rugged canyonlands and dune fields. Deserts are as ancient as the continents themselves, so the geologic evidence for an ancient arid land is not so remarkable. But the overwhelming knowledge of the dinosaur world outside Central Asia comes from much wetter environments. The dinosaur-bearing rocks of Wyoming and Montana represent swampy inlets and lush forests near vast inland seas. Similar wetlands are indicated in the dinosaur beds of Argentina, the rich coal deposits of western Europe, even the stream and river deposits of Antarctica and northern Alaska. By contrast, the Gobi's aridity, as represented by the red beds, is striking. Within these units are features of sedimentation, long curved lines called cross-beds that indicate the buildup and shifts of ancient sand dunes. Here are also featureless thick layers of red-brown or orange sand—like sand piles flattened by a child's beach shovel—that look like the products of violent and possibly destructive sandstorms.

Why should the arid conditions of the Gobi relate to the uniquely excellent preservation of the Gobi fossils? The whole area of inquiry relating to the question of death and burial and preservation of organisms that become fossils is called *taphonomy* (from the Greek *taphos*, meaning "burial"). I. A. Efremov, the commander of the Russian paleontological assault on the Gobi in the 1940s, was fascinated with this kind of detective work; he actually coined the term for the science. Taphonomically speaking, it seems that the creatures entombed in the Gobi red beds may have been covered rapidly in sand. Sealed off from scavengers, safe from further surface winds or flooding streams, the skeletons remained buried, inviolate, and beautifully articulated. There they lay in state for 80 million years until the wind scoured a sandstone hill enough to reveal a nubbin of bone—a tip of a skeletal iceberg—that may catch the eye of a prowling paleontologist.

Is this combination of moving sand and dinosaur graveyards merely fortuitous? The question has been pondered since the American Museum teams first visited the Flaming Cliffs seven decades ago. Remember what Osborn said about the discovery of the ostrichlike dinosaur *Oviraptor* on top of the "proto eggs" at Flaming Cliffs: "This immediately put the animal under suspicion of having been taken over by a sandstorm in the very act of robbing the dinosaur egg nest." Perhaps the American team was unduly influenced by their own familiarity with the terrific force of desert storms. Nonetheless, it is reasonable to at least consider the possibility that the sands so exquisitely preserving these creatures also brought death to them. Many subsequent explorers of the Gobi followed the intuitions of the 1920s expedition. When our team set out for the Gobi, we put these various arguments aside. "Let's find something interesting first and then worry about how it got there," we would have said. As the seasons rolled on and the discoveries accumulated, however, we were drawn back to these questions. We became more interested in Efremov's favorite science of taphonomy. And the wonders of Ukhaa Tolgod made us all instant taphonomists.

Actually, the investigation of the Gobi morgue has profited from important work by the Sino-Canadian team of paleontologists working in the Gobi of northern China. In 1993 a committee of authors, T. Jerzykiewicz, P. J. Currie, D. A. Eberth, P. A. Johnston, E. H. Koster, and J-J Zheng published an important paper interpreting the geology and taphonomy of the rich Cretaceous Bayan Mandahu locality. This impressive set of badlands consists of the same kind of red sandstones one finds at Flaming Cliffs, Khulsan, and Ukhaa Tolgod. In fact, Jerzykiewicz and coauthors made the convincing argument that both the rocks and the vertebrates they entombed were simply another example of the classic Djadokhta-like strata. Bayan Mandahu is embellished with protos, ankylosaurs, eggs, theropods, and various mammals and lizards that one would expect to find at Ukhaa Tolgod or Flaming Cliffs. In their observations of the fossils from this locality, the authors concluded that these animals were buried in rapidly moving sand soon after or even at the time of death. Moreover, they proposed that many individuals may have died trying to free themselves in a sandstorm. Several interesting clues were cited.

First, they observed that a large portion of the skeletons were preserved in a sandstone with a smooth unbroken texture and virtually no layering or bedding. This kind of rock is typically associated with deposits of massive sand movement, such as one might encounter in sandstorms.

Second, many of the dinosaurs, especially *Protoceratops,* seem to be in a pose that indicated a struggle for life. The protos were standing on their hind limbs with their snouts pointed upward and forelimbs tucked by their sides. Similarly, the large piles of ankylosaurs in the locality were positioned belly down, head up, limbs down. Jerzykiewicz and coauthors argued that it was highly unlikely that these creatures were simply burrowing dinosaurs at some point covered by shifting sand. Overwhelmingly, land burrowers dig their tunnels in a horizontal position with their heads down. It is highly inconvenient, nay, lethal, to have a snout up attempting to breathe in solid sand. Indeed, the dinosaur skeletons looked as if they may have been desperately "swimming" to the surface for air, much like skiers caught in an avalanche. Also noted were the famous "fighting dinosaurs" found by the Polish-Mongolian expedition at Tugrugeen, which looked for all the world like a proto and *Velociraptor* locked in mortal combat suddenly engulfed in a wall of sand.

Third, many of the skeletons found close together belonged to a single species, either *Protoceratops andrewsi* or an ankylosaur, such as *Pinacosaurus grangeri.* The cumbersome scientific term for this gathering is a "monospecific death assemblage." Such mass pileups have long been cited by taphonomists as an indication of catastrophe, a case of a disaster overtaking a crowd of social creatures belonging to the same species. Jerzykiewicz and coauthors noted that the aggregations of several *Protoceratops* and, in other places, juvenile *Pinacosaurus* were evidence of decimation at Bayan Mandahu. In one case, five medium-sized *Protoceratops* were found lying parallel to one another, oriented about 21 degrees to the horizontal, their bellies down, their heads upslope. These were probably buried by sandstorms on the lee side of dunes. Jerzykiewicz and colleagues also cited observations made by me and other paleontologists that localities like Tugrugeen had such mass death assemblages of *Protoceratops.*

Fourth, the skeletons of these dinosaurs are nearly complete, well articulated, and rarely show evidence of nicks or scars. This suggests that the animals were rapidly buried, essentially not subject to scavenging, and not broken up by disruptive forces, like moving water. The skeletons were simply buried where the animals died, not downstream from their place of death. Although Jerzykiewicz et al. did observe signs of scavenging on some juvenile *Pinacosaurus* skeletons (based on loose teeth near the skeletons, the scavengers were thought to be *Velociraptor*), such evidence was very rare. Most of the remains were smooth-surfaced and undisturbed.

This autopsy for the dinosaurs at Bayan Mandahu was published before our discovery of Ukhaa Tolgod. So when we began to find such an embarrassment of riches at our site we thought we had the makings of a good test case for the sand-death theory issued by Jerzykiewicz et al. Truthfully, we were a little skeptical of the idea. Lowell Dingus had seen some appreciable evidence of stream deposits, even some high-energy flooding, in the sequence of rocks at Ukhaa Tolgod. And some of these fluvial deposits had good fossils. As our field observations and fossil discoveries accumulated over the season we were admittedly led back to the idea that raging sandstorms were the *bête noire* for the critters of Ukhaa Tolgod. If anything, the site seemed to be an even more extreme example of a desert catastrophe. We had found some spectacular mass death assemblages—the crowd from Ankylosaur Flats, the plethora of protos from First Strike, the great swarm of oviraptorids in the region of the Camel's Humps, even the abundant oviraptorid nests from Xanadu. Although there were fossils from sandstones sandwiched by water-lain (fluvial) rocks, these samples were not nearly as rich as the samples from either sand dune facies (aeolian sandstones) or unstructured sands. Moreover, the "wet" part of the sequence was rich in mammals but extremely poor in larger skeletons, while the sand facies were rich in everything—dinosaurs, eggs, mammals, and lizards. Finally, the bone was extraordinarily pristine; those beautifully articulated skeletons showed little evidence of surface torment, including scavenging.

The strikingly repetitive nature of this carnage—from Tugrugeen, to Ukhaa Tolgod, to Bayan Mandahu, and many other sites—raises a very intriguing question. Was the storm portrayed in the opening of this chapter one massive disaster that wiped out all these creatures from various far-flung parts of the Gobi at a single moment in time? Our plotting of assemblages and rock sections at Ukhaa Tolgod gives us cause to decisively discard this idea. Not all the death assemblages occur at the same level in the rock sequence. In the section of rock above Ankylosaur Flats and Camel's Humps, a number of these mass concentrations of skeletons occur at higher and higher levels. Whatever catastrophe befell the denizens of Ukhaa Tolgod, it did not occur only once. Presumably, the same applies to Tugrugeen, Flaming Cliffs, Bayan Mandahu, and other localities. Repeated disasters seem to have been part of the harsh environment in which the Gobi creatures lived, died, and evolved.

How much time separates these events of mass death? We don't know. Maybe if Carl Swisher extracts some good paleomagnetic signals in the rock, we might be able to at least get a crude estimate of the passage of time. At this point, we can't say whether the events of death and entombment were a day, a week, a year, a millennium, or an age apart.

That is not the only element of uncertainty. Why is Ukhaa Tolgod even more richly concentrated in skeletons than other well-known sites? Earlier I made the point that many enticing red beds of the Cretaceous Gobi hardly have fossils at all. I speculated that the handful of rich localities represented widely separated refuges, oases, surrounded by seas of sand. In addition, any number of causes—drought, disease, flood, as well as sandstorm—could have obliterated concentrated populations of Cretaceous animals and plants. But even stretching the picture this far does not explain why Ukhaa Tolgod has such a distinctive mélange of fossils. Eggs are found in abundance at Flaming Cliffs, but the oviraptorids which laid those eggs were hardly around. The poignant picture of oviraptorids cov-

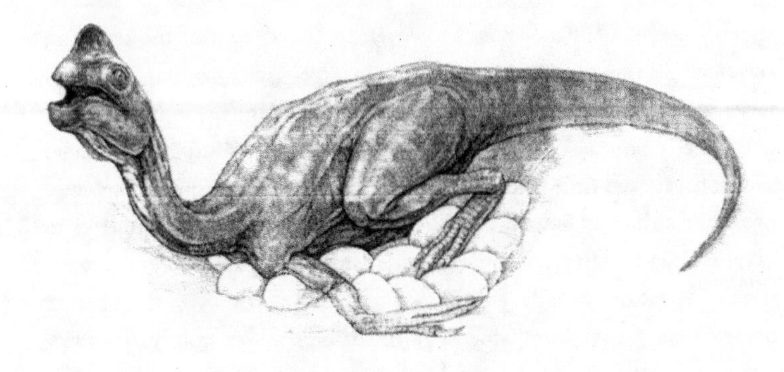

Oviraptor sitting on nest *(Mick Ellison, reprinted with permission, AMNH)*

ering their nests or hanging nearby the nests is illustrated by only one specimen from Flaming Cliffs—the remarkable 1923 find. By contrast, Ukhaa Tolgod is a mass of nests and oviraptorids; there may actually be hundreds of either there. And, as noted in the preceding chapter, there is even an example of a brooding parent on a nest. Clearly, this may have been a perpetual brooding ground for two or more taxa of oviraptorids, much like penguin colonies on Antarctica comprising several species. But why do we get so many of those beaked creatures at this particular spot? We don't know the answer to that one yet, although we're working on it.

Whatever the explanation, Ukhaa Tolgod casts a stunning new light on the Late Cretaceous world of the Gobi. The carnage here has a special character, and an illuminating one. Fortunately for us, the death and burial at this spot gave us embryos and huge masses of mammals and "colonies" of oviraptorids not found elsewhere. As for the poetic image of a good parent, that of an oviraptorid refusing to relinquish its nests in a raging storm, consider the example set by a living oviraptorid relative. Emperor penguins (*Aptenodytes forsteri*) plant their brooding colonies on the Antarctic continent as much as a hundred miles from the sea. The females lay their eggs and return to the coast before the onset of winter. The males, however, will stay to incubate the eggs. And there they stay for sixty-four days, despite the stress and mortality to embryos and adults that may come with the most ferocious winds on earth.

THE CRETACEOUS DISASTER

Natural phenomena involving lethal clouds or atmospheric upheaval have been implicated in more far-ranging catastrophes of the past than dust storms in the Cretaceous Gobi. Perhaps the most famous of these examples concerns a theory for dinosaur extinction itself. First off, it's important to discriminate what I've just been discussing—namely, mass death events at a particular time and place—from mass extinction events in which whole species or a great spectrum of species are completely eradicated. A desert storm may be a local disaster, but it certainly does not ensure the loss of species that lived over a broad range of the region or the continent.

On a bigger scale, however, such natural phenomena might be tied to mass extinction. And there is just about no bigger event of this nature than the kind of atmospheric turmoil that might come with the impact of an alien asteroid on the surface of the earth. Meteorites frequently fall to the earth, but most of these are golf-ball-sized objects that simply plant themselves in small depressions behind a desert sand dune or on the ice shield of Antarctica, where no one is there to witness the event. (There is currently an international expedition to search for meteorites in the Gobi Desert.) Sometimes meteorites, even slightly bigger ones, do encounter the earth in more populated areas. They go through the roofs of houses or the trunks of cars. The history of the earth shows us that much more destructive impacts are possible, as well. Meteor Crater in Arizona is 4,000 feet in diameter, 600 feet deep, and has a surrounding rim over 200 feet above the barren plain. It was apparently caused by an object that crashed into the earth anywhere between 5,000 and 50,000 years ago. But the devastation of the crash was not just confined to this crater. Large numbers of nickel-iron fragments, from gravel size to chunks weighing 1,400 pounds, have been found in a hundred-square-mile area. In addition, several thousand tons of nickel-iron droplets around the crater were condensed from a cloud of metallic vapors.

Even this spectacular crater does not represent the acme of extraterrestrial impact. Much larger craters, some as much as three miles in diameter, have been detected. These are usually camouflaged; many are very ancient, or have a very complex structure, or are under the sea. A particularly large crater seems to be somewhere between eastern Mexico and western Cuba. It was caused by a huge chunk from space. Doubtless it sent out an enormous cloud of noxious metallic vapors, a cloud possibly monstrous enough to enshroud the earth. Doubtless, it changed world weather patterns for years, perhaps even for decades. It is hard to imagine that this impact had no effects on a broad swath of animal and plant life in the seas and on land. Interestingly enough, the date of this Caribbean collision seems to be rather close to sixty-five million years before present, the critical date for the Cretaceous mass extinction event. Two members of our Gobi team, Lowell Dingus and Carl Swisher, have in fact been part of the research group responsible for studying and dating this event. The impact also corresponds with a worldwide surge—a spike—of a rare element, iridium, in rocks of the uppermost Cretaceous. Iridium is normally very rare in the crust of the earth, but it does make a much stronger showing in association with extraterrestrial impacts.

The theory that the (non-avian) dinosaurs and many other creatures were extinguished by the earth's collision with an asteroid is now more than a decade old. In recent years, the evidence for a coincident collision has accumulated. As noted, the dates for the collision have been refined, and the worldwide nature of this traumatic event has been further documented. There now seems less controversy that such a massive event did occur at the end of the Cretaceous, although a few detractors hold out.

Does that mean the long-standing mystery—what killed the dinosaurs?—has been solved? Well, not entirely. Even if such an impact did occur at sixty-five million years before present (give or take a half million), we cannot be sure it had the global impact ascribed to it. There is some evidence that a drastic decline in non-avian dinosaurs may have occurred well before the end of the Cretaceous. Moreover, this devastation was neither so overwhelming nor so rampant as the extermination at the end of the Permian. True, many families of mammals, birds, fishes, ammonites, belemnites, and bivalves (clams, oysters, and other dual-shelled species) went extinct. Over fifty percent of various marine planktonic groups were also erased. But numerous important lineages—many of the frogs, salamanders, turtles, lizards, mammals, crocodiles, birds, fishes, angiosperms, conifers, arthropods (insects, spiders, crabs, and sundry), gastropods (snails and kin), echinoderms, and nearly half the plankton species—went right on through the Cretaceous boundary. One of the real mysteries of Cretaceous extinction is therefore its taxonomic selectivity.

Why were the non-avian dinosaurs so persecuted? And if such an event was caused, as currently argued by many scientists, by a cataclysmic impact of a giant asteroid, why was this pattern of extinction so discriminating? Why did some dinosaurs, feathery, flying creatures of high metabolism that surely would not do well in clouds of metallic vapors, survive the event? Why did many other vertebrates, animals, plants, and marine organisms make it? The fact that this extinction event, like the Permo-Triassic extinction and other events before it, was selective really complicates the theory. It forces us to consider the subtleties of cause and effect relating to the survival of biological systems. Namely, we need to know how exactly such an event *selectively* snuffed out Cretaceous species. These are subtle connections for which we have few insights.

This is a shame. The asteroid impact theory has abiding superiority. It is the only explanation for Cretaceous extinction for which we have tangible evidence of a truly extraordinary event. We can say that the rock record supports the fact that a terrific impact did occur 65 million years ago. Some competing theories, those passionately defended by a variety of scientists, cannot be bolstered in this way. Problematically, all these theories make sense. Any number of things could have happened. Noxious clouds could have accumulated from a great worldwide pulse of volcanic eruptions. The drop in global temperature could have had ill effects on some of the fauna and flora. Extraordinary diseases could have swept through populations and even species. Changing sea levels and continental boundaries could have traumatized species adapted to a specific set of environmental conditions. Unfortunately, we don't have strong evidence that these phenomena were any more extreme at the end of the Cretaceous than at certain other times when mass extinction did not occur. In the case of some of these causes, such as widespread disease, we don't have any direct evidence at all. Finally, there is no way to establish *how* any of these proposed events *caused* extinction. The fossil record can tell us many things, just not everything.

THE AGE OF MAMMALS

Whatever its specific causes, the mass extinction event at the end of the Cretaceous made a gaping craterlike wound in the earth's biota. As I've noted, however, there were survivors. Birds may be the only dinosaurs to last out the Mesozoic but they indeed had company. As our stupendous samples of skulls at Ukhaa Tolgod showed, the Mesozoic world was writhing with lizards and mammals and other vermin that made it through the Cretaceous extinction event. The forests, fields, and deserts of the Mesozoic were teeming with these less conspicuous creatures. Yet the Mesozoic is famous as the "age of the dinosaurs." How odd this seems from a purely egocentric perspective. The Mesozoic was also the age of origin of our own branch of the tree of life, the mammals. One wonders whether, if dinosaurs were not so big and so grotesque, what we call the age of mammals—currently equated with the Cenozoic beginning "only" 65 million years ago—would be extended back to the early Mesozoic, when the first mammals appeared some 200 million years ago.

The earliest mammal-like forms, tiny shrewlike triconodonts, appeared some 200 million years ago during the Triassic Period, when prosauropods and iguanodontids were roaming the earth. The triconodonts were augmented by a number of other mammal-like clades during the succeeding Jurassic and Cretaceous periods. Many of these Mesozoic "experiments" waned and died out before or at the end of the Cretaceous. Several lineages, however, survived into the Cenozoic Era, a time interval popularly designated the "age of mammals." At least some of these survivors diversified into the modern mammals—the great range of kangaroos, koalas, primates, bats, whales, elephants, and aardvarks—that thrive today.

Mammal assemblages of the Mesozoic thus reveal a biological empire in transition. On one hand are forms like triconodonts that represent branches that terminated many millions of years ago. On the other hand are the precursors of the modern mammals, in some cases living nose to nose (or fang to claw) with the more archaic types. The Cretaceous Gobi preserves a pastiche of mammal and mammal-like species. This assemblage occurs too late to contain triconodonts (these are actually well represented in Early Cretaceous beds about a hundred miles northwest of the Flaming Cliffs), but it does produce some of the best-known specimens of archaic multituberculates, the first fossil mammal found by the C.A.E. at Flaming Cliffs. With their elongate, gnawing incisors, bladelike, nutcracking premolars, and broad, many-cusped molars, "multis"—as they are known in the paleontologists' parlance—were clearly filling the adaptive roles occupied later by modern rodents, rats, mice, chipmunks, and sundry. Multis thrived in the Mesozoic and even persisted in respectable numbers through the first 15 million years of the Cenozoic. The fall and eventual end of the multis show an interesting coincidence in time with the radiation of early rodents, the prototypes of mice and squirrels that undoubtedly represented their main competitors. The age of mammals is not without its own revolutions.

Inauspicious Antecedents

Multis represent the most abundant skulls and skeletons of mammals in the red beds of the Barun Goyot and Djadokhta. But there are other mammals there as well. The C.A.E., as well as the fiercely persistent "crawl team" under Zofia Kielan-Jaworowska, turned up the precious remains of mammals representing the roots of our own heritage. Our primary targets for Cretaceous mammals of the Gobi are these more "modern" mammals, like *Zalambdalestes* found at Tugrugeen and various sites, and the small "family" from Sugar Mountain.

There are three major clades of living mammals—the egg-laying monotremes, the marsupials, and the placentals. Monotremes have a very poor fossil record, with recent finds at last putting them back where they should be—in the Cretaceous of Australia and southern South America. (Remember that the two continents were broadly connected through a hospitable Antarctica during the Cretaceous.) Another of these branches, the marsupials, are diverse today in Australia and New Guinea but once flourished with great diversity in South America. Persistence of marsupials elsewhere, in North America for example, is today represented only by the opossum. But several kinds of opossum-like marsupials gathered in North America and the then connected western Europe during the Cretaceous and Early Cenozoic. In fact, the earliest record of the group—again a matter of isolated teeth or jaws with teeth—come from the Early Cretaceous of Utah. Records of early marsupials in Asia are sketchier, but there are now finds of bits and pieces of jaws and partial skulls emerging from Eocene strata in China. Under these circumstances, the possibility of a Late Cretaceous marsupial from the Gobi is very intriguing. This connection involves the form *Deltatheridium* first collected and described by the C.A.E. This rat-sized animal has triangular-shaped molar teeth very much like the living opossum. It also has an elongated robust upper canine, one good for biting on other creatures, including small mammals.

Originally, *Deltatheridium* was regarded simply as a rather primitive creature that anticipated the split between marsupials and the other modern group of mammals, the placentals. Zofia and her coauthor Percy Butler noted recently, however, that *Deltatheridium* indeed does have a very marsupial-like dentition. The allocation has stuck, at least for the last few years. *Deltatheridium* and its close relatives, genera like *Deltatheroides*, are now known from a few excellent specimens, comprising nearly complete skulls and partial skeletons. In addition to C.A.E., the Polish-Mongolian team as well as the Russians have had success in retrieving these specimens from various localities. We have desperately searched for *Deltatheridium* where the others had success—Flaming Cliffs, Khulsan, Kheerman Tsav, and an isolated, roasting gully at the bottom of a valley northwest of Altan Ula known as Ghurlin Tsav. We had no luck, though, at least through the 1994 season.

Some other recent discoveries of archaic Mesozoic mammals from Asia should be mentioned. These are again mostly teeth and jaws—but in one unusual case a nearly complete skull—of a variety of mammals from the deserts of Kazakhstan and Uzebekistan. The person behind these discoveries is Lev Nessov, a paleontologist from Saint Petersburg, who tragically died in 1995. Nessov conducted these explorations on a shoestring budget, taking third-class trains and actually walking from the stations to the field sites, he and his small crew taking their loads on their backs. It is a credit to Lev's energy and persistence that such important discoveries were made in spite of the difficulties of working in the eroding infrastructure of Russia and its neighboring countries.

The earliest placentals, sometimes also known as eutherians ("true beasts") in the scientific literature, come from the Early Cretaceous of North America, Mongolia, Kazakhstan, and Uzbekistan. These are again largely a matter of dental specimens, or unassociated (and therefore unidentifiable) parts of the ear regions and skeleton. The really good stuff—complete skulls and *associated* skeletons—doesn't turn up in abundance until the Late Cretaceous of the Gobi. These skulls were among the prizes of Andrews' 1920s expeditions. Joint Mongolian-Polish teams in the 1960s (and later Mongolian-Soviet teams) retrieved an impressive suite of placental skulls and skeletons from new Gobi sites, including the Tugrugeen beds. This collection was thoroughly studied and described in a series of excellent monographs by Kielan-Jaworowska.

My colleagues and I were most interested in building on this important work, but we knew this objective would be challenging. The placental skulls are very scarce and are among the tiniest in the Gobi collections. Larger ones are only about two inches in length while smaller ones barely measure three fourths of an inch, not much longer than the tip of a fountain pen. Even with the protective casing of a concretion, the delicate bones are easily damaged, and these miniatures are certainly elusive pickings for prospectors. Our effort got a big boost in 1991 when Jim Clark found the exquisite skull of *Zalambdalestes* at Tugrugeen.

It took several months of part-time preparation by my skilled artist Lorraine Meeker to relieve this five-inch skull of its surrounding rock matrix. It was not until fall of 1992, after our third season in the Gobi, that the skull was ready for study. Our probe of *Zalambdalestes* was aided by use of newfangled machinery. In collaboration with Dr. Tim Rowe at the University of Texas we put our small rat-sized *Zalambdalestes* skull under an industrial-strength CAT scan. This machine has an ion beam too intense for medical application but it will not damage inert materials like metals and fossilized skulls. The CAT scan made 1,600 high-resolution "slices" in cross section (one section every 20 microns). From the sections, a computer program generated an animated sequence. Viewing the animation was like traveling through the inside of the skull from one end to the other, examining details along the way and stopping en route to inspect eye-catching minutiae.

At such a point of interest we stopped to look at the carotid "plumbing" or, in more scientific terms, the carotid arterial system. These are the blood vessels routed through the back of the head that lead to the base of the brain and the eyeball. Of course, fossils do not preserve soft tissue in the form of nerves or blood vessels, but various holes and canals in the skull indicate the pathways of these structures. With the animated CAT-scanned images we could tell that the main pathway of the carotid ran in two branches on either side of the midline of the skull. This is a striking departure from the usual situation in modern placental mammals, in which the carotid crosses the base of the skull away from the midline and through the middle ear cavity. Just why the carotid varies in such position is not clear; possibly the vessel's position is related to the problem of packing a great deal of equipment in the form of nerves, blood vessels, small ear muscles, and middle ear bones (popularly known as the hammer, anvil, and stirrup) in the diminutive skulls of these mammals.

Interestingly, the middle route for the carotid seen in *Zalambdalestes* does occur in some rabbits and rodents. Could this be an indication of affinity? At this stage, it is not clear. The condition is scattered among certain mammals, including some archaic fossil forms. This midline route for the carotid could be a very primitive condition merely retained in rabbits, some rodents, and a few other mammals, but modified in most modern placentals. There is also suspicion that it might occur in other Mongolian species. We are anxious to settle this dilemma by casting a broader net of comparisons over mammals and by CAT-scanning our fine new skulls of other Mongolian genera, like *Kennalestes* and *Asioryctes*. These shrewlike forms are even much smaller than *Zalambdalestes*, but they should show details under our high-intensity scanner.

The Cretaceous placental mammal *Asioryctes (Ed Heck)*

LIFESTYLES OF THE SMALL AND INCONSPICUOUS

Anatomical data on *Zalambdalestes* and other forms dispel some myths about the stereotypic roles of the earliest mammals. The popular scenario for these Mesozoic animals is one that depicts swarms of stealthy, sharp-toothed shrews puncturing and consuming dinosaur eggs. Doubtless, some of these forms were capable of such habits, but there was a wide range of feeding preferences, as demonstrated by the seed-eating, nut-cracking multis, or the larger, possibly carnivorous forms like *Deltatheridium* (this beast could have consumed tiny *Asioryctes* or the abundant lizards, as well as dinosaur hatchlings known from the Gobi Cretaceous). The portrait of an ambulating, ground-dwelling shrew is also deficient in describing the range of movement for different Mesozoic species. A highly mobile ankle joint and grasping digits suggest that some multis were adept at climbing trees. New studies by Kielan-Jaworowska and the Russian anatomist Gambaryan show that the large bony orbits in the skulls of multis suggest very large eyes. The powerful jaw musculature, extending from the skull and inserting on the lower jaws to power the bite, is uniquely arranged, in ways not seen in other mammals. The limbs of these forms show some splaying—in a reptilian-like fashion—at least in the pelvic region of the skeleton. Long-limbed forms like *Zalambdalestes* were capable runners and leapers, and probably scampered energetically about, much like kangaroo rats in the desert night.

At the same time, what we do know of the anatomy of Mesozoic mammals suggests a rather narrow adaptive range in comparison to modern forms like kangaroos, whales, bats, and aardvarks. Our Mesozoic antecedents are typified by small size; even the largest of the multis are not larger than a groundhog.

In addition, there are indications that these Mesozoic mammals had a rather standardized and primitive sensory system, based on the study of endocasts, the filling of sandy matrix inside the hollow fossil skull that actually forms a cast of the brain. Endocasts of multituberculates and other forms show a relatively small cortical area, with few if any cortical folds or sulci, suggesting limited brain acuity. By comparison, think of the intricate folding of the surface of the human brain, which greatly increases the nerve tissue of the cortical region. The olfactory lobes (the smelling centers) in these archaic mammal brains are rather large, whereas the optic regions (the vision centers) are much smaller, and the opening for the passage of the optic nerve located in the braincase wall of the skull is also often small (the exception here being some of the large-eyed multis). The inferior colliculi—lobes near the back of the brain that represent hearing centers—are well developed. It appears that most of these mammals had a keen sense of smell and acute high-frequency hearing, much like that of living shrews and hedgehogs. And, like the latter, they had rather poor vision. Presumably, these mammals were most active at night, a time when hearing and smell are greatly exploited by living creatures with limited eyesight.

It seems surprising that such insights on lifestyles and adaptations can be pieced together from a few tiny bones. In post-Cretaceous rocks paleontologists are treated to much more dramatic evidence of mammalian history. During the Cenozoic appear fossil primates, whales, rhinos, camels, bears, and mammoths, the apotheoses of the glorious age of mammals. Inspecting these younger rocks, mammalian paleontologists can often walk upright, as evolution designed them to walk, instead of crawling on slopes on their hands and knees in search of a concretion with a flick of blue enamel the size of a match head embedded in the ghost outline of an almond-sized skull. Yet there is nothing so triumphant as finding one of these prizes and heading back to camp with a featherweight but profound discovery.

THE GREAT RADIATION

Of course mammals eventually came to dominate the landscape, at least in terms of sheer size and visibility. The Cenozoic—what we commonly call the age of mammals—begins with the mass extinction event some sixty-five million years ago. The non-avian dinosaurs disappeared, as did many other species. Yet mammals did not immediately sprout up and diversify to take their biological niches. For a few million years following the end of the Cretaceous there is a curious hiatus, a span of time in which there were no particularly large and diverse mammals. It is not until later in the Paleocene, the first epoch of the Cenozoic Era, that some of the larger herbivores and carnivores appear in the fossil record.

Early placental mammals
(Ed Heck, reprinted with permission, AMNH)

This early period, this "dark ages," is actually the late stage of one of the most fascinating pulses in the history of the vertebrates. Sometime in the Cretaceous began a flowering of the modern mammal groups, the placentals and marsupials. Our sense of this record is especially good in the case of the placental radiation. Over a relatively short span, the first members of the great orders of mammals branched off from one another. By the Late Paleocene many of these branches had sprung forth. The latest Cretaceous and the Paleocene saw the emergence of small progenitors of the insectivores, primates, carnivores, and various ungulate herbivores. By the end of the Paleocene—about ten or eleven million years after the Cretaceous extinction event—essentially all the modern orders arrived. These include the bats, the even-toed hoofed artiodactyls (camels, hippos, antelope, deer, cows), the odd-toed perissodactyls (horses, rhinos, and tapirs), the rodents, the lagomorphs (rabbits and pikas), and many others. The first whales appear only a few million years later.

These times of first appearance in the fossil record may not be precise. We cannot eliminate the possibility that some of the orders of mammals were older but are unknown for lack of proper fossils. Yet the later Cretaceous and the Paleocene safely brackets a ten- or fifteen-million-year span in which the great radiation of mammals took place. Major branches diverged from one another over a very short time. Rather than a stately tree of successive, widely spaced branches, this evolutionary history is a gloriously intricate bush—a starburst of evolution.

One might expect such radiations from time to time. They seem especially prevalent when a certain group of organisms takes advantage of the expansive range of habitats and conditions that are suddenly unoccupied after a mass extinction event. Thus these early bushlike histories are often called adaptive radiations. They typify not only the modern mammal orders but several other notable groups, birds for that matter, as well as certain kinds of fishes, insects, and marine invertebrates.

But such rapid bursts of speciation should have structure. Some branches should diverge earlier than others, even though the space between these events may be less than a million years. Are primates most closely related to bats or to tree shrews? Are rodents and rabbits close kin? Do all the wondrous array of herbivorous hoofed mammals take root in a

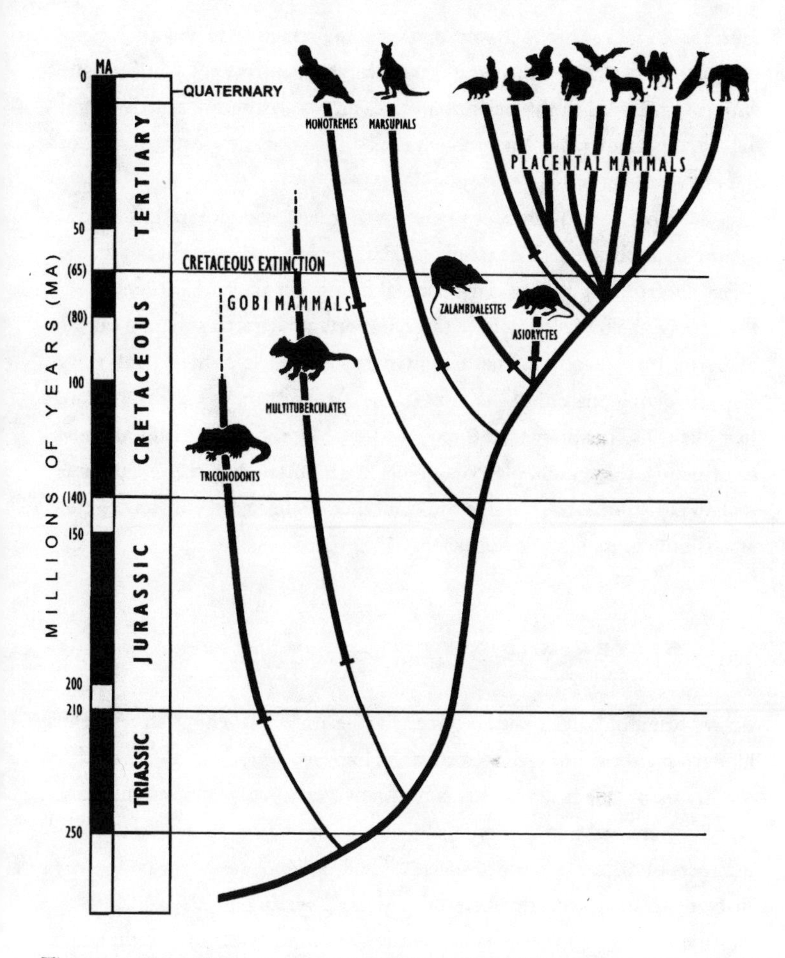

The mammalian radiation calibrated in millions of years (numbers) before present *(Ed Heck)*

single ancestor? It may seem odd that over a century of work since most of the major groups of mammals were first described we are still grappling with these questions. For some years I have pursued these problems, drawing not only on the evidence of early fossils but on data described in living mammals—ear anatomy, brain topography, even the DNA sequences found in certain genes. I have proposed some structure to this evolutionary history, not without sparking dispute and controversy. My dialogue with my critics is healthy. Science involves proposal, scrutiny, criticism, and debate, especially when the problem is as difficult as the great radiation of the placental mammals.

I am particularly anxious to see how our new evidence from the Gobi mammals applies here. The family from Sugar Mountain might not be the direct ancestors of a particular mammal order. But they are certainly near the base of the bush. They have those important characters that reveal the blueprint for a placental, one that gives us some sense of the evolutionary steps involving anatomical features in the early history of our diverse kin. Just like those maniraptoran theropods that Mark, Jim, and Luis are now scrutinizing, these Gobi placentals are a key to untangling the great mammalian radiation. And, just like the Gobi theropod project, our work in this area is only in its nascent stages.

THE BITTER END

As we attempt to draw more precisely the early branching structure of modern mammal lineages, we can stand back for a moment and marvel at the drama of this history. This is not, however, a story of awesome taxonomic diversity. Mammals are really not much of a group in terms of raw numbers of different kinds of species. Their 3,500 or so living species pale in comparison to the numbers of spiders, flowering plants, and insects. But mammals are marvelous in their range of lifestyles, architecture, and adaptations. In this way they parallel the history of dinosaurs, the high-profile land dwellers of the previous era. Mammals get bigger; they develop weird trunks and snouts and bony armor. They take to the trees and swing with alacrity, using their elongate limbs and tails. They develop wings and fly, and catch insects on the wing, navigating with an incredible sensory system known as echolocation. They become bulky, horned creatures, modern-day versions of Cretaceous ceratopsians. They return to the marine home of their ancestors and become the largest of all animals. They develop large brains and a bipedal gait, the use of tools, and the capacity to control and even destroy the biota with unprecedented power. And, like most other forms of life, they are extinguished in impressive amounts at various intervals. Extinction, then, is a phenomenon not reserved simply for dinosaurs and ancient sea creatures. It occurred many times in the last sixty-five million years of the Cenozoic—during the so-called age of mammals—and it is rather dramatically occurring now.

It seems likely that many mysteries of the geologic record, like the reasons for extinction, will persist. But before we disparage all the fanfare and argument built around such elusive problems, we can for a moment appreciate the power of our record of the history of the earth and its life. We know extinction, whatever its cause, is a reality because of this evidence. People, especially paleontologists, can accept extinction, renewal, diversification, and yet more extinction as intrinsic to our evolutionary legacy. We know that serpentine mosasaurs, segmented pill-buglike trilobites, winged pterosaurs, long-necked sauropods, and even bigger cockroaches than encountered in some bad hotel rooms in Hawaii are no longer around. If there is something that the fossil record teaches us that might be somehow attached to our "higher" sense of our place in the universe, it is that life in all its wondrous forms does not endure. Life even in its most baroque excesses of diversity and richness is fragile in the face of change. There is also nothing that really precludes the possibility of an ultimate extinction event, one wherein the planet reverts to some primeval lifeless state, or wherein the pitiful remnants of life persist but fail to rebound in any way that measures up to its halcyon history.

Indeed, E. O. Wilson's watershed book, *The Diversity of Life*, draws heavily on the lessons of the fossil record to remind us of the reality of extinction. Wilson also shows us how our own species is having its own asteroid-like impact on the bounty of life on this planet. Unlike various theories explaining past extinction events, this one unfortunately is much better founded. We have halved the amount of tropical rain forests since prehistoric times. We now have about the equivalent of a United States' worth of rain forest left in the world. Alarmingly, we are losing an area of rain forest annually that is about equal to the size of Florida. It doesn't take mathematical genius to calculate that, at the present rate of decimation, in less than fifty years we won't have any rain forest left. And our long reach even transcends the planet's collection of species. There is no question that we have significantly infused the atmosphere with carbon dioxide and chlorofluorocarbons, although, except for the case of the widening ozone hole, the jury is still out on how these contributions will permanently change our world climate. Perhaps not since a certain lineage of bacteria converted the atmosphere some two billion years ago has one small branch of life had such an enormous effect on its fellow species, even on the physical system, the air, and the water cycle of the planet we live on. In fifty years, when most of the world's ecosystems are devastated, and possibly thirty to forty percent of all its species are erased, we might unfortunately have our first tangible, and visceral, sense of what happens during a mass extinction event. Some poets, like Archibald MacLeish in "The End of the World," have vividly imagined it:

And there, there overhead, there, there hung over
Those thousands of white faces, those dazed eyes,
There in the starless dark the poise, the hover,
There with vast wings across canceled skies,
There in the sudden blackness the black pall
Of nothing, nothing, nothing—nothing at all.

CHAPTER 11

THE SECRET HISTORY OF LIFE

Having seen this life's dance performance of the utmost illusion and of the wealth of this world, which is mere driftwood; by giving birth to a mind of love and compassion, may we enter into the deeds of the Prince of Great Teachings!

From *The Secret Biography of the Eighth Bogdo Gegen* Anonymous. Late nineteenth–early twentieth century. Mongolia.

"Back to the Gobi!" was now our battle cry for one of the longest paleontological campaigns in history. With the onset of the 1995 season it had also become one of the biggest paleontological expeditions in history. Our team this year was a pastiche of regulars and new recruits, as well as a group of writers, photographers, and filmmakers from *National Geographic.* On July 1, 1995—1,833 days since I first struck out with Dashzeveg, Mark, and Malcolm for the Flaming Cliffs—we launched the sixth Mongolian Academy–American Museum Paleontological Expedition, leaving Ulaan Baatar with a remarkably ordered train of twelve vehicles and thirty people. Barsbold, who waved goodbye to us outside our compound next to the Natural History Museum, observed that it was possibly the largest assemblage of its kind ever to take on the Gobi. Some weeks later our bustling camp at Ukhaa Tolgod was visited by a giant Russian helicopter carrying U.S. Ambassador Donald Johnson and a party of eighteen. That night we watched on a tiny video screen a documentary with the original footage of the Andrews expeditions, appropriately titled *To the Ends of the Earth.* The viewing audience around the campfire was nearly the most populous settlement in the southern Gobi.

In 1995, Ukhaa Tolgod remained astoundingly generous with fossils. I was secretly worried that, with all the fanfare enveloping us, the locality might not deliver. But I was gloriously wrong. In the first three hours of work at the site we spotted over fifteen oviraptorid skeletons strung along one level near the top of the sandy butte next to Camel's Humps. One of these was another remarkable nesting oviraptorid, which we removed with much effort down a steep eighty-foot slope and into the back of the GAZ. At another site we found two oviraptorids—the smaller *Ingenia*—wrapped close together. The specimens, nicknamed Romeo and Juliet, were jacketed in huge amounts of plaster-encased burlap. The results were two 700-pound plaster monoliths—"grand pianos"—that made for an excruciating haul out.

But this season was not simply an oviraptorid raid. Mammals, too, still kept spilling out of the sediments. Over eighty skulls were recovered, and to our delight Jim and I each found the Holy Grail we had pursued for some years—skulls of the marsupial-like *Deltatheridium.* Jim also located another nice skeleton of *Mononykus* with a skull that showed details not as clearly preserved in the 1994 prize.

The fruits of our 1995 field labors cannot, however, be expanded upon yet. The smaller fossils arrived by air only in the fall and are currently being worked under Amy's sharp needle. The grand pianos arrived several months later by freighter. Meanwhile news of the nesting oviraptorid was published in *Nature* and released to the media before the end of the year. Additional funding from the National Science Foundation has bolstered our efforts to study the Cretaceous Gobi mammals, a complement to the grant currently funding the theropod project. Malcolm is also deeply involved in well-funded collaborative work on the Cenozoic mammals from Mongolia, and Dashzeveg has completed a series of manuscripts on various related themes. Colleagues, curators, postdoctoral fellows, and graduate students comprise a community of scientists in the museum as well as in the field working on the Gobi project.

Future goals are fairly clear to us. The evolutionary set problems concerning bird and mammal origins now involve several coinvestigators, including a postdoctoral fellow and mammal expert from Argentina, Guillermo Rougier, and our colleague John Wible from the University of Kentucky at Louisville. Mark also has invited Gao Keqin, a postdoctoral fellow, to sort out and describe the overwhelming sample of fossil lizards from Ukhaa Tolgod and elsewhere. Lowell and a small team of geologists plan to continue detailed mapping and sedimentology in the next field season in an attempt to disclose more about the original habitat at Flaming Cliffs, Tugrugeen, Khulsan, and Ukhaa Tolgod, and thus hope to answer some of the questions I raised in the previous chapter about life and death of these Cretaceous creatures. We still have our sights set on those small isolated outcrops west of burning Kheerman Tsav, and we may even probe farther in the Nemegt Formation, the units overlying our bountiful red beds. In future expeditions, who knows? Maybe we'll even find a skull to go with those amazing arms and claws of the mysterious *Deinocheirus.*

Thus a research machine is rolling along, one that I could never have imagined when I first heard the news of a Mongolian invitation in the winter of 1990. We are proud of our accomplishments, but the serendipity in all this—especially in the chance encounter with Ukhaa Tolgod—is rather arresting. In the spring of 1995, on the eve of our sixth season in the Gobi, I visited Zofia Kielan-Jaworowska in her laboratories at the Natural History Museum in Oslo, Norway (Zofia had some years before moved to Norway from Poland to take a distinguished position there). I showed her many slides of dinosaurs and mammals and sundry treasures from Ukhaa Tolgod, and had brought a batch of specimens, including some of the mammal skulls that so entranced her. A shocked Zofia fingered the specimens, exclaiming: "I could hit myself! I know our expedition was only ten kilometers from that place!"

In fact, they were closer yet. In 1995, Jim ran across some graffiti carved into a sandstone ledge that was dated "1971," accompanied by some scratchings that resembled closely those left by the Polish-Mongolian expedition on some of the canyon walls in Eldorado. There was no other explanation. This inscription was left in the banks of Gilvent Wash, only a mile or so from the fossil-stuffed pocket of Ukhaa Tolgod.

THE CIPHER OF EVOLUTION

The Gobi expeditions are my most vivid intersection with paleontology. But the drama of discovery goes on worldwide, carried out by many resourceful and energetic colleagues. China has now become a great storehouse of paleontological treasures. Work by many Chinese paleontologists has revealed cemeteries of huge dinosaur skeletons, fields of fossil eggs, more embryos, and exquisite fossil birds and mammals from the Mesozoic. These are being described in the handsome, newly rebuilt Institute of Vertebrate Paleontology and Paleoanthropology (IVPP) in Beijing. Teams are now stepping up work in the remote frontiers of Kazakhstan and Uzbekistan in search of dinosaurs and ancient mammals. Finds are turning up in the most far-flung of places—the steamy forests of southeast Asia, the empty wastelands of the Sahara, even on the ice-encased isles of Antarctica. *Giganotosaurus,* a tyrant more ponderous than *Tyrannosaurus* itself, was recently found in Argentina. And on the mammal front there continues to be a rash of findings related to the great radiation. These include new primates from the Eocene rocks of South America and archaic whales that actually ambulated on funny limbs in the Cenozoic of Pakistan. A monotreme-like form, *Gondwanotherium,* is now also known from the Cretaceous of Argentina. Within the last few months a wholly unprecedented array of dinosaurs, birds, and other vertebrates has turned up in the Cretaceous rocks of the isolated island of Madagascar.

With all this explosive information, this infusion of riches gained from this new age of exploration, we have plenty to work on. Indeed, some people ask me, why go back? Our team has incredible riches now from one Gobi locality, enough to keep a respectable body of paleontologists busy for years, or even generations. And indeed the enterprise is expensive and inconvenient, and on some days not even much fun. But it is hard to give up this life. The opening of an opportunity that came with Mongolian democracy isn't necessarily permanent; it's a window of opportunity, one that could close without warning in this chimeral world. In the meantime we're greedy. We have begun to know the Gobi better. We think we will find more Xanadus out there.

Some people also ask me the more general question: why all this effort for some bones stuck in a mound of red sand? Why is such a find worth the energy, danger, and discomfort in the field, not to mention the frustrating process of investigation, argument, debate, and criticism that comes with promulgating a scientific discovery? But I've never really thought twice about this occupation or about the need to somehow defend its impact on my life. I can't think of anything more pleasurable and fulfilling than the danger, discomfort, frustration, debate, and criticism that come with the work. Some may have another more probing question: well, you might enjoy the work, but is the work necessarily valuable? After all, do we really need a big pile of bones in a dark basement of a museum? Isn't paleontology a Victorian pastime that's served its purpose, and is carried on today by a bunch of dinosaur fanatics and romantics?

These questions come not just from the public at large but from learned and not always kindred scientists. They feel that some of the primary missions of paleontology—such as uncovering the pathway of evolution and phylogeny—are better served through the study of living organisms, particularly their genes. In some cases this may be true. There are many sticky problems that have not been successfully solved by the study of anatomy in either living or fossil organisms. Perhaps we'll do better with genes. For example, the subtle relationships of bacteria, with their rather muted variation in cell structure, have been most effectively studied through comparing their DNA. Just the same, for many other problems anatomy, especially when coupled with development or the ontogeny of organisms, has offered powerful insights into evolution, insights way beyond what we can recover from genes. Embryologists detected nearly a century ago that those three marvelous tiny ear bones in the mammalian head—the stapes, malleus, and incus—are the descendants of some small bones in the back of the jaw in non-mammalian vertebrates.

What is the importance of fossils in all this? Five years ago we had no *Mononykus*, no cluster of certain shrewlike placental mammals, and no embryonic oviraptorids. As documented by other paleontologists in other lands, we also had no whales that walked about on four limbs, a testament to the transition from land to water in the evolution of these aquatic beasts of the deep. As noted above, a recent finding of such a fossil from forty-million-year-old rocks in Pakistan is a shocking revelation that such a curious "walking whale" did indeed exist. Finally, the whole question of relationships of many living groups has little depth, richness, and fascination at all, little textured dimension, without the knowledge of the fossil record. Take birds, for instance. Among the living vertebrate groups there are two candidates for bird relatives. One of these is mammals and the other is that snaggle-toothed lineage of archosaurs, the crocodiles. It has been convincingly demonstrated that without information of extinct groups, like dinosaurs, synapsids, and ancient mammal-like forms, we would actually be led to the "wrong" conclusion, namely we would believe that birds and mammals were closely related. But when we include these critical fossils in the study we find that the similarities between birds and mammals—details of anatomical and gene structure—are lying to us; they probably arose independently in the two groups; in no way did they emerge in some immediate ancestor of birds and mammals. Suppose, though, we could come to the "correct" choice—that birds are more related to crocodiles—without using the fossil record. Isn't it far more meaningful to call upon the wealth of extinct lineages of dinosaurs and other fascinating groups to map the genealogy of birds? The *real* kin of birds, not just the closest kin that happen to be left alive, are other theropod dinosaurs. Not all branches of the tree of life produce foliage, but all branches, living and dead, must be accounted for in an appreciation of the structure and beauty of the whole tree.

I am sometimes frankly disappointed in colleagues who cave in to the criticisms of the limits of fossil evidence and the practice of paleontology. They start to develop an inferiority complex. There are attempts to recast paleontology into either something lavishly high-tech, or simply as a small part of the evidence for the big scientific problems in evolutionary biology. I certainly endorse the refashioning and reworking of science; in much of my work I draw heavily on the burgeoning information on gene sequences in vertebrates. But some of these concessions to technology are a bit overstretched. A couple of decades ago it became fashionable to call ourselves "paleobiologists," as if we were now after something grander and more literally vital than the traditions of the science allowed. One of my professors hated that fad. He was like the philosopher Nietzsche who, upon learning that the composer Wagner proclaimed that his operas and other works were not mere music, said, "*Not mere* music—no musician would say that."

Paleontology is an astoundingly rewarding science. It gives us much more than we dare hope. Earlier, I wrote of the odds against fossilization, the extraordinary luck involved in the preservation of the record of life, let alone our fortune in finding a Xanadu. It is hard to imagine our sense of existence—of who we are and where we come from—without this record. We would feel like strangers in a strange land with no past, no roots, no family tree that binds us to our native natural world. The legacy most treasured by Mongolians is the writings of their history, what they call *The Secret History of the Mongols.* Likewise, the secret history of life, our greatest legacy, is only cipherable from a few precious tidbits of the past, the fossils in the rock.

Paleontology, especially the recent flourish of investigations of climatic change and ice-age extinctions (where increasingly humans are implicated as major culprits), can also provide us with important lessons for mapping the future and help, perhaps, offer predictions that can be used to make decisions about our stewardship of nature, a sound policy for a future that better maintains the balance between humans and the thriving, evolving world around them. Currently our scientific knowledge of many of these things—and certainly the connection between science and the responsible actions of society—is poorly developed. But it will improve. It must improve.

View from the Sphinx

It's helpful to write these words after just returning from another rewarding but exhausting season in the Gobi. Our 1996 jaunt sustained the productivity of earlier seasons. Ukhaa Tolgod was again sprinkled with skeletons we could swear were not there before. Especially rewarding were some new dinosaur nests in pristine condition and several new aggregates of small mammal skeletons. About four miles west of the main locality we encountered a small gully carved into a miniature Ukhaa, and here too were protos, dromaeosaurs, and mammals. Lowell Dingus and his geologic team, along with a fine Mongolian geologist/paleontologist, Dr. Minchin, continued their obsessively detailed sampling and descriptions of those cross-bedded sandstones, with the hope of drawing a precise picture of a desert oasis from the Late Cretaceous. During our prolonged retreat from the Gobi that summer, we managed to relocate a famous site, Udan Sayr, worked many years earlier by Mongolian and Russian expeditions. The place was rich with fine fossils, including a superb skeleton of our "Komodo dragon," *Estesia*. It was well worth the trouble of getting there, and would be well worth a return in our upcoming season.

As in the case of the 1995 season, the fruits of our labors—including some real killers—have not been completely prepared, described, illustrated, and revealed to the scientific community and paleophiles the world over. Already, however, I must gather the energy to prepare for the next expedition. Life is a frantic, complicated, and telescoped bridge between field seasons. In 1996 we purchased some new Russian jeeps, and they might be supplemented by some imports for the 1997 season. We are also stocking up on declassified "spy satellite" images of the Gobi. These have greatly improved our mapping capabilities. Some of the images provide resolution of objects within ten feet of each other, and we are expecting images with even better focus, at six feet. These remote portraits were taken in the 1970s, and we can see no clear evidence, as expected, of any human activity at our beloved Ukhaa—no roads, camps, or big excavation sites. If the images were contemporaneous, just think what they would reveal with a six-foot resolving power. "Hey look! There's giant man, Jim Clark!" More important, some of the bigger oviraptorids and ankylosaurs conceivably could be picked up from outer space at this level of resolution.

But we also have one major logistical problem to solve for the upcoming season. In 1996 we rented that big lumbering Russian helicopter the ambassador had used so often. We put the helicopter to good use, skimming over the broiling badlands of Kheerman Tsav, landing in isolated pockets of fossil-producing rock that would have been supremely difficult objectives for our jeeps. Alas, I received the news some months after the

1996 season that our helicopter had crashed and burned. Fortunately, there were no fatalities—the flight crew was miraculously nursed back to health in one of those Mongolian hospitals—but the helicopter was simply a pile of incinerated metal on the steppe. So I am in desperate search for another, and hopefully more reliable, flying machine.

It is very difficult to solve these problems and gear up for planning something so remote to me. For ten months I am far away from this desert. I can no longer see its vast, incomprehensibly empty expanse, its naked ridges and spines of rock, and its bristling dinosaur graveyards. On a smothering winter's day looking out from my turret office at the American Museum of Natural History, seeing the blackened snow melt below the crush of a million tires and footprints on Seventy-seventh Street in Manhattan, it is easy to dream about those spectacular days of discovery in the Gobi.

But sometimes while I'm in the Gobi I dream of home. I feel lonely and cast adrift, out in the void between heated haze and blowing sand, alone in the immensity of a scorching inhospitable desert with a nearly constant idiot wind, where only nomads who have inherited the ways of centuries of their ancestors can carry on existence with grace and nobility. I feel that I am zigzagging through space and time, moving from Bayn Dzak to Eldorado, to Naran Bulak and back, traveling through rock sequences spanning 50 million years with no place to land and settle down. This realization produces a kind of melancholy that usually comes at the comfortable oasis of Naran Bulak, our sunny spring, at a time late in the season when the troops regroup to relax and recuperate. There the grass is uniquely welcoming, the water clear and cold. But it's 13,200 miles from this refuge to New York. Perhaps there is just enough comfort at Naran Bulak to want more. I long for the caress of loved ones, for a shower out of the wind, for a two-and-a-half-pound lobster, for ice in a glass, for the green, not of stubby, camel-chewed grass, but the green of a real tree with branches and leaves, for the privilege of not having to inspect the underside of my sleeping gear for scorpions and camel ticks.

In such moods it's good to take a walk. I drift away from our city of tents at Naran Bulak. I aim for the Sphinx, a sandstone monolith glowing white-hot in the setting sun. From the Sphinx I can sweep my arm like a sundial against the panorama. Moving clockwise, I rotate from the blue wisp of the Nemegt far to the northeast extending toward our valley of kings, Xanadu, and other pockets of Ukhaa Tolgod that are, from here, obliterated in the haze of the horizon. Then farther east and south are the darker shadows of Tost Mountain, blocking the entrance to the distant town of Gurvan Tes. To the southwest, the sun silhouettes the dinosaur-bearing beds of Tsagan Khushu, and due west a lonely pyramid of volcanic rock marks the junction for the path to the remote and invisible canyon-lands of Kheerman Tsav.

The sweep ends due north, my gaze fixed upon the hulking black beast of Altan Ula, its ridges sinking from crest to footings like rippling muscles. The mountain rises above several layers of canyons, ridges, and sand dunes, all foreshortened like stage sets in a window box. These encompass the chromatic mazes of the Altan Ula canyons and its fossil wonderlands. Finally there is Altan Ula itself—the mountain of gold. Its black ugliness belies any sense of that shimmering substance, but somewhere in those precipices are gold veins that griffins guard. Lost in the intricacies of ridge and canyon, pinnacle and crevice are also mountain sheep, snow leopards, and springs that trickle into carpets of moss and wildflowers, springs which have yet to provide the pleasure of drink to any human.

With the completion of this sweep, I am once again back in the embrace of the Gobi. I am one of its denizens, refashioned by wind and burning sun into a desert creature. I have an urge, like a roving beast, to leap across ten miles of rock and sand to the foot of the mountain of gold, drink from its virgin springs, and scramble to its highest crest. From there I can look into the blinding white of badlands and the huge empty valleys of blowing sand beyond. It seems that all my energies, my whole life, has been propelled by the obsession to probe some hidden valley, break open the rocks exposed on its cliff faces, and pull back the curtains to gaze on the immense scape of time with its cascade of prehistoric life. Like a future mariner among the galaxies, I feel the power that comes with exploration over these vast dimensions.

But those valleys I will come to know eventually. For now, I rotate slowly, almost due east. I begin walking a bearing of great familiarity, one that does not require GPS or Dashzeveg's bony finger to point the way, to my friends and the soft sound of laughter in camp.

NOTES

p. xiii. "It is surprising . . ." The Andrews American Museum of Natural History Expeditions to the Gobi Desert are chronicled in Andrews, R. C., 1932. *The New Conquest of Central Asia.* Natural History of Central Asia, Vol. I. The American Museum of Natural History, i–l + 678 pp. The quote is from p. 310.

PROLOGUE. A PALEONTOLOGICAL PARADISE

p. 8. . . . since the 1920s, when Roy Chapman Andrews . . . The results of these expeditions appear in numerous scientific and popular publications. The primary reference is Andrews, ibid.

p. 11. The birdlike dinosaurs . . . There is strong scientific consensus that birds are simply a subgroup of the dinosaurs derived from an ancestor related to theropod forms like *Tyrannosaurus* and *Velociraptor.* The theory has a long history of development, reviewed in Chapter 8.

p. 11. Some fossil eggs from the French Pyrenees . . . On p. 208 in *The New Conquest* . . . , Andrews related the exciting day of discovery of the dinosaur eggs from Flaming Cliffs:

"We saw a small sandstone ledge, beside which were lying three eggs, partly broken. The brown striated shell was so egglike that there could be no mistake. Granger finally said, 'No dinosaur eggs have ever been found, but the reptiles probably did lay eggs. These must be dinosaur eggs. They can't be anything else.' "

Contrary to Walter Granger's statement, dinosaur eggs had been found in France in the nineteenth century, although their discovery was not well publicized. The priest paleontologist Jean-Jacques Pouech (1859. *Bulletin de la Société Géologique de France,* Vol. 16, pp. 381–411) described bits of large fossil eggs but did not attribute them to dinosaurs. Philippe Matheran (1869. *Mémoires de l'Académie Impériale des Sciences, Belles Lettres et Arts de Marseille,* pp. 345–79) suspected that some large eggshell fragments either belonged to a giant bird or *Hypselosaurus,* a new dinosaur he had described. Paul Gervais, the influential zoologist and paleontologist, conducted the first detailed study of the fossil eggs from southern France but did not conclude that they necessarily belonged to dinosaurs (Gervais, P., 1877. "Structure des coquilles calcaires des oeufs et caractères que l'on peut en tirer,"

Journal de Zoologie, Vol. 6, pp. 88–96.) That these eggs belonged to dinosaurs is now generally accepted, but work at the sites in recent decades has failed to uncover an association between the eggs and skeletal remains. The identity of the dinosaurs that laid the eggs is still unknown. This history is reviewed by Buffetaut, E., and Leloeuff, J. L., 1994. "The Discovery of Dinosaur Eggshells in Nineteenth-Century France," pp. 31–34. In Carpenter, K., Hirsch, K. F., and Horner, J. A. (eds.), *Dinosaur Eggs and Babies.* Cambridge University Press, Cambridge.

p. 13. . . . the Sino-Swedish team . . . Bohlin, B., 1945. "The paleontological and geological researches in Mongolia and Kansu," *Reports of the Scientific Expeditions to the Northwest Provinces of China,* Publ. 26 (part 4), pp. 255–324.

p. 14. . . . the Russians in the 1940s. Efremov, I. A., 1956. *The Way of the Winds.* All-Union Pedagogical-Study Publisher, Ministry of Reserve Workers, Moscow, 360 pp. [In Russian.]

p. 14. . . . the Polish-Mongolian teams in the late 1960s and early 1970s . . . This series of expeditions is documented in numerous scientific papers. A popular account, but one that does not cover all expeditions, is Kielan-Jaworowska, Z., 1969. *Hunting for Dinosaurs.* The Maple Press Company, York, Pennsylvania, 177 pp.

p. 17. What follows is the story . . . A brief recounting of the expedition through the 1993 season appears in Novacek, M. J., Norell, M. A., McKenna, M. C., and Clark, J., 1994. "Fossils of the Flaming Cliffs," *Scientific American,* Vol. 271, No. 6, pp. 60–69.

CHAPTER 1. 1990—JOURNEY TO ELDORADO

p. 18. "Then in the early twenties . . ." Colbert, E. H., 1968. *Men and Dinosaurs.* E. P. Dutton, New York, 283 pp. The quote is from p. 202.

p. 22. . . . occupied himself with a book . . . Gould, S. J., 1989. *Wonderful Life.* W. W. Norton, New York, 337 pp.

p. 24. . . . Andrews' remarkable narrative . . . Andrews, op. cit.

p. 25. In 1919 he launched . . . Ibid., p. 4.

p. 25. . . . fossil eggs from Europe . . . Gervais, op. cit.

p. 27. "Almost as though led . . ." Andrews, op. cit., p. 162.

p. 27. "The tents were pitched . . ." Ibid., p. 162.

p. 28. "This immediately put the animal . . ." Osborn, H. F., 1924. "Three new Theropoda, *Protoceratops* Zone, Central Mongolia," *American Museum Novitates,* No. 144, p. 9.

p. 28. The fossil was named *Velociraptor* . . . First described by Osborn, ibid., pp. 1–3.

p. 29. "It was possibly the most valuable seven days . . ." Andrews, op. cit., p. 271.

p. 30. His writings capture . . . Efremov, op. cit.

p. 32. The collaboration of Polish and Mongolian scientists . . . Kielan-Jaworowska, op. cit.

p. 33. The Russians and Mongolians continued . . . Russian and Mongolian activities in later decades are highlighted in Lavas, J. R., 1993. *Dragons from the Dunes.* Academy Interprint, Auckland, 138 pp.

p. 33. Yet the closest . . . Currie, P. J. (ed.), 1993. "Results from the Sino-Canadian dinosaur project," *Canadian Journal of Earth Sciences,* Vol. 30, Nos. 10–11, pp. 1997–2272.

p. 36. A *Lonely Planet* travel book . . . Storey, R., 1993. *Mongolia, a Travel Survival Kit.* Lonely Planet Publications, Hawthorn, Australia, 235 pp.

p. 43. There are several species of ankylosaurs . . . Coombs, W., and Maryańská, T., 1990. "Ankylosauria," pp. 456–83. In Weishampel, D. B., Dodson, P., and Osmólska, H. (eds.), 1990. *The Dinosauria.* University of California Press, Berkeley, 733 pp.

p. 46. We named the lizard . . . Norell, M. A., McKenna, M. C., and Novacek, M. J., 1992. "*Estesia mongoliensis,* a new fossil varanoid from the Late Cretaceous Barun Goyot Formation of Mongolia," *American Museum Novitates,* No. 3045, pp. 1–24.

Chapter 2. Dinosaur Dream Time

p. 54. *Protoceratops* is related . . . The use of horns and frills in protoceratopsids and other neoceratopsians is discussed in Dodson, P., and Currie, P. J., 1990. "Neoceratopsia," pp. 593–618. In Weishampel et al. (eds.), op. cit.

p. 55. In *Velociraptor* the skull . . . The basic description of the dromaeosaur skeleton in Osborn, 1924, op. cit., has been expanded considerably by Ostrom, J. H., 1990. "Dromaeosauridae," pp. 269–79. In ibid. A vivid accounting is also given in Norell, M. A., Gaffney, E. S., and Dingus, L., 1995. *Discovering Dinosaurs.* Alfred A. Knopf, New York, i–xx + 204 pp.

p. 57. Nonetheless, the "fighting dinosaurs" . . . This remarkable discovery is related in Kielan-Jaworowska, Z., 1975. "Late Cretaceous dinosaurs and mammals from the Gobi desert," *American Scientist,* Vol. 63, pp. 150–59. Also see Lavas, op. cit.

p. 58. . . . a footprint, probably of a small lizard . . . Jerzykiewicz, T., Currie, P. J., Eberth, D. A., Johnston, P. A., Koster, E. H., and Zheng, J.-J., 1993. "Djadokhta Formation correlative strata in Chinese Inner Mongolia: an overview of the stratigraphy, sedimentary geology, and paleontology and comparisons with the type locality of the pre-Altai Gobi," pp. 2180–95. In Currie, op. cit.

p. 59. . . . the popular book . . . Crichton, M., 1990. *Jurassic Park.* Alfred A. Knopf, New York, 402 pp.

p. 61. These animals all shared . . . See Dodson and Currie, op. cit.

p. 63. Building such a classification . . . Biological classification developed over centuries, beginning with the writings of Aristotle. Two prominent works of this history cited here are Ray, J., 1693. *Synopsis Methodica Animalium Quadrupedum et Serpentini Generis.* 8° London, and Linnaeus, C., 1735. *Systema Naturae, sive Regna tria Naturae systematice proposita per Classes Ordines, Genera & Species.* Fol. Lugduni Batavorum. Subsequently, Linnaeus produced ten monumental editions of *Systema Naturae* between 1740 and 1759.

p. 64. There is a continuing debate . . . The so-called species problem—namely the question of defining species—is a major conundrum of comparative biology. The classic biological species definition (based on reproductive isolation) largely stems from Mayr, E., 1942. *Systematics and the Origin of Species.* Columbia University Press, New York. Opposing concepts, which emphasize the phylogenetic identity of species, are found in Simpson, G. G., 1944. *Tempo and Mode in Evolution.* Columbia University Press, New York. Another variation is presented by Hennig, W., 1966. *Phylogenetic Systematics.* University of Illinois Press, Urbana. The debate has taken on new dimensions exemplified in more recent papers by Queiroz, K. de, and Donoghue, M. J. 1988. "Phylogenetic systematics and the species problem," *Cladistics,* 4, pp. 317–38; and Nixon, K. C., and Wheeler, Q. D., 1992. "Extinction and the origin of species," pp. 119–43. In Novacek, M. J., and Wheeler, Q. D. (eds.) *Extinction and Phylogeny.* Columbia University Press, New York.

p. 66. According to some treatises . . . Weishampel et al. (eds.), op. cit.

p. 72. Relating species and their inclusive groups . . . Cladistics emerged from the studies of the German entomologist Willi Hennig in the 1950s (Hennig, W., 1950. *Grundzuge einer Theorie der phylogenetischen Systematik.* Deutscher Zentralverlag, Berlin) and was subsequently translated into English (Hennig, op. cit.). The American "school" of cladistics was promoted primarily by scientists at the American Museum of Natural History, represented by such works as Nelson, G., and Platnick, N. I., 1981. *Systematics and Biogeography—Cladistics and Vicariance.* Columbia University Press, New York; Eldredge, N., and Cracraft, J., 1980. *Phylogenetic Patterns and the Evolutionary Process.* Columbia University Press, New York; and Novacek and Wheeler (eds.), op. cit.

p. 72. The monophyly of Dinosauria . . . See also Norell et al., op. cit., and Gaffney, G., Dingus, L., and Smith, M., 1995. "Why cladistics?" *Natural History,* Vol. 104, No. 6, pp. 33–35.

p. 75. The sheer enormity of his coverage . . . Linnaeus, op. cit.

p. 76. But the actual investigation of the pattern . . . See also Novacek, M. J. "The meaning of systematics and the biodiversity crisis," pp. 101–8, in Eldredge, N. (ed.), 1992. *Systematics, Ecology, and the Biodiversity Crisis.* Columbia University Press, New York.

p. 79. The first three billion years of life . . . Many works deal with the early history of life. Among the best general treatments are those found in Schopf, J. W. (ed.), 1992. *Major Events in the History of Life.* Jones and Bartlett Publishers, Boston, pp. i–xv + 190 pp.; and Gould, S. J. (ed.), 1993. *The Book of Life.* W. W. Norton, New York, 256 pp.

p. 80. This baroque assemblage . . . Gould, *Wonderful Life,* op. cit.

p. 80. The Paleozoic Era . . . The earth calendar is under continual recalibration. An important technical reference is Harland, W. B., Armstrong, R. L., Cox, A. V., Craig, L. E., Smith, A. G., and Smith, D. G., 1989. *A Geologic Time Scale.* Cambridge University Press, Cambridge. More recent updates are in Berggren, W. A., Kent, D. V., Aubry, M.-P., and Hardenbol, J. (eds.), 1995. *Geochronology, Time Scales, and Global Stratigraphic Correlation.* Special Publication 54, SEPM (Society for Sedimentary Geology), Tulsa.

p. 81. . . . smallish, bipedal theropods like *Herrerasaurus* . . . See Sereno, P. C. 1995. "Roots of the family tree," *Natural History,* Vol. 104, No. 6, pp. 30–32.

p. 81. These include the famous Triassic animal *Coelophysis* . . Colbert, op. cit., pp. 139–42. Rowe, T., and Gauthier, J., 1990. "Ceratosauria," pp. 151–68. In Weishampel et al. (eds.), op. cit. Colbert, E. H., 1995. *The Little Dinosaurs of Ghost Ranch* Columbia University Press, New York. Schwartz, H. L., and Gillette, D. D., 1994. "Geology and taphonomy of the *Coelophysis* quarry, Upper Triassic Chinle Formation, Ghost Ranch, New Mexico," *Journal of Paleontology,* Vol. 68, pp. 1118–30.

p. 86. . . . the origin of the human lineage . . . Human evolution is the subject of a massive literature. One outstanding recent review is Tattersall, I., 1995. *The Fossil Trail.* Oxford University Press, Oxford, i–xi + 276 pp.

p. 87. The Holocene, beginning about 10,000 years ago . . . In Harland et al., op. cit., p. 118, the Holocene/Pleistocene boundary is dated at 10,000 years before present and is correlated "with a climatic event about then to be standardized in a varve [lake bed] sequence in Sweden." In any case, the boundary is rather sketchy and restrictive.

p. 88. The names endure, however, because of a curious history . . . See William B. Berry, 1968. *Growth of a Prehistoric Time Scale.* W. H. Freeman, San Francisco, 158 pp.

p. 89. "Stonebarrow, Black Ven, Ware Cliffs . . ." Fowles, J., 1970. *The French Lieutenant's Woman.* Signet, New York, p. 42.

p. 90. This decay constant is the key . . . The standard equation for radioactive decay is

$$\mathbf{t} = (1/\lambda) \ln [(\mathbf{D}/\mathbf{P}) + 1]$$

where $\mathbf{t}$ = age; λ = decay constant (fraction of parent disintegrating per unit time), $\mathbf{D}$ = number of daughter atoms produced by radioactive decay in the specimen; $\mathbf{P}$ = number of parent atoms in the specimen; and ln = the natural log ($\log_e$). Thus the age of a rock can be determined by knowing the ratio of the new isotope to the original isotope and the decay constant for the element in question. The higher the ratio, the older the rock. Harland et al., op. cit., pp. 190–219, provide a technical explanation of the equation and list the different decay constants for isotopes of uranium, thorium, potassium, and other elements.

p. 91. On a less tragic note . . . This legendary anecdote about Alexander Agassiz may be apocryphal (I first heard it as an undergraduate from my ecology professor). At any rate, Agassiz's profound knowledge of sea anemones and related marine creatures is undisputed.

CHAPTER 3. 1991—THE GREAT GOBI CIRCUIT

p. 94. "The combination of characters . . ." Owen, Sir Richard, 1842. Report on British Fossil Reptiles. Part II. *Report of the British Association for the Advancement of Science,* Eleventh Meeting, Plymouth, July 1841, p. 103. Owen claimed to derive this name and definition of Dinosauria from observations made on every relevant specimen known by 1840.

p. 101. "Featurelessness is the steppes' . . ." Theroux, P., 1975. *The Great Railway Bazaar.* Penguin Books, New York, pp. 43–44.

p. 102. . . . we reached Hurrendoch . . . This locality has yielded some fascinating dinosaurs as described by Barsbold, R., and Perle, A., 1984. "On the first new find of a primitive ornithomimosaur from the Cretaceous of M. P. R.," *Paleontol. Zh.* 1984, pp. 121–23. [In Russian.]

p. 105. As Roy Chapman Andrews wrote . . . Andrews, R. C., 1953. *All About Dinosaurs.* Random House, p. 32.

p. 107. Hadrosaurs . . . The group is reviewed in Weishampel, D. B., and Horner, J. R., 1990. "Hadrosauridae," pp. 534–61. In Weishampel et al. (eds.), op. cit.

p. 110. The martyrs were hunted down . . . The infamous purge of the 1930s in Mongolia is well documented by both written accounts and grisly remains. A brief recounting is provided by Storey, op. cit.

p. 112. . . . "illusionary town full of mysteries." In Efremov, op. cit.

p. 117. At Altan Ula . . . A popular account of the discovery of the Dragon's Tomb by the Russian team can be found in Lavas, op. cit., pp. 85–89.

p. 118. Something, of course, had to feed . . . The original description of *Tarbosaurus* is in Maleev, E. A., 1955. "New carnivorous dinosaurs from the Upper Cretaceous of Mongolia," *Doklady Akademie Nauk, S.S.S.R.*, 104, pp. 779–83. [In Russian.]

p. 123. Where were those brilliant skeletons . . . ? *Oviraptor* was first found by the Andrews American Museum Expeditions, but the Russians, Poles, and Mongolians subsequently found even better specimens. The smaller taxa, *Conchoraptor* and *Ingenia,* were retrieved from Kheerman Tsav. A summary of the group is provided by Barsbold, R., Maryańská, T., and Osmólska, H., 1990. "Oviraptorosauria," pp. 249–58. In Weishampel et al. (eds.), op. cit. A more recent study is Smith, D., 1992. "The type specimen of *Oviraptor philoceratops,* a theropod dinosaur from the Upper Cretaceous of Mongolia," *Neus Jahrsbuch Geol. Palaont. Abh.*, 186, pp. 365–88.

p. 125. "had slipped behind the earth's sharp edge" . . . Bowles, P., 1949. *The Sheltering Sky.* Ecco Press ed., 1978, p. 180.

p. 128. *Zalambdalestes* has elongate front incisors . . . Novacek, M. J., 1994. "A pocketful of fossils," *Natural History,* Vol. 103, No. 4, pp. 40–43.

CHAPTER 4. THE TERRAIN OF EONS

p. 131. "I pace upon the battlements . . ." Yeats, W. B., 1928. *The Tower.* pp. 85–90. In Williams, O. (ed.), 1952. *A Little Treasury of Modern Poetry.* Scribner, New York, 843 pp.

p. 133. Fossilized remains of fruit . . . The Nemegt environment is described in Lavas, op. cit., pp. 130–31.

p. 134. Nonetheless, the highest biomass . . . Bourlière, F., 1963. "Observations on the ecology of some large African mammals," pp. 43–54. In Howell, F. C., and Bourlière, F. (eds.). *African Ecology and Human Evolution.* Aldine, Chicago, 666 pp.

p. 134. The contrast between these rock units . . . Gradziński, R., Kielan-Jaworowska, Z., Maryańská, T., 1977. "Upper Cretaceous Djadokhta, Barun Goyot and Nemegt formations of Mongolia, including remarks on previous subdivisions," *Acta Geologica Polonica,* Vol. 27, No. 3, pp. 281–318.

p. 135. The geologists with the Andrews expeditions . . . Berkey, C., and Morris, F. K.,

1927. *Geology of Mongolia.* Natural History of Central Asia, 2, 475 pp. American Museum of Natural History.
p. 135. Gradziński et al., op. cit.
p. 140. . . . fossils may be the actual inspirations for the griffins . . . Mayor, A., 1991. "Griffin Bones, Ancient Folklore, and Paleontology," *Cryptozoology,* Vol. 10, 1991, pp. 16–41.
p. 140. This is not the first time fossils have been linked to well-known legends . . . Some of these earlier speculations, from the Cyclops to the Klagenfurt "dragon," are mentioned in Mayor, op. cit.
p. 141. ". . . According to the Issedonian-Scythian folklore . . ." Ibid., pp. 34–35.
p. 142. The brilliant William Diller Matthew . . . Matthew, W. D., 1915. "Climate and Evolution," *Annals of the New York Academy of Sciences,* Vol. 24, pp. 171–318.
p. 143. . . . the roots of humankind . . . Both Osborn and Matthew, as well as many others, were convinced that humans originated in Central Asia. An emphatic statement to that effect is given by Matthew, ibid., p. 209: "All authorities are to-day agreed in placing the center of dispersal of the human race in Asia. Its more exact location may be differently interpreted but the consensus of modern opinion would place it probably in or about the great plateau of central Asia."
p. 143. Alfred Lothar Wegener . . . The most often cited of Wegener's works on his theory of moving continents is Wegener, A., 1915. *The Origin of Continents and Oceans,* which was reworked and expanded in three editions published in 1920, 1922, and 1929. The appearance of each edition provoked an onslaught of criticism.
p. 145. Apparently the earth's magnetic field . . . The technical aspects of plate tectonics, such as the evidence of paleomagnetism, has been "translated" in numerous general works. A particularly successful popularization is Miller, R., 1983. *Continents in Collision.* Time-Life Books, Alexandria, 176 pp.
p. 153. Geographic isolation . . . is a major driving force . . . The notion that new species emerge from isolated populations is the cornerstone of Mayr's influential work, *Systematics and the Origin of Species,* op. cit.
p. 157. A warming trend . . . Temperature curves for the Cretaceous come from several data sources, notably isotopes of oxygen preserved in the shells of marine organisms. The evidence is nicely summarized in Benton, M., 1993. "Dinosaur Summer," pp. 126–67. In Gould, S. J. (ed.), *The Book of Life.*
p. 157. . . . this northern corridor was eminently livable. Clemens, W. A., and Allison, C. W., 1985. "Late Cretaceous terrestrial vertebrate fauna, North Slope, Alaska" (abstract), *Geological Society of America* (Abstracts with Program), 17.

CHAPTER 5. 1992—THE BIG EXPEDITION

p. 160. "As I rode I became part of the world . . ." Dodwell, C., 1979. *Travels with Fortune.* W. H. Allen.
p. 163. . . . like an *Allergorhai horhai* . . . The Mongolian government actually requested that Andrews search for this two-foot-long sausage-like, headless, limbless monster. The *Allergorhai horhai* was allegedly extremely poisonous to the touch; it slithered through the sands of the western Gobi. Andrews seems to only half dismiss this creature as myth and, by his own accounting, he at least attempted to find it. No one he encountered ever stated that they had seen an *Allergorhai horhai* first hand, but many nomads claimed to know others who had seen it. Andrews officially admits failure on this front (*The New Conquest* . . . , p. 62):

"I report it here with the hope that future explorers of the Gobi may have better success than we did in running to earth the *Allergorhai horhai.*"

Neither our team, nor anyone else, has found one.

p. 167. Permian reptiles . . . A technical report on Peter Vaughn's studies in New Mexico is Vaughn, P. P., 1969. "Early Permian vertebrates from southern New Mexico and their paleozoogeographic significance," *Los Angeles County Museum Contributions to Science,* No. 166, pp. 1–22.
p. 175. . . . a brilliant new light . . . on . . . mammalian history. The statements by Osborn, Granger, Simpson, and other authorities concerning the significance of the Gobi Cretaceous mammals are quoted extensively in Andrews, op. cit., pp. 271–74.
p. 179. "A recent linguistic analysis . . ." Mayor, A., 1991. Op. cit., p. 19.
p. 181. This "turtle death pond" . . . The specimens are now under study by Eugene Gaffney, a curator of vertebrate paleontology at the American Museum of Natural History.

CHAPTER 6. DINOSAUR LIVES—FROM EGG TO OLD AGE

p. 184. "It is most difficult . . ." Darwin, C. 1859. *The Origin of Species.* Mentor, 318 pp.
p. 186. In combination with an unusually massive . . . pelvic girdle . . . Borsuk-Bialynicka, M., 1977. "A new camarasaurid sauropod *Opisthocoelicaudia skarzynskii,* gen. n. sp. n. from the Upper Cretaceous of Mongolia," *Palaeontologica Polonica,* Vol. 37, pp. 5–64.
p. 186. A number of sauropods . . . Nowinski, A. 1971. "*Nemegtosaurus mongoliensis* n. gen., n. sp. (Sauropoda) from the Uppermost Cretaceous of Mongolia," *Palaeontologica Polonica,* Vol. 25, pp. 57–81.
p. 187. . . . sauropods were the ultimate dinosaurs. A recent popular account of the gigantic sauropods is Gillette, D. D., 1994. *Seismosaurus, The Earth Shaker.* Columbia University Press, New York, 205 pp.
p. 188. . . . wrestled with sauropod classification. McIntosh, J. S., 1990. "Sauropoda," pp. 345–401. In Weishampel et al. (eds.), op. cit.
p. 189. . . . terrestrial, high-browsing lifestyle. Various theories for sauropod behavior are found in Riggs, E., 1904, "Structure and relationships of opisthocoelian dinosaurs." Pt. II. The Brachiosauridae. *Publications of the Field Columbian Museum of Geology,* Vol. 2, pp. 229–48. Coombs, W. P., Jr. 1975. "Sauropod habits and habitats," *Palaeogeography, Palaeoclimatology, Palaeoecology,* Vol. 17, pp. 1–33; Dodson, P., 1990. "Sauropod paleoecology," pp. 402–7. In Weishampel et al. (eds.), op. cit.
p. 189. The landlubber lifestyle . . . Dodson, P., Behrensmayer, A. K., Bakker, R. T., and McIntosh, J. S., 1980. "Taphonomy and paleoecology of the Upper Jurassic Morrison Formation," *Paleobiology,* Vol. 6, pp. 208–32.
p. 190. The question of sauropod posture . . . Dodson, P., 1990. op. cit., p. 404.
p. 191. A special net of blood vessels . . . Ibid., p. 405.
p. 192. . . . gizzard stones. Gillette, D. D., 1995. "True Grit," *Natural History,* Vol. 104, No. 6, pp. 41–43.
p. 193. Unlike the . . . paintings of Charles Knight . . . The influential dinosaur reconstructions of Charles Knight appear in works of Colbert and many other popular books on dinosaurs. Despite a few inaccuracies, they provide a riveting view on the prehistoric past. The artist's own methods are described in Knight, C. R., 1947. *Animal Drawing: Anatomy and Action for Artists.* McGraw-Hill, New York.
p. 193. . . . using the massive hind legs and tail as a tripod. Osborn, H. F., 1899. "A skeleton of *Diplodocus,*" Memoirs, American Museum of Natural History, Vol. 1, pp. 191–214.
p. 195. But the evidence that dinosaurs were endothermic . . . The subject has attracted a voluminous literature. For reviews, see Farlow, J. O., 1990. Part II. Dinosaur energetics and thermal biology," pp. 43–55. In Weishampel et al. (eds.), op. cit; and Norell et al., op. cit., pp. 52–55.
p. 196. Careful analyses . . . Ricqlès, A. de, 1990. "Zonal 'growth rings' in dinosaurs," *Mod-*

ern Geology, Vol. 15, pp. 19–48. Reid, R. E. H., 1987. "Bone and dinosaurian 'endothermy,'" *Modern Geology,* Vol. 11, pp. 133–54.

p. 196. Finally, some late-breaking CAT-scan studies . . . Ruben, J. A., Leitch, A., and Hillenius, W., 1995.* Abstract, Annual Meeting of the Society of Vertebrate Paleontology. *Journal of Vertebrate Paleontology,* 50.

p. 197. . . . the emerging studies of isotopes . . . Barrick, E., and Shawers, W. J., 1995. "Oxygen isotope variability in juvenile dinosaurs (*Hypacrosaurus*): evidence for thermoregulation," *Paleobiology,* Vol. 214, pp. 552–60. This interesting analysis is based on the fluctuation of oxygen isotopes in fossil bone phosphate. Such a fluctuation is related to body temperature. In endotherms, where body temperature is held roughly constant, isotope variation should be low. This seems to be the case in the dinosaur *Hypacrosaurus.*

p. 197. . . . warm-blooded birds are a type of dinosaur. This analogy, however, is eroded by a new problem. A recent study of fossil bird bone suggests some of these species were ectothermic, weighing against the notion that their non-bird dinosaur relatives were necessarily endothermic; see Chinsamy, A., Chiappe, L. M., and Dodson, P., 1995. "Mesozoic avian bone microstructure: physiological implications," *Paleobiology,* Vol. 21, pp. 561–74.

p. 197. . . . passively warm-blooded . . . Despite the general recognition of at least passive (or inertial) endothermy in sauropods (Farlow, op. cit.), this has been disputed by Weaver, J. C., 1983. "The improbable endotherm: the energetics of the sauropod dinosaur *Brachiosaurus,*" *Paleobiology,* Vol. 9, pp. 173–82.

p. 198. . . . sauropod species were highly gregarious . . . Dodson et al., op. cit.

p. 199. . . . ingenious work on fossil trackways . . . Lockley, M., 1995. "Track records," *Natural History,* Vol. 104, No. 6, pp. 46–51.

p. 199. . . . sauropods migrated annually . . . Dodson, op. cit.

p. 200. . . . rich ceratopsid samples . . . Sampson, S. D., 1995. "Horns, herds, and hierarchies," *Natural History,* Vol. 104, No. 6, pp. 36–40.

p. 203. These crests have excited some paleontologists . . . Hopson, J. A., 1975. "The evolution of cranial display structures in hadrosaurian dinosaurs," *Paleobiology,* Vol. 1, pp. 21–43. Weishampel, D. B., and Horner, J. R., 1990. "Hadrosauridae," pp. 534–61. In Weishampel et al., op. cit.

p. 203. The idea was refined . . . Ibid.

p. 204. Both the Nemegt and Barun Goyot . . . Maryańská, T., and Osmólska, H., 1974. "Pachycephalosauria, a new suborder of ornithischian dinosaurs," *Palaeontologica Polonica,* Vol. 30, pp. 45–102.

p. 205. . . . pachycephalosaurs had extraordinarily thick skulls. Ibid., and Maryańská, T., 1990. "Pachycephalosauria," pp. 564–77. In Weishampel et al., op. cit.

p. 205. . . . skull roofing bones . . . The first to suggest that the thick skulls of pacycephalosaurs were a combat device was Colbert, E. H., 1955. *Evolution of the Vertebrates.* Wiley, New York, 479 pp. See also Galton, P. M., and Sues, H.-D., 1983. "New data on pachycephalosaurid dinosaurs (Reptilia, Ornithischia) from North America," *Canadian Journal of Earth Sciences,* Vol. 20, pp. 462–73.

p. 208. Instead, they suggest that these animals were primarily scavengers . . . Horner, J. A., and Lessem, D., 1993. *The Complete T. Rex.* Simon & Schuster, New York, 239 pp.

p. 208. The famous skeleton of *Tyrannosaurus* . . . Norell et al., op. cit., and Norell, M. A., personal communication.

p. 208. . . . murderous manner of assault . . . Molnar, R. E., and Farlow, J. O., 1990. "Carnosaur paleobiology," pp. 210–24. In Weishampel et al., op. cit.

*Respiratory turbinates and the metabolic status of some theropod dinosaurs and *Archaeopteryx*

p. 209. . . . designed to disembowel . . . Auffenberg, W., 1990. *The Behavioral Ecology of the Komodo Dragon.* University Presses of Florida, Gainesville, 406 pp.
p. 210. It has not escaped the notice of some paleontologists . . . Molnar and Farlow, op. cit.
p. 212. . . . new studies of trackways . . . Lockley, op. cit.
p. 212. Sometimes the clues left by fossils . . . Kielan-Jaworowska, 1969, *Hunting for Dinosaurs,* pp. 139–43.
p. 213. The Altan Ula arms . . . Osmólska, H., and Roniewicz, E., 1970. "Deinocheiridae, a new family of theropod dinosaurs," *Palaeontologica Polonica,* Vol. 21, pp. 5–19. Also see Norman, D. B., 1990. "Problematic Theropoda: 'coelurosaurs,'" pp. 280–305. In Weishampel et al., op. cit., and Lavas, op. cit., pp. 94–95.
p. 214. . . . petrified genitals of a giant antediluvian human . . . The specimen in question, now lost, is thought to be the distal femur of the theropod *Megalosaurus.* It was first described as belonging to a giant human by Robert Plot in 1676 and was named *Scrotum humanum* by R. Brooks in 1763. This history is reviewed by Norman, D., 1994. *The Illustrated Encyclopedia of Dinosaurs.* Crescent Books, New York, 208 pp.
p. 214. . . . Richard Owen coined the word "dinosaur." Owen, op. cit.
p. 214. The first alleged dinosaur eggs . . . Pouech, op. cit.; Matheran, op. cit.; Gervais, op. cit.
p. 214. With the retrieval of the Gobi fossil eggs . . . Andrews, op. cit., p. 208.
p. 214. Eggs were collected and identified . . . A brief history of dinosaur egg discoveries is provided in Lessem, D., 1992. *Kings of Creation.* Simon and Schuster, New York, 367 pp.
p. 214. But this chain of findings . . . Ibid.
p. 215. . . . Jack Horner and his team . . . Ibid., and Horner, J. R., 1982. "Evidence for colonial nesting and 'site fidelity' among ornithischian dinosaurs." *Nature,* Vol. 297, pp. 675–76.
p. 215. Horner worked this locality . . . Ibid.
p. 215. Add this to the hundreds of eggs . . . Egg shapes and sizes are discussed in Norell, M. A., 1991. *All You Need to Know About Dinosaurs.* Sterling Publishing, New York, 96 pp. See also Norell et al., op. cit.
p. 216. Why this comparative scarcity of eggs . . . ? Carpenter, K., 1982. "Baby dinosaurs from the Late Cretaceous Lance and Hell Creek formations and a description of a new species of theropod," *Contributions to Geology of the University of Wyoming,* Vol. 20, pp. 123–34.
p. 217. Embryos of dinosaurs . . . Norell, M. A., 1995. "Origins of the feathered nest," *Natural History,* Vol. 104, No. 6, pp. 58–61.
p. 219. Humans and other mammals and birds grow . . . Many aspects of growth and ontogeny are reviewed in Gould, S. J., 1977. *Ontogeny and Phylogeny.* Belknap Press, Harvard University, Cambridge, Massachusetts, 501 pp.
p. 220. . . . estimate growth rates in dinosaurs . . . Case, T. J., 1978. "Speculations on the growth rate and reproduction of some dinosaurs," *Paleobiology,* Vol. 4, pp. 320–28.
p. 220. These growth estimates, however, are very shaky. For criticisms, see Norell, 1991, op. cit., and Norell et al., 1995, op. cit.
p. 221. There seem to be many *unperceived* hazards . . . Darwin, op. cit.
p. 222. In the early years of evolutionary biology . . . An influential school of thought opted for a strongly directional form of evolution called orthogenesis. This perspective was further modified to embrace a notion of evolution as moving resolutely toward an ideal state, a concept called aristogenesis. The latter view is emphatically represented in Osborn, H. F., 1934. "Aristogenesis, the creative principle in the origin of species," *American Naturalist,* Vol. 68, pp. 193–235.
p. 223. We now view such presumptions . . . A review of early concepts of aristogenesis and

the problem of studying the evolution of adaptation is in Novacek, M. J., 1996. "Paleontological data and the study of adaptation." In Lauder, G. (ed.), *Adaptation.* Academic Press, San Diego (In press.) Also see Lauder, G., 1981. "Form and function: Structural analysis in evolutionary morphology," *Paleobiology,* Vol. 7, pp. 430–42.

CHAPTER 7. 1993—XANADU

p. 229. . . . that most unusual avian theropod . . . Altangerel Perle, Norell, M. A., Chiappe, L. M., and Clark, J. M., 1993. "Flightless bird from the Cretaceous of Mongolia," *Nature,* Vol. 362, pp. 623–26. Norell, M. A., Chiappe, L. M., and Clark, J., 1993. "A new limb on the avian family tree," *Natural History,* Vol. 102, No. 9, pp. 38–43.

p. 236. Osborn originally thought that *Saurornithoides* . . . Osborn, H. F., 1924. "Three new Theropoda . . . ," op. cit., pp. 3–7. Ostrom, J. H., 1969. "Osteology of *Deinonychus antirrhopus,* an unusual theropod from the Lower Cretaceous of Montana," *Peabody Museum of Natural History Bulletin,* 30, pp. 1–165. Barsbold, R., 1974. "Saurornithoididae, a new family of small theropod dinosaurs from central Asia and North America," *Palaeontologica Polonica,* Vol. 30, pp. 5–22. Osmólska, H., and Barsbold, R., 1990. "Troodontidae." pp. 259–68. In Weishampel et al. (eds.), op. cit.

p. 237. Oviraptorids are . . . unusual theropods. Osborn, H. F., 1924, op. cit. Barsbold, R., et al., 1990. "Oviraptorosauria," op. cit. Smith, D., 1992, op. cit. Norell et al., 1995, op. cit., pp. 124–25.

CHAPTER 8. FLYING DINOSAURS AND HOPEFUL MONSTERS

p. 243. "I believe there were no flowers then . . ." Lawrence, D. H., "Humming-Bird," 1929. *Collected Poems; Birds, Beasts, and Flowers!* Penguin, Black Sparrow Press, Santa Rosa.

p. 244. One member of the coelurosaur radiation . . . Osmólska, H., Roniewicz, E., and Barsbold, R., 1972. "A new dinosaur, *Gallimimus bullatus* n. gen., n. sp. (Ornithomimidae) from the Upper Cretaceous of Mongolia," *Palaeontologica Polonica,* Vol. 27, pp. 103–43. Barsbold, R., and Osmólska, H. "Ornithomimosauria," pp. 225–44. In Weishampel et al. (eds.), op. cit., pp. 225–44.

p. 244. . . . maniraptorans share an unusual design . . . Norell et al., 1995, op. cit.

p. 245. . . . much of the . . . evolution of birds took place in the Mesozoic. Chiappe, L., 1995. "The diversity of early birds." *Natural History,* Vol. 104, No. 6, pp. 52–55.

p. 246. The Mesozoic diversity . . . Ibid.

p. 247. The skeleton was named *Archaeopteryx* . . One of the most important of all vertebrate fossils is the subject of a very large scientific literature. The highlights include: Huxley, T. H., 1868. "On the animals which are most nearly intermediate between birds and reptiles," *Annual Magazine of Natural History,* London (4), 2, pp. 66–75. Heilmann, G., 1916. *The Origin of Birds.* Witherby, London, 208 pp. Ostrom, J. H., 1974. "*Archaeopteryx* and the origin of flight," *Quarterly Review of Biology,* Vol. 49, pp. 27–47. Wellnhofer, P., 1988. "A new specimen of *Archaeopteryx,*" *Science,* Vol. 240, pp. 1790–92. Gauthier, J., 1986. "Saurischian monophyly and the origin of birds," pp. 1–55. In Padian, K. (ed.), *The Origin of Birds and the Evolution of Flight.* Memoirs, California Academy of Sciences, 8. A recent review of these and other papers is provided by Ostrom, J. H., 1994. "On the origins of birds and of avian flight," pp. 160–77. In Prothero, D. R., and Schoch, R. M. (eds.). *Major Features of Evolution.* Publication of the Paleontological Society, 1994, No. 7, 270 pp.

p. 249. Some workers . . . strenuously reject the bird-dinosaur hypothesis. Martin, L. D., 1991. "Mesozoic birds and the origin of birds." pp. 485–540. In Schultze, H. P., and Truebe, L. (eds.). *Origins of the Higher Groups of Tetrapods.* Comstock, Ithaca. Feduccia, A., and

Wild, R., 1993. "Birdlike characters in the Triassic archosaur *Megalancosaurus,*" *Naturwissenschaften*, 80, pp. 564–66.
p. 251. *Avimimus incertae sedis* . . . Kurzanov, S. M., 1981. ["On the unusual theropods from the Upper Cretaceous of Mongolia."] *Transactions of the Soviet-Mongolian Paleontological Expeditions,* Trudy 15, pp. 39–50. [In Russian with English summary.]
p. 251. Notable among these is *Harpymimus.* Barsbold and Perle, op. cit.
p. 252. . . . it couldn't fly very well. Ostrom, J. H., 1975. "Bird flight; how did it begin?" *American Scientist,* Vol. 67, pp. 46–56. Bakker, R. T., 1975. "Dinosaur renaissance," *Scientific American,* Vol. 232, 58–78.
pp. 252–53. Under this view, *Archaeopteryx* could fly . . . Ruben, J., 1991. "Reptilian physiology and the flight capacity of *Archaeopteryx,*" *Evolution,* Vol. 45, pp. 1–17.
p. 253. . . . some remarkable results from detailed histology . . . Chinsamy, A., Chiappe, L. M., and Dodson, P., 1994. "Growth rings in Mesozoic birds," *Nature,* Vol. 368, pp. 196–97.
p. 254. . . . multiple . . . purposes for feathers . . . Randolph, S., 1994. "The relative timing of the origin of flight and endothermy: evidence from the comparative biology of birds and mammals," *Zoological Journal of the Linnaean Society,* Vol. 112, pp. 389–97.
p. 255. Traditionally, ranks of the Linnaean hierarchy . . . Linnaeus, 1735–59. *Systema Naturae.*
p. 257. The approach was clarified . . . Hennig, op. cit.
p. 257. The rules of classification . . . Queiroz, K. de, and Gauthier, J., 1992. "Phylogenetic taxonomy," *Annual Review of Ecology and Systematics,* Vol. 23, pp. 449–80.
p. 258. It is easy to see how this new system . . . Ibid.
p. 259. This is *Mononykus* . . . Altangerel Perle et al., op. cit.; Norell et al., 1993, *Natural History.* Perle, A., Chiappe, L. M., Barsbold, R., Clark, J. M., Norell, M. A., 1994. "Skeletal morphology of *Mononykus olecranus* from the late Cretaceous of Mongolia. *American Museum Novitates* 3105:1–29.
p. 261. The story of *Mononykus* . . . Ibid.
p. 262. . . . *Protoceratops* growth stages . . . Dodson, P., 1976. "Quantitative aspects of relative growth and sexual dimorphism in *Protoceratops,*" *Journal of Paleontology,* Vol. 50, pp. 929–40.
p. 263. An ontogeny is effectively a history . . . A general review of ontogeny and evolution is provided by Gould, *Ontogeny and Phylogeny.* There are, however, strongly contrasting views, notably Nelson, G. G., 1978. "Ontogeny, phylogeny, and the biogenetic law," *Systematic Zoology,* Vol. 27, pp. 324–45. Also see Nelson and Platnick, op. cit.
p. 264. There are special cases . . . Paedomorphosis and neoteny are reviewed and defined in Gould, *Ontogeny and Phylogeny.*
p. 266. The drama of these experimental manipulations . . . Ibid.
p. 267. Yet the fossil history of horses . . . The horse record is one of the outstanding documents of evolutionary and phylogenetic patterns. Classic treatments are Matthew, W. D., 1926. "The evolution of the horse; a record and its interpretation," *Quarterly Review of Biology,* 1, pp. 139–85; and Simpson, G. G., 1953. *The Major Features of Evolution.* Columbia University Press, New York. For a modern review, see MacFadden, B. J., 1992. "Interpreting extinctions from the fossil record: methods, assumptions, and case examples using horses (family Equidae)," pp. 17–45. In Novacek, M. J., and Wheeler, Q. D. (eds.), *Extinction and Phylogeny.*

CHAPTER 9. 1994—BACK TO THE BONANZA

p. 269. "Bones picked an age ago . . ." Muir, E., "The Road." Williams, O. (ed.), op. cit., p. 251.

p. 270. But the embryo in the half shell . . . The results of this discovery were reported by Norell, M. A., Clark, J. M., Dashzeveg, D., Barsbold, R., Chiappe, L., Davidson, A. R., McKenna, M. C., Perle, A., and Novacek, M. J., 1994. "A theropod dinosaur embryo and the affinities of the Flaming Cliffs dinosaur eggs," *Science,* Vol. 266, pp. 779–82. A popular account was published by Clark, J. M., 1995. "An egg thief exonerated," *Natural History,* Vol. 104, No. 6, pp. 56–57.
p. 271. The discovery of the true identity of the dinosaur embryo . . . Ibid.
p. 272. His "big mamma (or papa)" oviraptorid . . . Norell, M. A., Clark, J. M., Chiappe, L. M., and Dashzeveg, D. M., 1995. "A nesting dinosaur," *Nature,* Vol. 378, pp. 774–76.
p. 278. . . . our Komodo-like lizard . . . Norell et al., 1992. *Estesia* . . .
p. 280. Many specimens of *Shamosuchus* . . First described in Mook, C. C., 1924. "A new crocodilian from Mongolia," *American Museum Novitates,* No. 117, pp. 1–5.
p. 280. Gazin had been working . . . Gazin, C. L., 1956. "Paleocene mammalian faunas of the Bison Basin in south-central Wyoming," *Smithsonian Miscellaneous Collections,* Vol. 131, No. 6, pp. 1–57.
p. 280. This is a very odd observation . . . The offending statement is Gazin, ibid., p. 12. "The only nonmammalian specimens encountered during the collecting were four fragmentary dentaries, two portions of a maxillae, and a premaxilla of a lizard."
p. 283. This "family" appears to represent . . . Early finds of Cretaceous mammals from Mongolia were described in many papers, including Gregory and Simpson, 1926, op. cit., and a large series by Zofia Kielan-Jaworowska. *Asioryctes* is compared with other Cretaceous placental mammals in Kielan-Jaworowska, Z., 1984. "Evolution of the therian mammals in the Late Cretaceous of Asia. Part VII. Synopsis," *Palaeontologica Polonica,* No. 46, pp. 173–83.
p. 284. The *Mononykus* skull . . . Altangerel et al., 1993, op. cit., Norell et al., 1993, *Natural History,* op. cit., and Perle et al., 1994, op. cit. A full description of the new specimen of *Mononykus* with the skull has not yet been published.
p. 285. On the Chinese side . . . Jerzykiewicz, et al., op. cit.
p. 287. In the late fall of 1994 . . . The oviraptorid embryo appeared in Norell et al., 1994, *Science,* op. cit.
p. 287. I labored on a paper . . . Dashzeveg, D., Novacek, M. J., Norell, M. A., Clark, J. M., Chiappe, L. M., Davidson, A., McKenna, M. C., Dingus, L., Swisher, C., and Perle, A., 1995. "Extraordinary preservation in a new vertebrate assemblage from the Late Cretaceous of Mongolia," *Nature,* 374, pp. 446–49.

CHAPTER 10. DISASTERS, VICTIMS, AND SURVIVORS

p. 298. Reconstructions by . . . Shuvalov, V. F., 1982. "Paleogeography and history of the development of lacustrine systems of Mongolia in Jurassic and Cretaceous." In Martinson, G. G. (ed.). *Mezozoiske ozernye basseiny Mongolii.* Nauka, Leningrad, pp. 18–80 [In Russian].
p. 299. In the memoirs of the C.A.E. . . . Andrews, *The New Conquest* . . . , p. 308.
p. 301. The whole area of inquiry . . . Efremov, J. A., 1940. "Taphonomy, a new branch of paleontology," *Pan-American Geologist,* Vol. 74, pp. 81–93.
p. 302. Remember what Osborn said . . . Osborn, 1924. "Three new Theropoda," op. cit., p. 9.
p. 302. Actually, the investigation of the Gobi morgue . . . Jerzykiewicz et al., 1993, op. cit.
p. 303. Also noted were the famous "fighting dinosaurs" . . . Kielan-Jaworowska, 1975, *American Scientist,* op cit. Also see Lavas, 1993, op. cit.
p. 304. As our . . . fossil discoveries accumulated . . . Dashzeveg et al., 1995, op. cit.
pp. 307–10. The Cretaceous Disaster. Perhaps the most intensely discussed of all issues regarding the history of the Mesozoic was its termination. A few important papers include:

Alvarez, L. W., 1983. "Experimental evidence that an asteroid impact led to the extinction of many species 65 million years ago," *Proceedings of the National Academy of Science,* Vol. 80, pp. 627–42. Archibald, D. A., 1991. "Survivorship patterns of non-marine vertebrates across the Cretaceous-Tertiary (K-T) boundary in the western U.S.," pp. 1–2. In Kielan-Jaworowska, Z., Heintz, N., and Nakrem, H. A. (eds.). *Fifth Symposium on Mesozoic Terrestrial Ecosystems and Biota. Extended Abstracts.* Contributions from the Paleontological Museum, University of Oslo, No. 364. Clemens, W. A., 1986. "The evolution of the terrestrial vertebrate fauna during the Cretaceous-Tertiary transition," pp. 63–86. In Elliot, D. K. (ed.) *The Dynamics of Extinction.* Wiley Interscience, New York. Florentin, J.-M., Maurrasse, R., Sen, G., 1991. "Impacts, tsunamis, and the Haitian Cretaceous-Tertiary boundary layer," *Science,* Vol. 252, pp. 1690–93. Glen, W., 1990. "What killed the dinosaurs?" *American Scientist,* Vol. 78(4), pp. 354–70. Hunter, J., 1994. "Lack of a high body count at the K-T boundary," *Journal of Paleontology,* Vol. 68, No. 5, p. 1158. Sheehan, P. M., Fatovsky, D. E., Hoffman, R. G., Berghaus, C. B., and Gabriel, D. L., 1991. "Sudden extinction of the dinosaurs: Latest Cretaceous, Upper Great Plains, U.S.A.," *Science,* Vol. 254, pp. 835–39. Stanley, S. M., 1987. *Extinctions.* Scientific American Books. Williams, M. E., 1994. "Catastrophic versus noncatastrophic extinction of the dinosaurs: testing, falsifiability, and the burden of proof," *Journal of Paleontology,* Vol. 68, pp. 183–90.

p. 310. The earliest mammal-like . . . Mesozoic mammal history is massively documented. Of historical interest is Simpson's review in Simpson, G. G., 1971. "Concluding remarks: Mesozoic mammals revisited," pp. 181–98. In Kermack, D. M., and Kermack, K. A. (eds.). "Early Mammals," *Linnaean Society Zoological Journal,* Vol. 50, Suppl. 1. Two primary volumes of collected papers are Lillegraven, J. A., Kielan-Jaworowska, Z., and Clemens, W. A. (eds.), 1979. *Mesozoic Mammals. The First Two-Thirds of Mammalian History.* University of California Press, Berkeley; and Szalay, F. S., Novacek, M. J., and McKenna, M. C. (eds.), 1993. *Mammal Phylogeny (Vol. 1): Mesozoic Differentiation, Multituberculates, Monotremes, Early Therians, and Marsupials.* Springer Verlag, New York, 249 pp.

p. 310. . . . forms like triconodonts . . . Crompton, W. A., and Jenkins, F. A., 1979. "Origin of mammals." In Lillegraven et al. (eds.), op. cit., pp. 59–73. Crompton, A. W., and Luo, Z., 1993. "Relationships of the Liassic mammals *Sinoconodon, Morganucodon oehleri,* and *Dinnetherium."* pp. 30–44. In Szalay et al. (eds.), op. cit.

p. 311. . . . archaic multituberculates . . . Kielan-Jaworowska, Z., 1971. "Results of the Polish-Mongolian paleontological expeditions. Part III. Skull structure and affinities of the Multituberculata," *Palaeontologica Polonica,* Vol. 25, 5–41. Desui, M., 1993. "Cranial morphology and multituberculate relationships," pp. 63–74. In Szalay et al. (eds.), op. cit. Simmons, N., 1993. "Phylogeny of the Multituberculata," pp. 146–64. In Szalay et al. (eds.), op. cit.

pp. 311–14. Inauspicious Antecedents. The lines leading to the modern groups of mammals are well dissected. The Gobi mammals are covered in Gregory and Simpson, 1926, op. cit.; Kielan-Jaworowska, Z., Bown, T. M., and Lillegraven, J. A., 1979. "Eutheria." In Lillegraven et al. (eds.), op. cit. Kielan-Jaworowska, 1984, *Palaeontologica Polonica,* op. cit. More general patterns are treated in Rowe, T., 1993. "Phylogenetic systematics and the early history of mammals," pp. 129–45. In Szalay et al. (eds.), op. cit. Wible, J. R., and Hopson, J. A., 1993. "Basicranial evidence for early mammal phylogeny," ibid., pp. 45–62.

p. 311. There are three major clades of living mammals . . . This breakthrough in classification stems from the brilliant paper by Blainville, H. M. D., 1816. "Prodome d'une nouvelle distribution systématique de règne animal," *Bulletin Societé Philomatique,* 1816, pp. 67–81.

p. 311. Monotremes have a . . . poor fossil record . . . But it is getting better. See Archer, M., Flannery, F., Ritchie, A., and Molnar, R. E., 1985. "First Mesozoic mammal from Aus-

tralia: An early Cretaceous monotreme," *Nature,* Vol. 318, pp. 363–66. Pascual, R., Archer, M., Jaureguizar, E. O., Prado, J. L., Godthelp, H., and Hand, S. J., 1992. "First discovery of monotremes in South America," *Nature,* Vol. 356, pp. 704–6.
p. 312. Originally, *Deltatheridium* was regarded . . . Butler, P., and Kielan-Jaworowska, Z., 1976. "Is *Deltatheridium* a marsupial?" *Nature,* Vol. 245, pp. 105–6.
p. 313. Nessov conducted these explorations . . . Before his untimely death, Lev Nessov was greatly expanding our knowledge of Mesozoic mammal history in Asia. Among his papers are Nessov, L. A., 1985. "New Cretaceous mammals of the Kizylkum desert," *Vest. Leningradskogo University Geology and Geografik,* Vol. 17, pp. 1–18. [In Russian.] Nessov, L. A., and Kielan-Jaworowska, Z., 1991. "Evolution of Cretaceous Asian therian mammals," pp. 51–52. In Kielan-Jaworowska, et al. (eds.). *Fifth Symposium on Mesozoic Terrestrial Ecosystems . . . ,* op. cit.
p. 313. . . . the exquisite skull of *Zalambdalestes* at Tugrugeen. This analysis is described in Novacek, M., 1994. "A pocketful of fossils," *Natural History,* Vol. 103, No. 4, pp. 40–43.
pp. 315–17. Lifestyles of the Small and Inconspicuous. Also described in ibid. Another reference is Kielan-Jaworowska, Z., and Gambaryan, P. P., 1994. "Postcranial anatomy and habits of Asian multituberculate mammals," *Fossils and Strata,* Vol. 36, pp. 1–92.
pp. 317–20. The Great Radiation. Interest in the great radiation of placental mammals is rooted in the classic studies of Gregory, W. K. (1910). "The orders of mammals," *Bulletin of the American Museum of Natural History,* Vol. 27, pp. 1–524; and Simpson, G. G., 1945. "The principles of classification and a classification of mammals," ibid., Vol. 85, pp. 1–350. This interest was rekindled in the late 1970s and early 1980s with the publications of McKenna, M. C., 1975. "Toward a phylogenetic classification of the Mammalia," pp. 21–46. In Luckett, W. P., and Szalay, F. S. (eds.). *Phylogeny of the Primates.* Plenum Press, New York; and Novacek, M. J., 1982. "Information for molecular studies from anatomical and fossil evidence on higher eutherian phylogeny," pp. 3–41. In Goodman, M. (ed.). *Macromolecular Sequences in Systematic and Evolutionary Biology.* Plenum Press, New York. More recent reviews include McKenna, M. C., 1987. "Molecular and morphological analysis of high-level mammalian interrelationships," pp. 55–93. In Patterson, C. (ed.). *Molecules and Morphology in Evolution: Conflict or Compromise?* Cambridge University Press, Cambridge. Novacek, M. J., 1990. "Morphology, paleontology, and the higher clades of mammals," pp. 507–43. In Genoways, H. H. (ed.). *Current Mammalogy* (Vol. 2); Novacek, M. J., 1992a. "Mammalian phylogeny: shaking the tree," *Nature,* Vol. 356, pp. 121–25; Novacek, M. J., 1992b. "Fossils, topologies, missing data, and the higher level phylogeny of eutherian mammals," *Systematic Biology,* Vol. 41, pp. 58–73. Also numerous papers in Szalay, F. S., Novacek, M. J., and McKenna, M. C. (eds.), 1993. *Mammal Phylogeny (Vol. 2): Placentals.* Springer Verlag, New York. 321 pp.
pp. 319–20. . . . DNA sequences in certain genes. The question of the radiation of placental mammals has attracted an impressive spate of work on DNA in living mammals. Some notable papers are Adkins, R. M., and Honeycutt, R. L. (1991). "Molecular phylogeny of the superorder Archonta," *Proceedings of the National Academy of Sciences,* Vol. 88, pp. 1–5. Bailey, W. J., Slightom, J. L., and Goodman, M., 1992. "Rejection of the 'flying primate' hypothesis by phylogenetic evidence from the e-globin gene," *Science,* Vol. 256, pp. 86–89.
p. 320. Mammals get bigger . . . The Cenozoic history of mammals is covered in myriad volumes. Notable references are Carroll, R. L., 1988. *Vertebrate Paleontology and Evolution.* W. H. Freeman and Co., New York; and Prothero, D. R., 1994. "Mammalian Evolution," pp. 238–70. In Prothero, D. R., and Schoch, R. M. (eds.), op. cit.
p. 321. Indeed, E. O. Wilson's watershed book . . . Wilson, E. O., 1992. *The Diversity of Life.* Belknap Press, Harvard University, Cambridge, Massachusetts, 424 pp.

p. 322. "And there, there overhead . . ." MacLeish, A. *The End of the World,* p. 343. In Williams, O. (ed.), op. cit.

CHAPTER 11. THE SECRET HISTORY OF LIFE

p. 323. "Having seen this life's dance . . ." A number of texts are of importance to Mongolian traditions. This one derives from legends concerning the Bogdo Gegen—"the living Buddha"—of Urga (the ancient name for Ulaan Baatar).

p. 324. Meanwhile news of the nesting oviraptorid . . . Norell, M. A., et al., 1995, *Nature,* op. cit.

p. 325. In fact, they were closer yet. . . . The Polish-Mongolian team prospected the exposures of the eastern Nemegt Valley but concentrated on the more impressive canyonlands to the west. They did conduct reconnaissance of a canyon area known as Zos, which lies only a few miles northwest of Ukhaa Tolgod (Dashzeveg and Kielan-Jaworowska, personal communication).

p. 326. But the drama of discovery goes on worldwide . . . I have only highlighted the range of exciting new discoveries in vertebrate paleontology. These are regularly summarized in the *McGraw-Hill Encyclopedia of Science and Technology; Science and the Future, Encyclopedia Britannica; Geotimes;* and *Journal of Vertebrate Paleontology, Abstracts to the Annual Meetings of the Society of Vertebrate Paleontology.*

p. 327. Perhaps we'll do better with genes. . . . One limitation of gene studies is that the fossil record is virtually impervious to any kind of gene exploration. Only extraordinary preservation allows for the extraction of DNA in fossils. Some examples are DeSalle, R., Gatesy, J., Wheeler, W., and Grimaldi, D., 1992. "DNA sequences from a fossil termite in Oligo-Miocene amber and their phylogenetic implications," *Science,* Vol. 257, pp. 1933–36. Golenberg, E. M., Giannasi, D. E., Clegg, M. T., Smiley, C. J., Durbin, M., Henderson, D., and Zurawski, G., 1990. "Chloroplast DNA sequence from a Miocene Magnolia species," *Nature,* Vol. 344, pp. 656–58.

p. 327. Embryologists detected nearly a century ago . . . Gaupp, E., 1913. "Die Reichertsche Theorie (Hammer-, Amboss- und Kieferfrage)," *Arch. f. Anat. u. Entwicklungsgeschichte,* 1912, pp. 1–416.

p. 327. What is the importance of fossils . . . ? The answer to this question is more than simply posturing on principle; it is derived from some analytical treatments. For example, the importance of fossils in recovering the evolutionary branching of birds, mammals, and other amniotes was resoundingly demonstrated in a paper by Gauthier, J., Kluge, A. G., and Rowe, T., 1988. "Amniote phylogeny and the importance of fossils," *Cladistics,* Vol. 4, pp. 105–209. Other treatments include Donoghue, M., Doyle, J., Gauthier, J., Kluge, A., and Rowe T., 1989. "The importance of fossils in phylogeny reconstruction," *Annual Reviews of Ecology and Systematics,* Vol. 20, pp. 431–60; and Novacek, M. J., 1992. "Fossils as critical data for phylogeny," pp. 46–88. In Novacek, M. J., and Wheeler, Q. C. (eds.). *Extinction and Phylogeny.* Columbia University Press, New York. There are also examples where the sequence of occurrence of fossils remarkably predicts the sequence of evolutionary divergence based independently on cladistic study, but not all fossil records are this good—see Norell, M. A., and Novacek, M. J., 1992. "The fossil record and evolution: Comparing cladistic and paleontologic evidence for vertebrate history," *Science,* Vol. 255, pp. 1690–93.

p. 328. . . . the refashioning and reworking of science . . . The struggles, successes, and failures of scientists have been keenly observed by philosophers. For example, the tumult that led to the rise of cladistics is examined by Hull, D. L., 1988. *Science as a Process.* University of Chicago Press, Chicago.

p. 329. "*Not mere* music . . ." Nietzsche, F. 1888. *The Case of Wagner.* Reprinted, Vintage Books, Random House, New York, 1967, p. 177.

p. 329. The legacy most treasured by the Mongolians is the writings of their history . . . One of the world's great libraries of sacred and traditional texts is in the basement of the municipal library in Ulaan Baatar. An interesting narrative of early twentieth-century Mongolia and its traditions is Haslund, H., 1934. *In Secret Mongolia.* The Mystic Travellers Series, Adventures Unlimited Press (First Printing 1995), U.S.A., 366 pp.

Select Reading List

Andrews, R. C., 1932. *The New Conquest of Central Asia.* Natural History of Central Asia, Vol. I. The American Museum of Natural History, i–l + 678 pp.

Berry, W. B. N., 1968. *Growth of a Prehistoric Time Scale.* W. H. Freeman, San Francisco, 158 pp.

Carpenter, K., Hirsch, K. F., and Horner, J. A. (eds.), 1994. *Dinosaur Eggs and Babies.* Cambridge University Press, Cambridge.

Carroll, R. L., 1988. *Vertebrate Paleontology and Evolution.* W. H. Freeman and Co., New York.

Colbert, E. H., 1968. *Men and Dinosaurs.* E. P. Dutton, New York, 283 pp.

———, 1995. *The Little Dinosaurs of Ghost Ranch.* Columbia University Press, New York.

Currie, P. J. (ed.), 1993. "Results from the Sino-Canadian dinosaur project," *Canadian Journal of Earth Sciences,* Vol. 30, Nos. 10–11, pp. 1997–2272.

Efremov, I. A., 1956. *The Way of the Winds.* All-Union Pedagogical-Study Publisher, Ministry of Reserve Workers, Moscow, 360 pp. [In Russian.]

Eldredge, N. (ed.), 1992. *Systematics, Ecology, and the Biodiversity Crisis.* Columbia University Press, New York.

———, and Cracraft, J., 1980. *Phylogenetic Patterns and the Evolutionary Process.* Columbia University Press, New York.

Gillette, D. D., 1994. *Seismosaurus, The Earth Shaker.* Columbia University Press, New York, 205 pp.

Goldensohn, E. (ed.), 1995. *Natural History,* Vol. 104, No. 6, pp. 1–88. (The "dinosaur issue"—summarizes much of the current work as well as many of the results of the Mongolian-American Museum Gobi Expedition.)

Gould, S. J., 1977. *Ontogeny and Phylogeny.* Belknap Press, Harvard University, Cambridge, Massachusetts, 501 pp.

———, 1989. *Wonderful Life.* W. W. Norton, New York, 337 pp.

——— (ed.), 1993. *The Book of Life.* W. W. Norton, New York, 256 pp.

Hennig, W., 1966. *Phylogenetic Systematics.* University of Illinois Press, Urbana.

Kielan-Jaworowska, Z., 1969. *Hunting for Dinosaurs.* The Maple Press Company, York, Pennsylvania, 177 pp.

———, 1975. "Late Cretaceous dinosaurs and mammals from the Gobi desert," *American Scientist,* Vol. 63, pp. 150–59.

Lavas, J. R., 1993. *Dragons from the Dunes.* Academy Interprint, Auckland, 138 pp.

Lessem, D., 1992. *Kings of Creation.* Simon and Schuster, New York, 367 pp.

Mayr, E., 1942. *Systematics and the Origin of Species.* Columbia University Press, New York.

Miller, R., 1983. *Continents in Collision.* Time-Life Books, Alexandria, 176 pp.

Nelson, G., and Platnick, N. I., 1981. *Systematics and Biogeography—Cladistics and Vicariance* Columbia University Press, New York.

Norell, M. A., 1991. *All You Need to Know About Dinosaurs.* Sterling Publishing, New York, 96 pp.

———, Gaffney, E. S., and Dingus, L., 1995. *Discovering Dinosaurs* Alfred A. Knopf, New York, i–xx + 204 pp.

Norman, D., 1994. *The Illustrated Encyclopedia of Dinosaurs.* Crescent Books, New York, 208 pp.

Novacek, M. J., Norell, M. A., McKenna, M. C., and Clark, J., 1994. "Fossils of the Flaming Cliffs," *Scientific American,* Vol. 271, No. 6, pp. 60–69.

——— and Wheeler, Q. D. (eds.), 1992. *Extinction and Phylogeny.* Columbia University Press, New York.

Schopf, J. W. (ed.), 1992. *Major Events in the History of Life.* Jones and Bartlett Publishers, Boston, pp. i–xv + 190 pp.

Simpson, G. G., 1944. *Tempo and Mode in Evolution.* Columbia University Press, New York.

Storey, R., 1993. *Mongolia, a Travel Survival Kit.* Lonely Planet Publications, Hawthorn, Australia, 235 pp.

Szalay, F. S., Novacek, M. J., and McKenna, M. C. (eds.), 1993. *Mammal Phylogeny (Vol. 1): Mesozoic Differentiation, Multituberculates, Monotremes, Early Therians, and Marsupials.* Springer Verlag, New York, 249 pp.

———, Novacek, M. J., and McKenna, M. C. (eds.), 1993. *Mammal Phylogeny (Vol. 2): Placentals.* Springer Verlag, New York. 321 pp.

Tattersall, I., 1995. *The Fossil Trail.* Oxford University Press, Oxford, i–xi + 276 pp.

Ternes, A. (ed.), 1994. *Natural History,* Vol. 103, No. 4, pp. 1–107. (The "mammal issue"—with many articles reviewing aspects of the history of mammals, including coverage of the Gobi fossil mammals.)

Weishampel, D. B., Dodson, P., and Osmólska, H. (eds.), 1990. *The Dinosauria.* University of California Press, Berkeley, 733 pp.

Wilson, E. O., 1992. *The Diversity of Life.* Belknap Press, Harvard University, Cambridge, Massachusetts, 424 pp.

Index

Aardvarks, 259, 260, 316
Aarvaheer, 274
Agassiz, Alexander, 91, 335
Alexander, Kevin, 161, 228
Algae, 85
All About Dinosaurs, 60
Allergorhai horhai, 337
Allosaurus, 193
Almas, 177, 179
Almas Canyon, 178, 179
Almas Pass, 177
Altai Mountains, 140
Altan Ula, 31, 32, 45, 115–16, 117, 119–20, 133, 212, 213, 332, 336
  1992 expedition to, 164
  1994 expedition to, 286
Ameral, Bill, 227, 271
America-Asiatic Association, 25
American Museum of Natural History, 5–9, 14–16, 24, 25, 34–36, 95, 227
  Barosaurus at, 188, 193, 194
  dinosaur halls at, 59, 115, 117, 188
  expeditions from. *See* Flaming Cliffs; Gobi Desert
  Tyrannosaurus rex at, 208
Amino acids, 79
Ammonites, 85, 86, 149
Amphibians, 73, 82
Amphioxus, 73
Andrews, George, 24
Andrews, Roy Chapman (or team of), 8, 12, 13, 21–30, 60, 174–75, 333, 337
  desert storms and, 299
  dinosaur eggs and, 214, 216
  documentary about, 324
  evolution and, 142, 143, 154, 158
  quoted, 29, 105
  Mononykus and, 261
  Oviraptor and, 240
  See also Central Asiatic Expeditions (C.A.E.)
Angiosperms, 85
Animalia, 73
Ankylosaur Flats, 236, 240, 277, 281, 304, 305
Ankylosaurs (ids), 11, 31, 32, 70, 86, 102, 112, 135, 290, 303, 330, 333
  evolution and, 67
  found in 1993, 232, 234, 235
  plate tectonics and, 154, 155
  Protoceratops and, 52–53
  species of, 43–45
Anteaters, 44, 259, 260

Apatosaurus, 67, 84, 85
Aptenodytes forsteri, 306
Araucariaceae, 133
Archaeopteryx, 244, 247–54, 255, 258, 259, 341, 342
Archeological evidence, 29
Archosaurs, 73, 81, 84, 244, 248, 328
Ardèche Valley, 13
Arimaspeans, 179
Aristogenesis, 340–41
Armadillos, 45
Arthropods, 138
Artiodactyls, 318
Arts Bogd, 21, 126, 127
Asiatic dinosaur rush, 18
Asioryctes, 283, 314, 315
Asteroid impact theory, 309
Auffenberg, Walter, 209
Aves, 245, 258
Avialae, 230, 244, 245, 250
Avimimus, 250–51
Awash, 13

Bacteria, 78, 79
Badlands, 4, 5, 12, 13, 31, 40, 41, 116, 141, 170, 227, 286, 332. *See also* Canyonlands; Eldorado
Bagà Bogd, 126
Bagaceratops, 61, 62, 70, 262
Baishin Tsav, 109, 111, 112
Baker, James, 19
Bakker, Bob, 189, 198–99, 252
Barosaurus, 84, 188, 189, 190, 193, 194, 199
Barsbold, Rinchen
  background/description of, 17, 33, 130
  Harpymimus and, 251
  Ingenia and, 239
  Saurornithoides and, 236
Barun Goyot Formation, 42, 111–12, 134, 136–37, 244, 298, 311, 333
  "bone-heads" in, 204
  vs. Nemegt Formation, 140
  sauropods in, 190
Bats, 12, 316, 318
Bayandalay, 36, 37
Bayan Mandahu, 58, 285, 302, 303
Bayersaichan (Bayer), 22, 37, 39
Bayn Dzak, 27, 31, 32, 116, 135
Bayn Shireh, 31
BBC team, 160, 162, 163–64, 172
Bears, 316
Bedford, Duke of, 24
Beetles, 84–85
Behrensmayer, Anna K., 189, 198–99
Berkeley Geochronologic Laboratory, 90
Berkey, Charles, 25, 27, 135
Berry, William, 88
Biological classification, 63–76, 188, 252, 254–58, 345
Biomass, 134, 198
Birds, 28, 65, 69, 73, 246, 334
  crocodiles as relatives of, 328
  cuckoo behavior in, 270
  dinosaurs and, 12, 69, 84, 149, 196, 197, 218–19, 243–49, 284, 328, 333. *See also Mononykus*; Theropods, birds and
  diversity of skeletons of, 12
  early/earliest, 11, 16–17, 247
  evolution of, 82, 86, 243–58, 333. *See also specific geologic times*
  "mantling" by, 252
  modern, 174, 248, 252, 258, 259
  plate tectonics and, 159
  as predators, 252, 253–54
  "proto," 253
  synapsids vs., 168
  wishbone in, 248, 249
Bison Basin, 280, 343
Bivalves, 149
"Blank series," 42, 134, 135, 136
Blashford-Snell, John, 179
Bogd, 22
Bogdo Gegen, 346
Boldsuch, Tsagaany, 274
Bone histology, 195–96
Borsuk-Bialynicka, 186
Botanical surveys, 29

Bothrocaryum, 133
Bowles, Paul, 125
Brachiopods, 81
Brachiosaurids, 151, 152, 186
Brachiosaurus, 84, 187, 193, 199
British Museum of Natural History, 24
Brontosaurus, 66–67
"Brown hills," 225
Bryozoans, 81
Buddhism/Buddhists, 109, 110, 111. *See also* Bogdo Gegen
Bugin Tsav, 133, 180–81, 229
Bulgan, 20, 21, 127, 176
Burgess Shale, 80
Butler, Percy, 312

Caballo Mountains, 168, 169, 170, 177
Caecilians, 73
Camarasaurids, 186, 190, 192, 338
Cambrian, 80, 82, 87, 88
Camel's Humps, 235, 240, 277, 281, 284, 304, 305, 324
Camels, 316, 318
Canadian team. *See* Sino-Canadian team
Canyonlands, 109, 121–22, 133, 346
"Carbon-dated," 90
Carbon dioxide, 321
Carnivores, 317, 318
Carnosaurs, 68, 193
Carpenter, Jim, 273
Carpenter, Kenneth, 216
Case, Ted, 220
CAT scans, 15, 182, 196, 215, 313–14
Cave paintings, 13
Cells, 79
Cenozoic, 26, 29, 82, 86–88, 281, 310, 317, 326
  birds and, 159, 250
  mammals and, 310, 311, 316, 324–25, 344
  marsupials and, 312
  plate tectonics and, 159
Central Asiatic Expeditions (C.A.E.), 24, 26–30, 143, 175, 214–15, 261, 299, 311, 312
Central Field, 277, 295–301
Centrosaurinae, 200–1
Ceratopsians, 28, 54–55, 62, 70–71
  Cretaceous and, 86
  frills in, 54, 55, 61, 62, 70, 200
  head crests in, 204
  horns in 54, 62, 200–1
Ceratosaurs, 81
Cetaceans, 24
Champsosaur, 106
Chiappe, Luis
  background and description of, 227
  birds and, 246, 250
  burn injury of, 230–31
Chinsamy, Anusuya, 253
Chlorofluorocarbons, 322
Chloroplasts, 79
Choibalsan, 110
Chordates, 73, 263
Cladistics, 72, 257, 334, 346, 347
Cladograms, 71, 72, 74, 246, 252
Clark, Jim, 330
  almas and, 179
  background and description of, 100
  birds and, 250
  Zalambdalestes and, 128
Clarke, Philip, 160
Clemens, William, 157–58
Climate, 142, 151–52, 156–57, 322, 329
  body temperature and, 195–96
Climate and Evolution, 142
Coelophysis, 81, 83, 173, 335
Coelurosaurs, 244, 246, 247, 248, 249
Colbert, Edwin H. (Ned), 18, 81, 206–7
"Cold-bloodedness." *See* Ectotherms
Colville River, 157
Compsognathus, 247
Computers, 63–64, 96
Conchoraptor, 239, 336
Conifers, 40, 84, 85, 133, 158
Conrad, Fred, 100
Continental drift, 143–46
Coombs, Walter, 189

Copralites, 58
Coral reefs, 80, 149
Corythosaurus, 154, 202–3
Cretaceous, 40, 55, 78–79, 83, 85–86
  ankylosaurs of, 44–45
  Argentina, Spain, and China in, 245, 246
  Australia and South America in, 312
  birds of, 245–46, 253
  different meanings of name of, 88
  dinosaurs of, 9, 10, 28, 31, 107, 109
  Early, 100, 157, 311
  Late. *See* Late Cretaceous
  marsupials of, 312
  mass extinction at end of, 221, 222, 307, 308, 309, 310–11, 318
  microscopic fossils from, 121
  North America and Patagonia in, 246
  plate tectonics and, 152–53, 155–59
  platypus of, 250
  sauropods of, 186
  theropods of, 249
  vertebrates of, 11
Cretaceous Gobi, 40, 90–91, 140, 204–5, 251, 278–79, 305–6, 338
  as desert, 298
  dinosaur hatchlings in, 315
  disasters in, 307–10, 343–44
  fossil preservation in, 301
  hadrosaurs in, 117
  lizards in, 29, 278, 281
  mammals in, 15, 28–29, 86, 137, 175, 311, 324
  placentals in, 313
  plate tectonics and, 158–59
  turtles in, 281, 338
"Cretaceous Komodo," 102. *See also* *Estesia*; Komodo dragon
Cretaceous Nemegt, 133, 134, 137
"Cretaceous sheep," 53
Crichton, Michael, 59
Crick, Francis, 26
Crinoids, 81
Crocodiles, 53, 73, 168, 195, 219–20, 244, 280, 328
Cuckoo, 270
Currie, P. J., 302
Cycads, 84–85
Cyclops, 141, 337
Cynomorium sangaricum, 42

Dabs, 39
Dalan Dzadgad, 4, 19, 100, 101, 161, 171, 230
Darwin, Charles, 184, 221, 222–23
Dashzeveg, Demberelyin
  background and description of, 4–5
  Cenozoic mammals and, 324–25
  canyon named for, 8
  leadership by, 14, 95
  Polish-Mongolian team and, 32, 33
  search for, 123–24
  son of, 22
Datang, 34
Daus, 38–39, 40, 224
Davasambu, 35
Davidson, Amy (background and description of), 227, 228
Davs, 39
Deinocheirus, 213, 325
Deinonychus, 55, 155, 213
Deltatheridium, 312, 315, 324
Deserts, 81, 295–301
Desert storms, 299–301. *See also* Sandstorms
Devonian, 80, 82, 88
Diatoms, 267
Dieter-Sues, Hans, 206–7
Differentiation, 266
Dimetrodon, 80, 83, 167–69, 170
Dingus, Lowell
  background/description of, 100, 114–15, 119
  and mass extinction of dinosaurs, 308
  and original habitat of Gobi, 298, 325, 330
Dinomania, 59
Dinosaur embryos, 215–16, 217, 220–21, 245, 340
  birds and, 261–64

found in 1993, 234, 235, 244, 269–70
found in 1994, 278
Dinosaur evolution, 60, 67–76, 81–91
Dimetrodon and, 168
plate tectonics and, 150–59
Dinosaur hatchlings, 234, 235, 262, 270, 315
"Dinosaur museums," 59
Dinosaur skeletons/skulls, 16, 28–29, 42, 43, 233–35, 241, 303
biological classification and, 70
complete/nearly complete, 304
from Gobi vs. other places, 10–11
juvenile, 157
large vs. small, 11
number of, compared with eggs, 216
smaller, 11, 134
See also Jaws, dinosaur; Skeletons/skulls
Dinosauria, 72–74, 83, 94, 257, 334, 335, 336
Dinosaurs
age of, 310
aggression in, 207–13
armor-plated/armored, 11, 43, 70, 187. *See also* Ankylosaurs; Sauropods
baby, 214, 217–19, 220. *See also* Dinosaur hatchlings
beaked, 27, 61, 63, 262. *See also* Dinosaurs, parrot-beaked
biological classification and, 63–76
birdlike, 11, 17, 261, 333
birds and. *See* Birds, dinosaurs and
body temperature of, 194–98, 339
"bone-head," 204
brooding by, 218–19
cannibalism by, 173
colors of, 185
Dimetrodon and kin vs., 168
duck-billed, 31, 70, 181, 215, 216. *See also* Hadrosaurs; *Saurolophus*
environments of, 15
extinction of, 10, 221–23, 307–11, 334, 344
"fighting," 57, 127, 130, 135, 164, 303
"flat-topped," 205
flesh-eating, 67, 68
flying, 244
food/digestion of, 192, 195, 199
frilled, 53, 54, 55, 200–1. *See also* Ceratopsians; *Protoceratops*/"protos"/protoceratopsids
gigantism in, 186–94
growth in, 220–21, 262
head-butting by, 206–7
head crests in, 201–4
hearing of, 203–4
herds of, 198–200
horned, 54, 200–1. *See also* Ceratopsians
identifying fossils of, 61–66, 73
juvenile, 216–17, 219–20, 261–62
knowledge about, before Gobi expeditions, 25
life spans of, 220–21
lifestyles/behavior of, 185, 186, 190–94, 198–223
long-necked, 11, 40, 69. *See also* Sauropods
mating hierarchy among, 199, 204
metabolism of, 194–98
migration of, 199
naming, 64–76, 254–58
non-bird/non-avian, 86, 250, 253, 262, 308, 309, 339
ostrich-like, 244. *See also* Ornithomimids
pack hunting by, 57, 212–13
parrot-beaked, 27, 61, 71. *See also* Ceratopsians; *Psittacosaurus*
pecking order among, 199–200
plate-backed, 67, 70. *See also* *Stegosaurus*
popularity of, 59, 60
posture of, 198
as predators, 207–13, 252, 253
rhinolike, 70. *See also Triceratops*
sense of smell in, 207
size of, 186–94, 221
"smartest," 237
snake-necked, 69. *See also* Sauropods

social relationships of, 198–207
spiked, 31, 32, 43, 200–1. *See also* Ankylosaurs; Hadrosaurs
tanklike, 11, 31. *See also* Ankylosaurs
toothless, 28
very primitive early, 69
vision of, 203–4, 207, 237
Diplodocids, 151, 186, 187, 192, 193, 338
Diversity of Life, The, 321
Djadokhta Formation, 127, 134, 135, 136, 190, 244, 298, 311, 334
Barun Goyot Formation vs., 136, 137
Nemegt Formation vs., 139–40
DNA, 26, 64, 65, 78, 79, 138, 319, 327, 345, 346
Dodson, Peter, 66, 189, 198, 203, 204
Dolphins, 24
"Dragon," 14, 141, 167–68. *See also* Komodo dragon
"Dragons' Tomb," 117, 336
Dromaeosauridae, 245
Dromaeosaurs(ids), 32, 55, 56, 114, 236, 270, 271, 276, 288, 330, 334
birds and, 250
"higher intelligence" of, 173–74
pack hunting by, 57
plate tectonics and, 154, 155
See also Velociraptor
"Dune dwellers," 29
Dung, 215. *See also* Feces

Eagles, 22
Ear/hearing, 203–4, 314, 316, 319, 327
Earth calendar, 77–78, 82, 87–88, 273
Earthquakes, 144, 147
Eberth, D. A., 298, 302
Echidna, 250, 257
Echolocation, 320
Ectotherms, 195, 196, 253
Ederingian Nuruu, 32
Edmontosaurus, 117, 203
Efremov, Ivan A., 30, 112, 117, 134, 302
"Egg Mountain," 215, 216, 218
Eggs, 8, 13, 17, 19, 27–28, 46, 108, 213–20, 333, 340, 343
after 1920s expeditions, 20
at Bayan Mandahu, 303
cleidodic, 217
first discovery of, 8, 11, 25, 333–34
at Flaming Cliffs, 8, 11, 17, 214, 271, 306, 333
found in 1920s, 8, 29
found in 1991, 112
found in 1993, 234, 241, 271
found in 1994, 278, 286
Gobipteryx, 245, 286
and growth curves, 220–21
robber/raider of, 28, 271, 302. *See also Oviraptor*/oviraptorids
See also Dinosaur embryos
Eldorado, 15, 18–48, 111–15, 116, 137, 164, 175, 326, 333
Electron microscopes, 63
Elephants, 141, 192
Empedocles, 141
Enantiornis, 245
Endemics, and plate tectonics, 156
Endotherms, 195, 196, 197, 220–21, 252, 338, 399, 340
Energy, 79
Enzymes, 79
Eocene, 326
Erdenandalay, 166
Erdensu, 99
Erlian, 35, 99, 161
Erlikosaurus, 109, 156
Estes, Richard, 46
Estesia, 46–48, 102, 278, 281, 286, 290, 291, 329a, 333
Eukaryotes, 79–80
Eutherians, 313
Evolution, 10, 26, 77–91, 142–43, 218, 281, 340–41
climate and, 142–43, 151–52, 156–57
convergent, 259–60
of dinosaurs. *See* Dinosaur evolution
ear bones and, 327
and extinction of dinosaurs, 221–23, 310–11

geographic isolation and, 153, 337
Gobi as bounteous cradle of, 15
of mammals, 316, 318
ontogeny and, 263–68, 327
plate tectonics and, 148–59
systematics and, 76
transitions in, 12, 252
See also Geology; Humans

Feathers, 247, 252–53, 254, 259
Feces, 58, 215. *See also* Dung
First Strike, 240, 277, 304
Fish(es), 42, 73, 82, 219, 259
Flaming Cliffs, 6, 14, 16, 23–30, 49–53, 57–58, 167, 175, 287
dinosaur eggs discovered in, 8, 11, 17, 214, 271, 306, 333
and folklore about griffins, 141
fossils missing from, 58
future goals regarding, 325
1923 expedition to, 237, 261, 271, 306
1991 expedition to, 112
1992 expedition to, 166–67, 171–73
1994 expedition to, 274–76
Polish-Mongolian expeditions to, 32
Soviet expedition to, 31
See also Djadokhta Formation
"Float," 116
Flowers, 84–85, 152, 158
Flynn, John, 92–93
Footprints, 58, 69, 212, 334
Foraminifera, 149
Forbidden City, 26
Forests, 84, 85, 158, 198
Fossil preservation, 58, 77, 90, 267
in Gobi, 10–11, 17, 138–39, 301
at Tugrugeen, 127
at Ukhaa Tolgod, 278, 284
Fossils, 132, 198, 199, 346
birds and, 246, 247, 253, 254
"chunks of skin" as, 117
as "dragon bones," 140
from Gobi vs. other places, 10–11
griffins and, 140, 141, 179, 337
"killer," 120, 129
"living," 250
microscopic, 121
myths/legends and, 140–41, 337
and naming strata, 134–37
preparing, 227
systematics and, 76
taxonomic turnover and, 221
tools for removing, 104–5
French Lieutenant's Woman, 89
Fruit, 133

Gaffney, Gene, 280
Gallimimus, 244, 341
Galton, Peter, 206–7
Gambaryan, 315
Gastropods, 149
Gauthier, Jacques, 248, 257
Gazin, C. Lewis, 280
Genes, 265–67, 327, 346
Genghis Khan, 102
Genotypes, 265, 267
Geological Institute, 5, 95
Geology, 25, 29, 77–91, 134–37, 139, 143–59
Gervais, Paul, 214, 333
Ghurlin Tsav, 312
Giganotosaurus, 69, 326
Gila monster, 46
Gilbent Uul, 40
Gill slits, 73, 263
Gilmore, Charles, 214, 215
Gilvent, 40, 224, 226, 231, 277, 326
Ginkgoes, 84
Gizzard stones, 192
Glen Rose Limestone, 190, 212
Global positioning systems (GPS), 15, 20, 95–97, 101, 103, 115
Glossopteris, 144
Glyptodonts, 45
Gobi Desert, 3, 12, 16, 245, 331–32
almas and, 179
biological Garden of Eden in, 134
changes since 1920s in, 19

compared with other deserts, 120, 295–96
compared with other sources of fossils, 10–11, 12, 14, 17, 132–34, 137
constellations seen in, 101, 125
Cretaceous. *See* Cretaceous Gobi
drought in, 298
elevations of, 19–20
future goals regarding, 325
geological description of, 15
gold mines in, 140
Jurassic, 39
key fossils from, 267–68, 280
maps of, 19, 20, 22, 109, 276–77, 329
Monument Valley in, 121
1920s expeditions to, 8–9, 12, 13, 19, 158, 214, 275, 280, 302. *See also* Central Asiatic Expeditions (C.A.E.)
1922 expedition to, 19, 26–28
1923 expedition to, 28–29, 261
1930 expedition to, 30
1940s expeditions to, 30–31
1960s expeditions to, 14, 15, 32, 186. *See also* Eldorado
1970s expeditions to, 14, 32, 33
1980s expeditions to, 33
1990s expeditions to, 15–17
1990 expedition to, 3–9, 14, 19–23, 33–48, 94. *See also* Eldorado
1991 expedition to, 94–130, 237–38, 299–301
1992 expedition to, 160–83, 280
1993 expedition to, 224–42. *See also* Xanadu
1994 expedition to, 269–88
1995 expedition to, 323–26
1996 expedition to, 329–30
nomads of, 4, 19, 37, 140, 331
"outback" of, 31
plate tectonics and, 148, 154, 155–56
preservation of fossils in, 10–11, 17, 138–40, 301
roads and trails in and to, 3–4, 19, 37, 101, 109, 274
storms in. *See* Desert storms
Triassic, 39
wildlife in and near, 108
See also Flaming Cliffs; Nemegt Valley; Ukhaa Tolgod; *specific teams*
Gobipteryx, 245, 258, 286
Gold, 32, 140
Gondwanotherium, 326
Gould, Stephen Jay, 22, 80
Goyo, 42
GPS. *See* Global positioning systems (GPS)
Granger, Walter, 8, 25, 27–29, 175, 275, 333
Granger's Flats, 277
Great Wall, 34, 35, 110
Gregory, William King, 175
Griffins, 140, 141, 179
Growth, 219–21, 262, 263–64
Growth of a Prehistoric Time Scale, 88
Guchin Us, 22
Gurvan Saichan, 3, 20, 22, 23, 27, 164, 231
Gurvan Tes, 111, 235, 236, 332
Gymnosperms, 84, 158

Hadrosaurines, 203
Hadrosaurs, 31, 32, 70, 86, 107–9, 116–19, 203–4, 215, 247
head crests of, 201–4
plate tectonics and, 154, 155, 157
Haloxylon, 27
Harpymimus, 102, 251
Hedgehogs, 316
Heilmann, Gerhard, 248, 249
Hell Creek, 114, 214
Helodermatids, 46
Hennig, Willi, 257, 334
Herbivores, 117–18, 192, 207, 272, 317, 318
Herodotus, 179
Herrerasaurus, 81, 335
Hesperornis, 246
Hesperornithiforms, 248
Hippos, 318
Holocene, 87, 335

Homalocephale, 205, 206
Homer, 141
Homo sapiens, 78–79, 87, 219
"Hopeful monsters," 266–67, 341–42
Hopson, James, 203
Horner, Jack, 208, 215, 218
Horses, 267, 318, 342
Hugin Djavchalant, 108
Humans, 13, 26, 87, 142, 143, 335
  embryos of, 263–64
  growth in, 219
  Scrotum humanum and, 66, 340
Hurrendoch 100, 102, 106, 251, 336
Hutchison, Howard, 281
Huxley, Thomas Henry, 248
Hymenoptera, 85
Hypacrosaurus, 339
Hypselosaurus, 220
Hypsilophodontids, 154, 215

Iberomesornis, 247
Ice ages, 87, 152, 329. *See also* Polar ice
Ichabodcraniosaurus, 114
Ichthyornis, 246
Iguanodontids, 102, 215, 310
  plate tectonics and, 151, 154, 159
Incertae sedis, 250–51
Indricotherium, 26
Ingenia, 239, 241, 277, 324, 336
Insectivores, 318
Insects, 64, 85, 138, 149, 152
Institute of Vertebrate Paleontology and Paleoanthropology (IVPP), 326
Invertebrates, 256
Iridium, 308
Isotopes, 90, 156, 196, 197, 335, 339

Japanese team, 241
Jaws
  dinosaur, 28, 63, 118, 123
  lizard, 42
  mammal, 231, 233, 235
Jerzykiewicz, T., 288, 302–4
Jim's Pocket, 277
Johnson, Donald, 324
Johnston, P. A., 302
Jones, Indiana, 24
Jornada del Muerto, 169, 170
Junior synonyms, 66, 67
Jurassic, 39, 82, 83–84, 151–52
  birds and, 244, 246, 247
  mammals and, 310
  sauropods and, 186, 192
Jurassic Park, 11, 55, 59

Kalgan, 34, 110
Kangaroos, 316
Kannemeyerıa, 150
Kazakhstan, 312, 313
Kelly, Jeanne, 227
Kennalestes, 314
Keqin, Gao, 325
Khara Khutuul, 107–8
Kheerman Tsav, 32, 120, 121–26, 133, 175, 245, 286, 325, 332
  1991 expedition to, 237–38
  1992 expedition to, 181–82
  1994 expedition to, 286
  1996 expedition to, 330
Khoobor, 129
Khulsan, 16, 42, 111, 115, 116, 164, 175, 286–88, 325
Khulsan dragon, 45–48, 112. *See also* Komodo dragon
Kielan-Jaworowska, Zofia, 32–33, 212, 283, 311–15, 325–26
Klagenfurt, 141
Knight, Charles, 193, 338
Komodo dragon, 14, 46, 52, 209–10, 329a
Komodo Island, 211–12
Koster, E. H., 302
Kozlowzki, Roman, 32

Lagomorpha/lagomorphs, 128, 318
Lakes. *See* Ponds/lakes/streams
Lama/lamaism, 109, 111
LANDSAT, 15, 276, 285

Landslide Butte, 215
Langdon, George, 161, 171, 172, 179
Lanns, John, 228, 230, 273
Late Cretaceous, 10, 32, 49–52, 57, 78, 186, 249, 289–94, 306, 330, 333
  placentals/marsupials of, 312, 313
  plate tectonics and, 157, 158–59
Lawson, Doug, 149
Lessem, Don, 208
Life
  hierarchy of, 75–76, 219, 255, 257
  history of, 222–23, 254, 329, 335, 346–47
  See also Biological classification
Limestones, 78, 85, 247
Linnaeas, Carolus/Linnean, 63, 75–76, 92, 255–58, 334
Lithology, 41
Lizards, 42, 46–52, 73, 168, 195, 219–20, 278, 303, 310, 334
  as "dragon," 14, 46, 329a. *See also* Komodo dragon
Lizard skeletons/skulls, 17, 29, 32, 46–48, 114, 123, 129, 233–40
  at Kheerman Tsav, 182
  in Nemegt Valley, 135
  at Tugrugeen, 229
  at Udan Sayr, 330
  at Ukhaa Tolgod, 272, 278, 281, 288, 310, 325
Lockley, Martin, 199, 212
Lynch, John, 160

McIntosh, Jack, 188, 198–99
McKenna, Malcolm
  Cenozoic mammals and, 324
  description and background of, 7
  eggs and, 108
  leadership by, 14, 95
  quoted on "first time," 12
McKenna, Priscilla, role of, 97
Mader, Bryn, 261
Maiasaura, 215
Mamenchisaurus, 190
Mammalia, 257, 345
Mammals, 11–12, 30, 65, 73, 136, 168, 174–75, 310–22, 344, 345
  age of, 310–11, 316, 317, 321
  evolution of, 81, 82, 83, 90, 142, 149, 151. *See also specific geologic times*
  extinction of, 45, 320–22
  finger bones of, 42
  ground-hugging, 86
  growth in, 219
  large land, 26, 29
  lifestyles of, 315–17, 320
  modern, 12, 15, 53, 310–12, 320, 344, 345
  radiation of, 318–20, 326, 345
  shrewlike. *See* Shrewlike (forms)
  teeth of, vs. dinosaur teeth, 63
  See also Marsupials; Multituberculates/multis; Placentals
Mammal skeletons/skulls, 11, 17, 25, 90, 121, 128, 129, 175, 231–35
  found in 1920s, 29
  found in 1960s and 1970s, 32
  found in 1993, 233–34, 240, 272
  found in 1994, 282–83
  found in 1995, 324
  found in 1996, 330
  at Kheerman Tsav, 182
  number of, from Gobi, 137
  small/tiny, 11, 15
  at Tugrugeen, 174, 175, 229
  at Ukhaa Tolgod, 272, 288, 305, 310, 330
  See also Jaws, mammal; Skeletons/skulls
Mammoths, 316
Mandal Obo, 166, 275
Maniraptora, 69, 236, 238, 244–45, 250, 255, 256
Maps, 19, 20, 22, 95, 109, 276–77
Marine organisms, 80, 86, 149, 156
Marine reptiles, 86, 144
Mark's Egg Site, 277
Marsupials, 257, 259, 311–12, 318
Maryańská, 239
Matthew, William Diller, 142–43, 158

Mayor, Adrienne, 140, 141, 179
Meeker, Lorraine, 313
Megalosaurus, 340
Men and Dinosaurs, 18
Mesosaurus, 144
Mesozoic, 81, 83–86, 128, 222, 273, 282, 310, 315–17
  birds of, 174–75, 245, 246, 247, 248, 341, 342
  lizards of, 278, 310
  mammals of, 174–75, 287, 310–22, 345
  placentals of, 283
  plate tectonics and, 151
  turtles of, 281
Metabolism, 79, 194–98
Metamorphosis, 263, 264
Meteorites, 307–8
Mid-oceanic ridges, 144–45, 146
Mike's *Mononykus*, 277
Minchin, Dr., 330
Mitochondria, 79
Molnar, Ralph, 210
Mongolia, 4, 7–8, 36–37, 98–99
  American Museum of Natural History and, 14, 15, 16, 34–36
  democracy in, 34, 102, 106, 113, 161, 327
  folklore/legends of, 141, 179, 337, 346. *See also* Almas; Griffins
  history of, 102, 110–11, 329, 347
  Inner (Nee), 21, 30, 33, 334
  maps of, 20, 109
  plate tectonics and, 156
  Soviet Union and, 14, 30–31, 102
  Western culture in, 228
Mongolian Academy of Sciences, 14, 32, 33, 35, 95, 270, 274
Mongolian food, 21, 97, 113, 161, 162, 163
Mongolian names, 22–23
Mongolian State Museum, 35
Mongolian team, 12, 14, 33, 109. *See also* Polish-Mongolian team
Mongolian Television and Radio, 162
Mononykus, 17, 229–30, 237, 241, 248, 259–61, 291, 343
  found on 1994 expedition, 284, 286
  found on 1995 expedition, 324
  as "hopeful monster," 268, 342
Monophyletic group, 72
Monotremes, 257, 311–12, 326, 345
Monument Valley, 121, 170, 171
Morrison Formation, 189–90, 192, 199
Moss animals, 81
Mountain building, 144, 153
Multituberculates/multis, 114, 174, 175, 229, 232–33, 272, 311, 315, 344
"Mummy," mud (of hadrosaur), 117
Municipal Museum, 31
Museums, 10
Mutations, 266–67

Nahas, Anthony, 160
Naran Bulak, 100, 115, 121, 126, 177, 179, 180, 181, 286, 331
National Science Foundation, 271, 324
Natural History Museum, 31, 57, 99, 117, 162. *See also* Municipal Museum
Natural selection, 267
Nemegt Formation, 112, 115, 118, 133, 134, 136, 186, 338
  vs. Barun Goyot and Djadokhta formations, 139–40, 336
  "bone-heads" in, 204
  future goals regarding, 325
  sauropods of, 190
  Soviet-Mongolian expedition to, 229
  theropods of, 244
Nemegt Valley, 14–17, 39, 40, 332
  1991 expedition to, 111–12
  1992 expedition to, 164, 176–79
  1993 expedition to, 224–42
  Polish-Mongolian expeditions to, 32, 134
  Soviet expedition to, 30–31
  two different environments of, 133–34, 135, 136, 140

Neoceratopsia, 70–71, 334
Neoteny, 264
Nessov, Lev, 312, 345
Nests, 11, 17, 28, 43, 218, 234, 270, 290, 304, 306. *See also* Eggs
New Conquest of Central Asia, The, 24, 214
Nomina dubia, 66, 67
Norell, Mark
  background and description of, 5
  bird ancestry and, 250
  eggs and, 217, 269
  leadership by, 14, 95
  phylogeny and, 218–19
  quoted on Cretaceous sheep, 53
North American Cordillera, 153
North Pole, 145, 148, 156, 157
Notochord, 73, 263
Novacek, Michael (author)
  background of, 167–71
  at Kheerman Tsav, 123–26
  as leader of expedition, 95
Nyssa sylvatica, 133
Nyssoidea, 133

Oases, 289–94, 297–98
Oceanic crust, 144–45, 146–48
Oceans, life in, 80, 85
Ontogeny, 262–68, 327
Opisthocoelicaudia, 186, 338
Opossum, 312
Ora, 209–12
Ordovician, 80, 82
Origin of Species, The, 184
Orlov, I., 30
Ornithischia, 67–68, 69–76, 81, 257
  plate tectonics and, 150
Ornithomimids, 244, 251, 336
Orthogenesis, 340
Osborn, Henry Fairfield
  background of, 24, 25, 26
  Central Asia and, 142, 143, 154, 158
  mammals and, 174–75
  natural selection and, 222
  sauropods and, 193
  Saurornithoides and, 236
Osmólska, Halska, 66, 239
Ostrom, John, 236, 247, 248, 252
Oviraptor/oviraptorids, 69, 122, 123, 237–40, 245, 250, 251, 306, 330, 336
  dromaeosaurs and, 270
  embryos of, 271. *See also* Dinosaur embryos
  found in 1923, 237
  found in 1991, 237–38
  found in 1993, 237, 270, 272
  found in 1994, 277–78, 284–85
  found in 1995, 324
  as "hopeful monster," 268
  as parents, 272, 278, 291, 306
  plate tectonics and, 155
  two species of, 238–39
  at Ukhaa Tolgod, 288–94, 304, 306
  wishbone in, 249
Oviraptor philoceratops, 28, 336
Owen, Richard, 94, 214, 336
Oxygen, 79, 80, 156, 337
Ozone hole, 322

Pachycephalosaurs, 156, 204–7, 339
Paedomorphosis, 264–65
Paleocene, 317–18
Paleolithic cave paintings, 13
Paleomagnetics, 145–46, 273, 337
Paleontological Institute, 5
Paleontology, 10–11, 60–66, 76, 91, 139, 149, 267, 301–7, 327
  author's background in, 167–71
  criticisms of, 328
  field, 103, 172–73
  geological time and, 77–91
  important lessons of, 329
  Mongolia and, 25
  tools of, 103–6
  vertebrate, 5, 11, 33, 57, 346
Paleozoic, 80, 82, 84, 335
Parasaurolophus, 202, 204
Pass, Jerry, 160
Patagopteryx, 246, 247

Pelecanimimus, 251
Penguins, 306–7
Perrissodactyls, 318
Perle, Altangerel
  background of, 95, 100
  description of, 101–2
  Harpymimus and, 251
Perle, Chimbald, 100
Permian, 80–81, 82, 83, 167–68, 170, 221, 308–9, 338
  meaning of name, 88–89
  plate tectonics and, 150, 152
Phenotypes, 265, 267
Photosynthesis, 79
Phylogenetic taxonomy, 257–58
Phylogeny, 218, 252, 262–63, 334, 344, 345, 346
Pinacosaurus, 43, 44, 46
Pinacosaurus grangeri, 303, 304
Pisanosaurus, 81
Placentals, 12, 128, 235, 257, 283, 311–22, 326
  first high-quality skeletons of, 12
  new species of, 283
Plankton, 81, 149, 267
Plants, 55, 58, 65, 82, 83–84, 116, 152, 192, 215. *See also* Flowers; Fruit; Trees; *specific types*
Plateosaurus, 69, 81, 150
Plate tectonics, 145–58, 337
Platypus, 250, 257
Pleistocene, 87, 335
Pliny the Elder, 91, 141
Plot, Richard, 66
Pneumatization, 239
Polar ice, 80, 87, 152, 156
Polish-Mongolian team, 12–15, 32–33, 57, 115–16, 129, 137, 325–26, 333, 346
  Barun Goyot Formation named by, 42
  "bone-heads" found by, 204
  "fighting dinosaurs" found by, 303
  Opisthocoelicaudia and, 186
  placentals found by, 313
  See also "Blank series"; Tugrugeen
Ponds/lakes/streams, 52, 53, 58, 133, 134, 139, 189, 190, 203, 275, 280
  fluvial, 136, 304
  Late Cretaceous, at Ukhaa Tolgod, 290, 304
Porpoise, 24
Precambrian, 79, 82, 87
Prenocephale, 206
Primates, 12, 86, 316, 318, 326
Prosauropods, 83, 310
Protein, 79
Protoavis, 249
Protoceratops/"protos"/protoceratopsids, 18–19, 28, 32, 42, 52–55, 57, 67, 135, 333, 334
  at Bayan Mandahu, 303, 304
  biological classification of, 70
  as "Cretaceous sheep," 53
  eggs of, 216, 234
  embryos of, 271
  found on 1993 expedition, 232, 234
  frills in, 54, 55, 61, 62, 63, 262
  growth in, 220, 221, 262
  at Hugin Djavchalant, 108
  identifying fossils of, 61–62
  and mythical griffins, 140, 141
  plate tectonics and, 154, 156, 158
  skulls of, 42, 70, 123, 127, 271
  "social tendencies" of, 54
  at Ukhaa Tolgod, 288, 290, 291, 304, 330
  See also Ceratopsians
Protoceratops andrewsi, 9, 27, 28, 61, 62, 65, 262, 303
Protoceratops kozlowskii, 62, 65
Psittacosaurus, 70, 102, 214–15
Pterosaurs, 73, 81, 84, 149, 244

Quaternary, 82, 86–87
Queiroz, Kevin de, 257
Quetzalcoatlus, 149

Rabbits, 128, 314
Radioactive decay, 90, 103, 136, 335

Rain forests, 134, 321
Ray, John, 63
Red hills, 17, 40–45, 47, 111, 225
Red rocks, 167
Reid, Robin, 196
Remote sensing, 103
Reptiles, 30, 81, 82, 86, 95, 144, 196, 338
Rhinoceros/rhino, 141, 316
Ribosomes, 79
Ricqlès, Armand de, 196
Riggs, Elmer, 189
Rocks, 89–91, 131–32, 135, 139
  Cambrian, 87
  Cenozoic, 26
  dating of, 89–91, 103, 135–36, 144–45. *See also* Paleomagnetics
  igneous, 77, 90, 147
  metamorphic, 77
  oceanic, 144–45
  red, 167
  sedimentary, 77–78, 132, 133, 139
  strata of, 134–37, 139, 279
  time-regressive formations of, 133
Rodents, 272, 311, 314, 318
Rougier, Guillermo, 325
Rowe, Tim, 15, 313
Ruben, John, 196
"Ruins, The," 112
Russian team, 12, 14, 33, 115–16, 118, 241, 280, 333. *See also* "Blank series"; Soviet team
Ryan, Michael, 200

Sahara, 13, 295, 296, 297, 326
Saichania, 43, 45
Salamanders, 264
Salt flat, 39
Sampson, Scott, 200, 215
San Andreas fault system, 77, 147
Sandblasts, 112–13, 114
Sand cliffs, 27
Sand dunes, 4, 27, 52, 53, 58, 115, 278, 289–97
Sandstones, 121, 175–76, 275, 303
  cross-beds in, 41, 58, 115, 276, 330
  field notes on, 41
  geological time and, 78
Sandstorms, 21, 28, 114, 303, 304. *See also* Desert storms
Sangre de Cristo Mountains, 171
Saskatchewan, 157
Satellites, 15, 19, 96, 97, 101, 330–31. *See also* LANDSAT; Remote sensing
Saurischia, 67–69, 257
Saurolophus, 117, 118, 180–81
  head crests/spikes in, 202–3, 204
  plate tectonics and, 154, 155
Sauropods, 11, 32, 67–69, 85–86, 109, 116, 186–94, 197–98, 212, 338
  armor plating of some, 187
  eggs of, 215
  endothermy in, 339
  migration/herding of, 199, 339
  plate tectonics and, 151, 154, 159
  size of, 84
  social relationships of, 198–99
Saurornithoides, 155, 236, 237
Saynshand, 100, 106, 107, 108
Scientific Commission of Mongolia, 30
Screen-washing technique, 121
Scrotum humanum, 66, 340
Scythians, 140, 179
Sea algae, 85
Sea "lilies," 81
Seals, 12
Secret History of the Mongols, The, 329
Seismosaurus, 84, 187, 338
Serengeti, 134, 139, 198, 203
Sereno, Paul, 81
Severy Mountain, 38
Sexual dimorphism, 54, 201, 203
Shabarrakh Usu, 27, 53
Shackleford, J. B., 27
Shales, 78, 116, 133, 138, 276
Shamosuchus djadochtensis, 280
Sheeregeen Gashoon, 32
Sheltering Sky, The, 125
Shoulder girdle, 212, 245

Shrewlike (forms), 53, 83, 86, 112, 233, 235, 282, 310, 314
Shrews, 315
Shuvalov, 298
Siberia, 21
Sibling species, 65
Silts/siltstones, 41, 78
Silurian, 80, 82
Simpson, George Gaylor, 175
Sino-Canadian team, 33, 58, 285, 302, 333
Sino-Swedish team, 13–14, 30, 333
Sinornis, 245, 247
Skeletons/skulls, 138–39
  at Bayan Mandahu, 303
  CAT scans of. *See* CAT scans
  complete/nearly complete ones, 43, 45, 57, 134, 137
  at Dragons' Tomb, 117
  found in 1993, 231–34, 240, 271
  found in 1994, 275–77, 282–86
  found in 1996, 330
  giant, 180–81. *See also* Sauropods
  Gobi as richest site of, 17
  in Nemegt Formation, 134
  See also Dinosaur skeletons/skulls; Mammal skeletons/skulls
Skin, fossils as "chunks of," 117
Slays, 78
Smith, Evan, 277
Smithsonian Natural History Museum, 189, 257
Snakes, 73, 86, 168, 178
Sodnam, Dr., 33
Solar technology, 97
Solnhofen, 247
South Pole, 145
Soviet Academy of Sciences, 30
Soviet team, 30–31, 137. *See also* Russian team
Soviet-Mongolian team, 229, 259, 313
Sphinx, 331
SPOT satellite, 15
Starfish, 81
Stegoceras, 205, 206
Stegosaurus, 67, 70
Streams. *See* Ponds/lakes/streams
Sugar Mountain, 281–84, 311, 320
Supersaurus, 187
Swamp, 79
Swedish team. *See* Sino-Swedish team
Swisher, Carl, 90, 273, 305, 308
Synapsids, 80–81, 83, 150, 168, 328
Syntarsus, 150
Systema Naturae, 75, 257, 335
Systematics, 76, 91, 335

Talarurus, 43
Tanke, Darren, 200
Taphonomy, 302, 303, 335, 338
Tapirs, 318
Tarbosaurus, 11, 16, 31, 32, 44, 69, 118–19, 181, 207
  original description of, 336
  plate tectonics and, 155
  tooth/teeth of, 116, 118
Tarbosaurus bataar, 64
Tarchia, 43, 44
Tatal Gol, 227
Taxonomy, 65–76, 92, 221, 256–58
  sauropods and, 188
Technology, 13, 15–16, 90, 97, 103, 328–29. *See also specific types*
Temujin, 102
Tendagaru Formation, 190
Tertiary, 82, 86, 87, 121, 174
Tetrapods, 30, 73, 168, 263
Therizinosaurids, 244
Theropods, 11, 15, 17, 32, 69, 173, 174, 235, 236, 237, 303
  at Baishin Tsav, 109
  biological classification of, 67
  birds and, 11, 60, 159, 230, 236, 243–58, 328
  body temperature of, 196
  branches of, 244
  cladogram for, 246
  Cretaceous, 86, 249
  diversity of skeletons of, 12
  grace of, 198
  limbs of, 43, 46

maniratoran, 238, 249, 255. *See also* Maniraptora
pack hunting by, 212
plate tectonics and, 156, 157
Scrotum humanum and, 66
specialized, 245, 251
at Tugrugeen, 174
at Ukhaa Tolgod, 287
wishbone in, 249
See also *Estesia; Harpymimus; Mononykus; Velociraptor*
Time (geological), 77–91
Torosaurus latus, 70
Tost Mountain, 332
To the Ends of the Earth, 324
Trans Altai Gobi, 32
Trans-Siberian Express, 34–35
Trees, 84, 116, 133, 158, 203, 248, 249, 289. *See also* Conifers
Triassic, 39, 81, 83, 84, 148, 150–51, 173, 248, 249, 310
Triceratops, 44, 55, 70
Triconodonts, 83, 310, 311, 344
Trilobites, 149
Tritylodon, 29
Troodon/troodontids, 155, 157, 236, 238, 241, 245, 250, 288
Tsagan Khushu, 32, 331
Tugrik, 126
Tugrugeen, 16, 32, 57, 126–28, 164, 172–76, 229–31, 304
"fighting dinosaurs" at, 303
future goals regarding, 325
Mononykus at, 259
placentals at, 313
sand dunes at, 296, 297
Tule River, 274
Turtles, 53, 73, 168, 181, 195, 280, 281, 338
Tuul Gol, 99–100
Tyrannosaurus, 11, 31, 44, 64, 69, 86, 119, 155, 207–10, 248

Udan Sayr, 330
Udanoceratops, 61
Ukhaa Tolgod, 17, 40, 269–71, 304–5
future goals regarding, 325
as hatchery, 278
key fossils from, 267–68, 280
Late Cretaceous, 289–94, 330
multituberculates at, 272
1993 expedition to, 225, 235, 237, 240, 241, 243–44
1994 expedition to, 276–88
1995 expedition to, 324
1996 expedition to, 330
Oviraptor at, 271
as richest locality of Mesozoic fossils, 287, 305–6
rock strata at, 279
sandstorms/desert storms at, 294, 305
Ulaan Baatar, 4, 5, 14, 15, 22, 25, 31, 35–37, 94, 99, 273–74, 345, 347
hotel bar in, 162
in 1930s, 110
railroad of, 100, 163
Ulaan Ushu, 285
Ultrasaurus, 187
Ural Mountains, 148
Urga, 22, 25, 346
Uzebekistan, 312, 313

Varanids, 46
Varanus komodoensis, 209–10
Vaughn, Peter, 167
Velociraptor, 11, 15, 32, 55–57, 69, 86, 173–74, 212–13, 236, 286, 333
limbs of, 56, 212
naming of, 28
plate tectonics and, 155, 156
"protos" as food for, 19, 52–53, 57
skeletons of, 56, 57, 128
skull of, 15, 29, 55, 128, 234, 334
teeth of, 52, 55, 128, 173
at Tugrugeen, 128, 164, 173–74, 229
at Ukhaa Tolgod, 290–94
Vertebrate paleontology, 5, 11, 33, 57, 346
Vertebrates, 11, 72, 73–74, 138

embryos of, 263–64
evolution of, 143, 317
growth in, 219
invertebrates vs., 256
Nemegt Formation and, 134, 136
Permian, 168
plate tectonics and, 150–51, 154
pneumatization in, 239
range of, 137
turtles and, 280–81
Volcanoes, 144, 147

"Warm-bloodedness." *See* Endotherms
Watson, James, 26
Watts, Tim, 160
Way of the Winds, The, 30
Wegener, Alfred Lothar, 143–44, 146, 148, 150, 337
Weishampel, David, 66, 203, 204
Wellenhofer, Peter, 247, 248
Werscheck, Barbara, 33, 161, 269
Wetlands, 301
Whales, 12, 24, 191, 194, 316, 318
fishes and, 259
limbs on archaic, 326, 328
Wible, John, 325
Wilford, John Noble, 100, 273
Wilson, E. O., 321
Wonderful Life, 22
Wrists, 244, 247–48

Xanadu, 235–42, 273–78, 304, 332, 341
Xerophytic plants, 55

Young, Neil, 284

Zaks, 27
Zalambdalestes, 112, 128–29, 164, 182, 276, 311, 313–16, 336, 345
Zheng, J-J, 302
Zofia's Hills, 277
Zoological surveys/zoology, 25, 29
Zos Canyon, 280, 346

Look for the hour-long documentary on the Gobi expedition entitled *The Dinosaur Hunters*, part of the National Geographic "Explorer" series.

About the Author

Michael Novacek is Senior Vice-President and Provost of Science, as well as Curator of Vertebrate Paleontology, at the American Museum of Natural History in New York City. Over the past six years he has led the first western expeditions to the Gobi Desert in sixty years, unearthing a series of startling and historic fossil finds. His search for dinosaurs and other fossils has also taken him to remote areas of Baja California, Mexico, the Andes of Chile, the Arabian peninsula, and the American West.